Springer Series on

ATOMIC, OPTICAL, AND PLASMA PHYSICS

The Springer Series on Atomic, Optical, and Plasma Physics covers in a comprehensive manner theory and experiment in the entire field of atoms and molecules and their interaction with electromagnetic radiation. Books in the series provide a rich source of new ideas and techniques with wide applications in fields such as chemistry, materials science, astrophysics, surface science, plasma technology, advanced optics, aeronomy, and engineering. Laser physics is a particular connecting theme that has provided much of the continuing impetus for new developments in the field, such as quantum computation and Bose-Einstein condensation. The purpose of the series is to cover the gap between standard undergraduate textbooks and the research literature with emphasis on the fundamental ideas, methods, techniques, and results in the field.

Please view available titles in *Springer Series on Atomic, Optical, and Plasma Physics* on series homepage http://www.springer.com/series/411

Man Mohan

Editor

New Trends in Atomic and Molecular Physics

Advanced Technological Applications

 Springer

Editor
Man Mohan
Department of Physics and Astrophysics
University of Delhi
Delhi, India

ISSN 1615-5653   Springer Series on Atomic, Optical, and Plasma Physics
ISBN 978-3-642-38166-9   ISBN 978-3-642-38167-6 (eBook)
DOI 10.1007/978-3-642-38167-6
Springer Heidelberg New York Dordrecht London

Library of Congress Control Number: 2013944561

Printed on acid-free paper

Springer is part of Springer Science+Business Media (www.springer.com)

Preface

In recent years, the field of Atomic, Molecular and Optical physics has undergone a revolutionary change due to tremendous developments of cutting edge Scientific Technology and high performance computers. The research in this field has grown very widely and covers a broad spectrum of topics including: Structure and Dynamics of atoms and molecules, high-precision spectroscopy, cold atoms, Bose-Einstein condensates, Optical Lattices, free electron lasers etc. This field is also of great interest for applications in several neighboring areas such as astrophysics, plasma physics, condensed matter physics, biology, medicine and nano-technologies.

This present book presents the latest research from around the world and covers the topics on: Atomic Structure, Collision Physics, Photo Excitation and ionization Processes, Ultra-Cold atoms, Bose-Einstein Condensate and state-of-art technological applications in the field of Astronomy, Astrophysics, future energy source from Fusion, Biology and Nano-technology.

Under the topic of Atomic Structures, Biemont from University of Liege, Belgium in Chap. 1, presented the progress realized regarding the radiative properties (transition probabilities, lifetimes, branching fractions) of the heavy atoms and ions ($Z > 37$) of the fifth and sixth rows of the periodic table and of the lanthanides. Their knowledge is vital and strongly needed in astrophysics in relation with the problematic of stellar abundance determination, cosmochronology and nucleosynthesis. In Chap. 2, Man Mohan et al. described the present status of atomic structure calculations using several different international computer codes like CIV3, GRASP etc. and have shown the importance of correlations and relativistic effects for multi-electron systems.

In Chap. 3, Nicholas Guise et al. from NIST (USA) described highly advanced design of compact permanent-magnet Penning traps with novel unitary architectures for capture of highly charged ions extracted from an EBIT source. They demonstrated confinement within two trap geometries for various ion species. In their experiment storage times exceed 1 second in a room-temperature apparatus, enabling spectroscopic and charge-transfer studies relevant to astrophysics and metrology. Paul Mokler et al. from Max-Planck Institute, Heidelberg, Germany in Chap. 4, demonstrated that K-L inter-shell higher-order electronic recombination for highly charged ions ($18 < Z < 36$) can dominate first-order dielectronic recombination. These electron-correlation effects stress the importance of multi-electron processes and have to be included into plasma models. In Chap. 5, Aarti Dasgupta et al. from Naval Research Laboratory (USA) described the world's most powerful X-ray source, the wire-array Z-pinch, which offers a promising route for producing energy through controlled thermonuclear fusion and in which high atomic number materials are investigated as an intense X-ray radiation source on the Z facility at the US Sandia National Laboratories. In their studies, spectroscopic analysis is the key tool to understand the conditions and X-ray performance of high dense plasmas.

Further work under extreme conditions is presented by F.B. Rosmej, F. Petit-demange of Ecole Polytech., Lab., France. They studied (Chap. 6) Auger electron heating of dense matter driven by irradiation of solids with intense XUV/X-ray Free Electron Laser radiation which permits studies of dielectronic capture channels that are usually closed in standard atomic physics investigations (like, e.g., in accelerators, EBITS, Tokamaks, astrophysics). After disintegration of crystalline order dense strongly coupled plasma is formed where dielectronic capture coupled to excited states is identified as the primary population source of autoionizing hole states. Opacity is a fundamental quantity for plasmas and gives a measure of radiation transport. For explaining this, Sultana N. Nahar the Ohio State University, USA has illustrated in Chap. 7, the necessity for high precision atomic calculations for the radiative processes of photoexcitation. On the collision side with charged particles the situation is reviewed by A. Bhatia from NASA USA in Chap. 8 and R. Srivastava et al., IIT Roorkee in Chap. 9. R. Karazija et al. in Chap. 10, describe the conditions and regularities for the formation of a narrow group of intense lines in the emission and photoexcitation spectra of free ions. In Chap. 11, Henrik Hartman discusses fluorescence process behind emission lines prominent in spectra of many low-density types of plasma.

Advanced technology in the field of AMO has led to laser-cooling of alkali atoms near to zero degree Kelvin resulting in Bose Einstein condensation (BEC). Bose Einstein condensation physics has emerged as an active frontier at the cross section of AMO and condensed matter physics. In this direction, the subject of Chap. 12, by Indu Satija et al., USA, deals with the Topological Insulators, new exotic state of matter with importance at both fundamental and technological frontiers. Their particular focus is the detection of these states in the ultra-cold atom laboratories. Chapter 13 describes the investigation of coherently prepared atomic medium under external perturbation and its prospective application in developing an atomic frequency offset reference. Vermuri et al., in Chap. 14, discuss and predicted a ratchet-like motion for the light which can be achieved by utilizing phase-displaced inputs and a linear gradient for index of refraction across the waveguide array showing BEC application in communication engineering.

Chapters from 15 to 19 emphasize on the high technological applications of Atomic Physics in different fields from Astronomy, Astrophysics, Biology and Nanotechnology. In Chap. 15 from Anil Pradhan et al. from Ohio State Univ., USA, describes an astonishing connection between the physics of K-shell X-ray transitions in heavy elements (high-Z), and potential cancer therapy and diagnostics (theranostics) using high-Z nanoparticles embedded in cancerous tumors. The methodology is termed Resonant Nano-Plasma Theranostics (RNPT), and relies on the use of monochromatic X-rays and nanobiotechnolgy of targeting and delivery of high-Z (e.g. gold or platinum) nanomaterials to malignant cells. Irradiation by monochromatic X-rays is far more efficient than conventional broadband X-ray devices, such as CT scanners for imaging or Linear Accelerators for therapy. In Chap. 16, R. Schuch group from Atomic Physics Center, Stockholm, Sweden have reviewed the work on the transmission characteristics of slow and highly charged ions (HCI) through thin foils and also presented the preliminary results of pre-equilibrium energy loss and charge states of HCI after transmission through ultra-thin carbon nanosheets which are of the thickness of single and three molecular layers. Their work on interaction of energetic charged particles with matter is a subject of large interest because of its potential applications such as material modifications, radiation detectors, biological and medical treatments, and ion-implantation. In all, the book covers from basic to highly advanced topics having wide applications in several neighboring fields of science and technology.

This book will be useful for Scientists, Teachers, graduate and post graduate students, especially dealing in Atomic, Molecular and Optical Physics. The book has a wide scope with applications in Plasma Physics Astrophysics, Cold Collisions, Nano-technology and future energy sources like JET (UK) and ITER (international Thermonuclear Experimental Reactor) where several countries are co-operating for harnessing fusion power.

I am thankful to Professor Dinesh Singh, the Vice-Chancellor of University of Delhi, Dr. A.K. Maini, the Director, LASTEC, DRDO, New Delhi, Prof. R.P. Tandon, The Head of Dept. of Phys., Univ. of Delhi for their encouragement.

I am thankful to Prof. P.G. Burke, FRS and Prof. Alan Hibbert of Q. Univ. Belfast who introduced me the subject of multi-electrons atoms and ions. I am thankful to Prof. T.W. Hansch (Nobel Laureate) and Late Prof. H. Walther for giving me the opportunity to work in Max Planck Institute, Garching. I am also thankful to Prof. N.P. Bigelow the Head of Quantum Opt. and BEC Div., Univ. of Rochester, USA for the constant encouragement.

I am also thankful to Dr. A.K. Razdan (LASTEC), Dr. P.K. Jha, Dr. A.B. Bhattacherjee, Dr. A.K. Singh, Dr. Sanjay Tyagi, Dr. Vinay Singh, Dr. Nupur Verma, Prof. R. Ramnathan, Prof. R. Seth, Prof. R.K. Sharma, Dr. Rinku Sharma, Dr. N. Singh, Dr. Rachna Joshi, Dr. Kriti Batra, Dr. Alok Jha, Dr. Monica, Dr. Tarun Kumar, Jagjit Singh, Siddhartha Lahon, Priyanka Verma, Sunny Aggarwal, Sonam Mahajan and Mr. Manoj Malik for helping me in making this book. I am thankful to my wife Sneh Mohan, daughter Anu and son Anurag, daughter-in-law Skita of their constant patience and encouragement for finishing the project.

Convener and Organizing Committee   Man Mohan
Cchem MRSC (London)
Alexander Humboldt Fellow (Germany)
Senior Associate, International Centre for Theoretical Physics
(ICTP), Trieste (Italy)
Visiting Professor, France, Canada, Japan

Contents

Chapter 1
Recent Investigations of Radiative Lifetimes and Transition Probabilities in Heavy Elements $(37 \leq Z \leq 92)$

Émile Biémont

Abstract We present a review of the progress realized regarding the radiative properties (transition probabilities, lifetimes, branching fractions) of the heavy atoms and ions ($Z \geq 37$) of the fifth and sixth rows of the periodic table and of the lanthanides. In all cases, the discussion is limited to the first three ionization stages. The present review is motivated by the recent developments in astrophysics, laser physics or plasma physics. Further progress in the domain considered here is essentially hindered by the poor knowledge of the atomic spectra particularly concerning the energy levels.

1.1 Introduction

There is a growing interest in atomic data of heavy ions in different fields of physics including plasma diagnostics in fusion research, absorption spectroscopy in environmental studies, and investigations of the chemical composition of astrophysical objects. As part of an on-going effort to meet this demand, we report, in the present paper, on the progress realized during the past few years concerning the radiative parameters determination of three groups of heavy elements: the fifth and the sixth rows of the Mendeleev table and also the lanthanides ions (LI).

In order to clarify the relative importance of the **r**- and **s**-processes for the production of heavy elements in the Galaxy, the astrophysicists need accurate atomic data (transition probabilities, oscillator strengths, radiative lifetimes, branching fractions (BF), hyperfine structure constants, ...) particularly for the heavy elements belonging to the sixth row ($55 \leq Z \leq 86$) of the Mendeleev table. In astrophysics, radiative data are strongly needed for transitions originating from the ground level or from low-excitation levels which are predominantly populated in cold stars. There is also an increasing demand for weaker high-excitation lines currently identified on the high-resolution astrophysical spectra now available. In addition, opacity calculations require data for a huge number of lines emitted in all the spectral ranges.

É. Biémont (✉)
IPNAS (Bât. B15), Université de Liège, Sart Tilman, 4000 Liège, Belgium
e-mail: e.biemont@ulg.ac.be

M. Mohan (ed.), *New Trends in Atomic and Molecular Physics*,
Springer Series on Atomic, Optical, and Plasma Physics 76,
DOI 10.1007/978-3-642-38167-6_1, © Springer-Verlag Berlin Heidelberg 2013

An element like tungsten, which is important for thermonuclear fusion research because it is used as a plasma facing material in Tokamak devices, belongs also to this group. Due to its high melting point and thermal conductivity, and its low tritium retention and erosion rate under plasma loading, tungsten is indeed a very attractive element for Tokamaks. The International Thermonuclear Experimental Reactor (ITER), which will be the next step toward the realization of fusion, will use tungsten, together with beryllium and carbon-fiber reinforced composites, as plasma facing materials. Atomic data for tungsten ions are therefore urgently needed.

The heavy refractory elements of the fifth row, frequently difficult to produce in the laboratory, require also further investigations, many gaps subsisting concerning the available atomic data. As an example, ruthenium plays an important role in stellar nucleosynthesis. When the stars are in the 'asymptotic giant branch (AGB)' phase, they experience thermal pulses, generating a rich nucleosynthesis by the s-process. The convective envelope of the star may penetrate the region where s-process elements have been produced, and may bring them to the stellar surface, where they become observable. A few s-process elements are particularly interesting with that respect because they provide information on the time-scales involved in the process and ruthenium is among these elements.

The investigation of the atomic structure of the lanthanide ions is often prevented by the complexity of the configurations concerned involving an open 4f-shell, by the high density of low-energy levels and, consequently, by the huge number of transitions generally appearing in the visible spectral range. Their knowledge however is vital and strongly needed in astrophysics in relation with the problematics of stellar abundance determination, cosmochronology and nucleosynthesis, these elements appearing generally overabundant in chemically peculiar (CP) stars. Outside astrophysics, the lighting-research community is interested in LI, the rare-earths salts being used in many commercial metal-halide high-intensity discharge lamps.

In astrophysics, strong magnetic fields have been detected in hot stars (e.g. O, A and B types). Detailed investigations of these fields require the knowledge of accurate Landé g-factors. Many of these g-factors are unknown or inadequately known, particularly for the heavy elements or ions of the periodic table. Available data for many levels (even of low excitation energy) are still lacking or, when they exist, their accuracy frequently suffers from the limitations of the "old" laboratory analyses.

Due to their complexity, but also to their low cosmic abundances, most of the heavy elements have generally attracted less interest of the spectroscopists during the last decades but the analysis of a large number of high-resolution astrophysical spectra, obtained in recent years from the ground or from space (e.g. with the Hubble Space Telescope), has modified somewhat the situation and has stimulated further theoretical and laboratory investigations of these ions.

For these reasons, we present here an update of the results obtained concerning the radiative properties (oscillator strengths, A-values, lifetimes, BFs, Landé factors, …) of the three groups of elements just mentioned (more precisely, the first three ionization stages only have been considered). The present paper is an update of recent reviews on similar topics [5–7].

1.2 Experimental Techniques

1.2.1 Measurement of Lifetimes for Metastable States Using Heavy Ion Storage Rings

The optical observation, at a heavy-ion storage ring, of the light emitted from metastable levels of singly ionized elements allows atomic lifetime measurements in the millisecond up to the second range. We briefly consider here some lifetime measurements performed at the Stockholm heavy-ion storage ring (the CRYRING), a sophisticated experimental device which was operated by the Manne Siegbahn Laboratory (MSL) up to 2009 and which is updated by the DESIREE (Double ElectroStatic Ion Ring ExpEriment) device. This work has been done in collaboration with Prof. S. Mannervik and his team at Stockholm University. The measurements made at CRYRING have been extensively discussed in a review paper [63] and the details will not be repeated here. This technique was interesting because it did allow, with an accuracy reaching typically 1 % or so, the measurement of long lifetimes (up to several tens of seconds) of low-lying metastable states frequently important in astrophysics. In addition, these long lifetimes, very sensitive to small configuration interaction (CI) effects, provided the opportunity to test the theoretical models used in atomic structure calculations for heavy elements or ions.

Among the difficulties encountered when using this technique (for a discussion, see e.g. Refs. [54, 62]), let us mention, besides the technical difficulties associated with a very sophisticated experimental device, that only a limited number of levels were accessible through time-consuming measurements and also the fact that some corrections, like those associated with repopulation effects, needed to be carefully taken into account. During the measurements, there was a possibility of laser probing of the populations [64]. In this approach, an allowed transition was induced from the metastable level to an upper level. This upper level, whose lifetime reached generally a few ns, was decaying back to a metastable or another low-lying level, the intensity of the fluorescence decay being proportional to the population of the metastable state.

In the framework of a well-established collaboration between Stockholm, Liège and Mons universities, the CRYRING measurements have been applied with success to Sr II [12], Xe II [93] (fifth group), Ba II [43] (sixth group), La II [29] and Nd II [20] (LI). When the comparison was possible, an excellent agreement theory-experiment was observed.

The investigation of the radiative decay of the metastable level 5d $^4D_{7/2}$ in Xe$^+$ has appeared particularly interesting [93]. Lifetime measurements of this level in a storage ring are difficult since magnetic mixing of the metastable with a short-lived level quenches its population. Theoretically, we found that the decay is heavily dominated by an M2 transition and not by M1/E2 transitions (as usually expected). Decay rates were determined at different magnetic field strengths B in order to allow a nonlinear extrapolation to $B = 0$ and the experimental measurement ($\tau = 2.4 \pm 0.8$ s) was found in agreement with the calculated MCDF value (2.32 s), but much smaller than previous estimations.

A similar effect was found in Kr II [13] where the radiative lifetime of the metastable $4s^2 4p^4 (^3P) 4d\ ^4D_{7/2}$ level shows also an unusual situation regarding the importance of an M2 depopulation channel. While the first order M1 and E2 channels were expected to contribute in a dominant way to the decay, the experimental result ($\tau = 0.57 \pm 0.03$ s) was far too short to be due to these channels and, only if second order contributions to the decay branches (including essentially the M2 contribution), were taken into account in the calculations, could the unexpected short lifetime be explained.

For the discussion, let us mention that in the lighter homologous Ar^+ (not belonging to the groups of elements considered in the present paper), a laser probing investigation of the $3s^2 3p^4 (^1D) 3d\ ^2G_{7/2,9/2}$ metastable doublet states [58, 92] has allowed us to establish the unexpected and extraordinary strong contribution of an electric octupole (E3) transition to the ground state, in addition to the M1 decay channels to the 3d $^{2,4}F$ states and the E2 contributions to the 4s ^2P, ^2D states. It is interesting to mention that a similar effect (unexpected importance of an E3 contribution) was neither found in Kr II nor in Xe II.

1.2.2 Allowed Transitions Investigated with Laser-Induced Fluorescence Spectroscopy

A second technique, widely used for lifetime measurements in heavy ions, is the time-resolved laser-induced fluorescence (TR-LIF) spectroscopy. For the description of the method, see e.g. the numerous references quoted in [5–7]. The lifetime of an excited atomic state i, τ_i, can be measured by observing the fluorescence decay from this level using the following formula:

$$I(t) = I(0)e^{-\dfrac{t}{\tau_i}} \tag{1.1}$$

where $I(0)$ is the initial intensity and $I(t)$ the intensity at time t.

A very direct way of investigating atomic lifetimes is to use short-pulse laser excitation and subsequent fast time-resolved detection of the fluorescence light emitted. This technique has been described in many papers (see e.g. [5, 7]) and, consequently, only the essential characteristics will be recalled here. Slight modifications and adaptation of the basic technique were needed for experiments according to the ions considered. The experimental results discussed in the present paper have been obtained at the Lund Laser Centre (LLC) in Sweden (in the framework of collaborations with Prof. S. Svanberg and his group) and at the Jilin University at Changchun (China) (in coll. with Prof. Z.K. Jiang and Prof. Z.W. Dai and their collaborators).

A laser-produced plasma is used as an ion source. This plasma is obtained by focusing, on a solid target placed in a vacuum chamber, the green beam of a Continuum Surelite Nd:YAG laser, which has a pulse duration of 10 ns and an energy of about 5–10 mJ. A small plume of plasma is produced after each pulse. The expanding plasma is crossed by an excitation beam at about 1–2 cm above the target.

The delay between the plasma production and the excitation pulse can be varied by externally triggering the lasers used in the experiment with a digital delay generator. The plasma density and temperature in the observed region can be adjusted by changing the plasma production laser pulse energy, the size of the focus point, the distance above the target surface, and the delay times between the ablation and excitation pulses. Three types of laser excitation have been considered in our work, i.e. one-step excitation, two-step excitation and two-photon excitation. A detailed description can be found, for example, in the papers quoted in Refs. [5–7] and in the Tables 1.1 and 1.3.

In relation with the main characteristics of the TR-LIF technique, the following points must be underlined:

- Many levels are accessible in a selective way by using different excitation schemes and different dye lasers.
- The method has appeared efficient and useful for investigating many neutral, singly ionized and doubly ionized heavy elements. We were able to produce also trebly ionized atoms but only in the case of cerium.
- The lifetimes accessible, with an accuracy of typically a few percent, are ranging roughly from 1 ns to several hundreds of nanoseconds.

In order to get the relevant transition probabilities or oscillator strengths, the experimental lifetimes, however, must be combined with BF measurements (using e.g. the FTS method) or calculations (adopting e.g. the HFR or MCDF approaches: see Sect. 1.3).

The TR-LIF method, combined with BF determination, has been extensively used in many atoms and ions of the fifth and sixth periods and of the lanthanide group. The results are discussed in the following sections.

1.2.3 Study of Intermediate Charges with the Beam-Foil Spectroscopy

The beam-foil spectroscopy (BFS) was one of the rare methods allowing the investigation of the atomic structure of moderately charged ions. The experimental device of Liège University was used for that purpose. The experimental details can be found in the relevant publications (see e.g. [21]) and, consequently, we just recall here the main characteristics of this method.

A beam of ions was produced by a Van de Graaff accelerator equipped with a conventional radio-frequency source. The beam was analyzed by a magnet and focused inside a target chamber. Beams with energies up to 2 MeV could be produced. Inside the chamber, the beam was excited and ionized by passing through a very thin (about 20 μg/cm^2) carbon foil. Just after the foil, the light, emitted by the excited ions, was observed at right angle by a Seya-Namioka-type spectrometer equipped with an $R = 1$ m concave 1 200 l/inch grating blazed for normal incidence at 110.0 nm. The entrance slit of the spectrometer had a width of 120 μm and was situated at 15 mm from the axis of the 10 mm diameter ion beam.

The light was detected by a thin, back-illuminated, liquid nitrogen cooled CCD detector specially developed for far UV measurements. The CCD worked under vacuum and was cooled down by liquid nitrogen to -90 °C for noise reduction. The CCD, which replaces the exit slit of the spectrometer, was tilted to an angle of 125° relatively to the spectrometer exit arm axis in order to be tangential to the Rowland circle. Under that geometry, it has a dispersion of 0.02 nm/pixel and detects light over a 20 nm wide region with a fairly constant resolution giving a line width (FWHM) of about 0.12 nm. The whole system was working under vacuum (10^{-5} Torr). The CCD images were transferred to a networked computer and analyzed by a specially written software.

A problem affecting the beam-foil spectra is the rather low spectral resolution just mentioned. At that resolution, we find many lines blended. In addition, beam-foil excitation is non-selective; this results in the production of a range of ion charge states (the spectra of neighboring ions often overlap each other) and in the excitation of a plethora of individually weak spectral lines that arise from the multitude of excited levels. The use of a CCD camera represents a major improvement over spectral scanning techniques and assures both better data accumulation (counting statistics) and constancy of conditions across a selected spectral range.

We have investigated with the BFS the spectrum emitted by the trebly ionized lanthanum (La^{3+}) [21]. This work has emphasized the difficulties inherent to the investigation of the fourth spectrum of the rare-earths which could eventually be observed in some hot stars. Some xenon ions (Xe V, VI, VII and VIII) have also been studied using the BFS technique combined with relevant calculations [15, 17, 19, 39, 40].

1.3 Theoretical Approaches

1.3.1 General Characteristics

In the lanthanides, strong configuration interactions are expected to occur between the low-lying configurations for the first ionization stages along the isoelectronic sequences. A second characteristic of the LI is the sudden collapse of the 4f orbital occurring at lanthanum. This collapse is enhanced when the number of 4f electrons is increasing. These considerations and the fact that a huge number of levels arise from the $4f^N$ configurations make the analysis of the lanthanide spectra extremely difficult and time consuming.

The radiative properties of the LI have been rather little investigated in the past in relation with these complex electronic structures (unfilled 4f shell) and the fact that the laboratory analyses are still extremely fragmentary or even missing for many ions. Low motivation resulted also from the rather small cosmic abundances of these elements in astrophysics compared, e.g., to the high abundances of the iron-group elements.

Theoretical investigations of the elements or ions of the fifth and six rows of the periodic table are also extremely demanding having in mind the strong constraint resulting from the simultaneous consideration, in the calculations, of both configuration interaction and relativistic effects.

1.3.2 The Relativistic Hartree-Fock Method

For heavy systems, accurate calculations of atomic structures require that both intravalence and core-valence correlation are considered in a detailed way. Simultaneous treatment of both types of effects, within a CI scheme, is very complex and reliable only if extensive configuration mixing is considered. However, in practice, the number of interacting configurations that can be introduced in Cowan's codes [27] is severely limited by the computer capabilities. In particular, the inclusion of core-polarization (CPOL) effects through the consideration of a huge number of configurations with open inner shells is essentially prevented by computational reasons. As the CPOL effects are expected to be important in lowly charged heavy atoms, an approach in which most of the intravalence correlation is represented within a CI scheme while core-valence correlation for systems with more than one valence electron is described by a CPOL model potential with a core penetration corrective term can be considered as an alternative. This approach was suggested by Migdalek and Baylis [66] and adopted in our calculations.

Cowan's suite of computer codes [27], based on the HFR approximation, was modified accordingly (see e.g. [5, 81]). This approach, although based on the Schrödinger equation, does include the most important relativistic effects such as the mass-velocity corrections and the Darwin contribution.

For an atom with n valence electrons, the one-particle operator of the potential can be expressed as:

$$V_{P1} = -\frac{1}{2}\alpha_d \sum_{i=1}^{n} \frac{r_i^2}{(r_i^2 + r_c^2)^3} \tag{1.2}$$

where α_d is the static dipole polarizability of the ionic core for which numerical values are available in the literature [38] and r_c a cut-off radius corresponding to the HFR expectation value of $\langle r \rangle$ for the outermost core orbital of a given ion. Additional corrections, which include the core-penetration effects [44, 45], were considered as described in Ref. [5].

The HFR + CPOL method has appeared successful and efficient for describing the atomic structure of many heavy atoms and ions. In particular, the theoretical lifetime values agree, in most cases, within a few (generally <20) percent with the experimental results as they were measured with the TR-LIF method. Larger discrepancies were observed however for a number of weak transitions (but in this case, the experimental results are also generally less accurate!). Some care must be exercised when the transitions are affected by substantial cancellation effects in the calculation of the line strengths [27]. As the experiment allows generally to obtain

BFs for a rather limited number of transitions, the above theoretical methodology has allowed to extend the BF determination to many transitions of astrophysical interest.

1.3.3 The Multiconfigurational Dirac-Fock Method

In some specific ions (for more details, see the publication lists given in the Tables 1.1 and 1.3), a fully relativistic technique i.e. the multiconfigurational Dirac-Fock (MCDF) method was used [42, 78]. This approach, which includes relativity in a more detailed way because it is based on a fully relativistic scheme, should a priori be more accurate than the HFR + CPOL approximation for heavy ions but this approach, however, was generally found less flexible. In addition, it consumes more computer time and its use is frequently hindered by difficult convergence problems. This is why, for most ions, the HFR + CPOL method was generally prioritized.

The list of specific ions, which were considered in the framework of the MCDF approach, includes Tc II [75], Sb I [47], Bi II [72], Pb II and Bi III [83]. Some lowly charged tungsten ions (i.e. W IV, V, VI) were also investigated with the MCDF approach.

1.4 The Results

1.4.1 The Fifth Row of the Periodic Table

The fifth row of the periodic table includes the elements Rb ($Z = 37$) up to Xe ($Z = 54$). We report, in Table 1.1, a summary of the results obtained so far along this group. 206 lifetimes have been measured by TR-LIF spectroscopy at the LLC (Sweden) or at the University of Jilin (China). For some ions (see the starred m-values in Table 1.1), only theoretical results (HFR + CPOL) have been obtained. The configurations involved are given in column 4 and the number of transitions (N), for which transition probabilities have been determined, appear in column 5. In most cases, only the strongest transitions, depopulating the levels for which the lifetimes have been measured, are given in the different publications where the details about the experiments and about the calculations can also be found (see the last column of Table 1.1).

Some short comments about the different ions are relevant here.

In Y II, the overall quality of the HFR + CPOL calculations has been assessed by comparisons with new and previous lifetime measurements. New measurements are concerning 5 levels of the lifetime 4d5p and 5s5p configurations in the energy range 32 048–44 569 cm^{-1}. A similar theoretical model applied to Y III leaded to results in good agreement with new laser measurements for two 5p levels [23].

Table 1.1 Fifth-row elements: number of measured lifetimes (m), configurations involved, number of depopulating transitions (N) and the corresponding references

Z	Ion	m	Conf.	N	Ref.
39	Y II	5	4d5p, 5s5p	84	[23]
	Y III	2	5p	182	[23]
40	Zr I	17	$4d^25s5p$, $4d^35p$, $4d5s^25p$	78	[61]
	Zr II	16	$4d^25p$	242	[59]
41	Nb II	17	$4d^35p$	107	[71]
	Nb III	20^*	$4d^25p$	76	[71]
42	Mo II	16	$4d^45p$	110	[57]
43	Tc II	6^*	$4d^55p$	20	[75]
44	Ru I	10	$4d^75p$, $4d^65s5p$, $4d^76p$	163	[37]
	Ru II	23	$4d^65p$	178	[77]
	Ru III	6^*	$4d^55p$	25	[77]
45	Rh II	17	$4d^75p$	113	[87]
46	Pd I	6	$4d^95p$	20	[100]
47	Ag II	–	$4d^95d$	7	[14]
		12^*	$4d^85s^2$, $4d^96s$, $4d^95d$	65	[25]
48	Cd I	11	5s5p, 5snd ($n = 6$–9), 5sns ($n = 7, 8$)	102	[99]
	Cd II	5	$4d^{10}5p$, $4d^{10}6s$, $4d^{10}5d$	10	[99]
50	Sn I	40	5pnp ($n = 10$–13, 15–19, 27, 31, 32)		[107]
			5pnf ($n = 4, 5, 9$–19, 22, 23), 5p8p		[107]
	Sn I	9	5p7p		[106]
51	Sb I	12	$5p^26s$, $5p^25d$	55	[47]
Total		$206 + 44^*$		1637	

In Zr I, for three levels, we confirm previous measurements while, for 14 other levels, the lifetimes have been measured for the first time [61]. In Zr II, for 12 levels, there were no previous results available [59].

We used the TR-LIF technique to measure 17 radiative lifetimes in Nb II [71] while BFs were measured from spectra recorded using FTS. In addition, transition probabilities (76 transitions in the range 143.0–314.0 nm) were calculated, for the first time, in Nb III using a HFR + CPOL method. The derived solar photospheric niobium abundance, $A_{\mathrm{Nb}} = 1.44 \pm 0.06$ (logarithmic scale), was found in agreement with the meteoritic value. The stellar Nb/Eu abundance ratio determined using synthetic spectra, including hyperfine broadening of the lines, for five metal-poor stars confirms that the **r**-process is a dominant production method for the n-capture elements in these stars.

The lifetimes measured by Hannaford and Lowe [46] in Mo II concern levels with energies in the range 45 800–50 700 cm^{-1}. Lifetimes for higher energy levels are missing although transitions from these levels are likely to be observed in astrophysics or in plasma physics. Consequently, we have reported lifetime measurements for levels with higher energies, i.e. in the range $48\,000$–$61\,000$ cm^{-1} [57].

In astrophysics, technetium is an **s**-process element with no stable isotope. Consequently, no experimental transition probabilities or lifetimes are available for this atom. Using three independent theoretical approaches (CA, HFR + CPOL, AUTOSTRUCTURE), oscillator strengths have been calculated for a set of Tc II transitions of astrophysical interest and the reliability of their absolute scale has been assessed [75]. The examination of the spectra emitted by some Ap stars has allowed the identification of Tc II transitions in HD 125 248. This Tc II detection should however await confirmation from spectral synthesis relying on dedicated model atmospheres.

In 1984, the solar photospheric abundance of ruthenium was determined [9] from 9 lines of Ru I using a one-dimensional (1D) model [48]. The result, $A_{Ru} = 1.84 \pm 0.10$, differed from the meteoritic result ($A_{Ru} = 1.76 \pm 0.03$) by 0.08 dex. The BFs were derived from arc measurements [26] affected by large systematic errors. New f values have been obtained for Ru I from LLC measurements (10 new lifetimes) [37] in the range 225.0–471.0 nm. A recent 3D model proposed by Asplund *et al.* [3] was used for the calculations. The new abundance value is now: $A_{Ru} = 1.72 \pm 0.12$ which is very close to the meteoritic result. The f values obtained for Ru II show that the lines of this ion are too weak to be observed in the photospheric spectrum [37, 77].

In singly ionized rhodium, no radiative data have been published so far in the literature despite the fact that several Rh II lines have been identified in different astrophysical spectra such as the solar spectrum [68] and the spectra of the HgMn type star χ Lupi [56], the super-rich mercury star HD 65 949, the HgMn star HD 175 640 and the peculiar Przybylski's star HD 101 065 [28]. For that reason, 17 radiative lifetimes of Rh II have been measured with the TR-LIF technique and combined with theoretical BFs (HFR + CPOL model) in order to deduce new oscillator strengths for a set of 113 Rh II transitions in the spectral range 153.0–417.6 nm [87].

Similar motivations, in particular the need of oscillator strengths for investigating the chemical composition of CP stars, for the modeling of stellar atmospheres and for the understanding of the buildup of the elements in nucleosynthesis, was a strong motivation to undertake new lifetime measurements and BF determination in Pd I [100], Sb I [47], Cd I and Cd II [99].

Radiative parameters have been obtained by laser-produced plasma and TR spectroscopy for transitions depopulating the levels belonging to the $4d^8 5s^2$, $4d^9 6s$ and $4d^9 5d$ configurations of Ag II [14, 25]. The light emitted by the plasma was analyzed by a grating monochromator coupled with a time-resolved optical multichannel analyzer system. Spectral response calibration of the experimental system was performed using a deuterium lamp in the wavelength range from 200 to 400 nm, and a standard tungsten lamp in the range from 350 to 600 nm. The transition probabilities were obtained from measured BFs and theoretical radiative lifetimes.

The interstellar gas-phase abundance of tin appears to be enriched with respect to that of the Sun [97]. In the stars, tin isotopes are produced by the p-, **s**- and **r**-processes and some transitions of Sn I have been identified in the photospheric solar

Table 1.2 Investigated xenon ions

Z	Ion	Conf.	Ref.
54	Xe II	5d	[29]
	Xe V	$5s5p^3$, $5s^25p5d$, $5s^25p6s$, $5s^25p^2$, $5s^25p6p$, $5s^25p4f$	[17]
	Xe VI	$5s^2nl$ (np, nf, nh, nk; $n \leq 8$), $5p^3$, $5s^2nl$ (ns, nd, ng, ni; $n \leq 8$), $5s5p^2$	[15]
	Xe VII	$5s5p$, $5p^2$, $5s5d$, $5s6s$, $5p5d$, $4f5p$, $5p5d$, $5s5f$	[19]
	Xe VIII	$5p$, $5d$	[19]
	Xe IX	$4d^96p$, $4d^94f$, $4d^95f$	[39, 40]

spectrum and in the spectra of some CP hot stars [1]. In recent years, with the increasing demand for an experimental realization of quantum logic devices, potential interest for new or improved lasers and other optical devices, the interest in accurate information on high Rydberg states of Sn I has also rapidly increased [55, 91]. For these reasons, using the TR-LIF technique in a Sn atomic beam, 40 natural radiative lifetimes have been measured for even-parity $5pnp$ ($J = 1, 2$) and $5pnf$ levels along the Rydberg series and for all the 5p8p perturbing states with energies in the range $52\,263.8$–$59\,099.9$ cm^{-1} [107]. A two-step laser excitation scheme was used in the experiment. Through an analysis of the energy levels structure by the multichannel quantum defect theory (MQDT), the channel admixture coefficients have been obtained and used to fit the theoretical lifetimes to the experimental ones in order to predict new values for the levels not measured. A generally good overall agreement between experimental and theoretical MQDT and HFR lifetimes has been achieved except for a few levels. Landé g-factors have been measured by the same method and also by the Zeeman quantum-beat technique for most of these even-parity levels [102]. Lifetimes for 9 levels and Landé g_J factors for eight levels of the 5p7p configuration have also been obtained by the same techniques. The results obtained were compared with HFR + CPOL results [106].

We summarize in Table 1.2 the results obtained so far for some xenon ions (Xe II, Xe V, Xe VI, Xe VII, Xe VIII and Xe IV).

1.4.2 The Sixth Row of the Periodic Table

The results obtained for the sixth row of the periodic table, i.e. for the elements Cs ($Z = 55$) up to Rn ($Z = 86$), are presented in Table 1.3. The presentation is similar to that of Table 1.1. Up to now, 167 lifetimes have been measured by TR-LIF spectroscopy for these atoms or ions. For the starred values in Table 1.3, only theoretical results (HFR + CPOL) have been obtained. In most cases, only the strongest transitions depopulating the levels for which the lifetimes have been measured are given in the different publications. The following comments are relevant here.

Hafnium ($Z = 72$) has been observed in the spectra of some stars like the barium star HD 202 109 [104], Sirius A [103] and the photosphere of δ Scuti [104]. Hafnium

Table 1.3 Sixth-row elements: number of measured lifetimes (m), configurations involved, number of the depopulating transitions (N) and the corresponding references

Z	Ion	m	Conf.	N	Ref.
72	Hf I	2	$5d6s^2 6p$	–	[60]
	Hf III	9	$5d6p$	55	[60]
73	Ta I	14	$5d^3 6s6p$, $5d^4 6p$	23	[32]
	Ta II	3	$5d^3 6p$	100	[85]
	Ta III	6	$5d^2 6p$	206	[36]
74	W I	141^*		143	[76]
	W II	9	$5d^3 6s6p$, $5d^4 6p$	6265	[70]
	W III	2	$5d^3 6p$	4826	[88]
75	Re I	11	$5d^4 6s^2 6p$, $5d^5 6s6p$	81	[74]
	Re II	7	$(5d + 6s)^5 6p$	45	[73]
76	Os I	12	$5d^6 6s6p$, $5d^7 6p$	129	[82]
	Os II	9	$5d^5 6s6p$, $5d^6 6p$	137	[82]
77	Ir I	9	$5d^7 6s6p$	206	[101]
	Ir II	4	$5d^7 6p$	223	[101]
78	Pt II	8	$5d^8 6p$	164	[84]
79	Au I	3	$5d^9 6s6p$	6	[33]
	Au II	1	$5d^9 6p$	63	[33]
	Au II	22^*	$5d^9 7s$, $5d^9 6d$	114	[18]
	Au III	60^*	$5d^8 6p$, $5d^7 6s6p$	146	[30]
80	Hg I	10	$6sns$ ($n = 7$–10), $6snd$ ($n = 6$–11)	–	[24]
81	Tl I	15	$6s^2 ns$ ($n = 7$–14), $6s^2 nd$ ($n = 6$–12)	48	[16]
82	Pb I	29	$6pns$ ($n = 7$–13), $6pnd$ ($n = 6$–13)	–	[53]
	Pb I	3	$6p7s$	136	[11]
	Pb II	1	$7s$	2	[83]
83	Bi II	42^*	$6pnp$ ($n = 7$–8), $6pnf$ ($n = 5, 6$), $6s6p^3$, $6pns$ ($n = 7$), $6pnd$ ($n = 6$)	43	[72]
	Bi III	–	$7s$	2	[83]
Total		$167 + 265^*$		13163	

is also a possible chronometer for stellar and galactic evolution [79]. In astrophysics, radiative data are needed for transitions originating from the ground term or from low excitation levels which are mostly populated in cold stars. This has motivated an investigation of hafnium atoms and ions [60]. Radiative lifetimes of nine odd levels in Hf III and of two odd levels in Hf I have been measured for the first time by laser spectroscopy and transition probabilities have been deduced for 55 transitions of Hf III.

During the past few years, there have been several analyses of the atomic structure of neutral tantalum (Ta I), an element with "moderate" complexity, the ground configuration being $5d^3 6s^2$, but little has been done concerning the transition probability determination while there is an obvious need in astrophysics [52, 95], the photospheric abundance of tantalum being still unknown [2]. Radiative lifetimes of 14 odd-parity Ta I levels, in the energy range $30\,664$–$45\,256$ cm^{-1}, have been measured [32] and new transition probabilities deduced for a set of strong lines depopulating the levels investigated experimentally.

Singly ionized tantalum, Ta II, has been looked for but not detected in the χ Lupi CP star [31]. The richness and complexity of the Ta II spectrum impose severe limitations in the determination of absolute transition probabilities which is affected by the fragmentary knowledge of the atomic structure of this ion even for low-lying configurations. 3 new lifetimes have been obtained experimentally in Ta II and compared with the calculated results obtained for the low-lying levels ($E <$ $44\,000$ cm^{-1}) [85].

Ta III suffers from the lack of accurate theoretical and experimental f-values or lifetimes, the only set of transition probabilities available in this ion being that reported by Azarov *et al.* [4]. We have realized the first lifetime measurements for six odd-parity levels belonging to the $5d^2 6p$ configuration. Weighted transition probabilities (gA values) have also been obtained for the strong lines depopulating the investigated levels [36].

Radiative data of tungsten ions are important in plasma physics. With its low yield and high threshold for sputtering [69], tungsten is widely used in fusion reactors as a divertor target. Tungsten is also important in astrophysics, neutral tungsten being identified in Ap stars (see e.g. [51]). More recently, W I lines have been detected and investigated in the spectrum of HD 221 170 (see e.g. [105]). The dominant species, except in cool stars, is in fact expected to be W II, the line of singly ionized tungsten at 203 nm having been observed in the UV spectrum of one Am star [90]. Some lifetime measurements are available in W II but radiative data are still lacking for many transitions. Till now, doubly ionized tungsten (W III) has not been identified in hot stars. In W III, the only work available is that of Schultz-Johanning *et al.* [94]. In our work, an extensive set of oscillator strengths has been calculated for W II [70] and W III [88] which represents a considerable extension of the available data, the accuracy of the new results being assessed through comparisons with the TR-LIF measurements performed for 9 levels of W II and 2 levels of W III. The uncertainty in the new oscillator strengths, not affected by cancellation effects, should not exceed 15 %. Transition probabilities for allowed and forbidden lines in W I, II and III have been discussed in two recent papers [76, 86]. For the E1 transitions, recommended values are proposed from a critical evaluation of the data available in the literature. For the M1 and E2 transitions, for which no data had been published, a new set of radiative rates has been obtained using a HFR + CPOL approach. The tables summarizing the compiled data are expected to be useful for plasma modeling in fusion reactors. An extension of these calculations to higher ionization stages (W IV, W V, W VI, W VII and W VIII) is in progress.

An investigation of Re (I and II) [73, 74] and Os (I and II) [82] ions was also carried out. For ten of the 11 odd-parity levels measured in Re I, there were no previous results available. Reliable semi-empirical transition probabilities have been deduced for 81 Re I and 45 Re II lines. For 9 levels of Os I and 4 levels of Os II, the lifetimes were measured for the first time. It was possible to deduce oscillator strengths for 129 transitions of Os I and 137 transitions of Os II appearing in the wavelength range 180.0–870.0 nm. These results have allowed us to revise the abundance of osmium in the solar photosphere ($A_{Os} = 1.25 \pm 0.11$). The newly derived oscillator strengths have been applied as well to derive the osmium abundance in the carbon-rich metal-poor star HD 187 861 [82].

Metal-poor halo stars with greatly enhanced abundances of the heavy neutron-capture elements have been identified and have stimulated much interest in recent years [96]. It has been emphasized recently [49, 50] that accurate abundance values of iridium in stars is of great significance not only in radioactive cosmochronology but also for putting constraints on the structure and nucleosynthetic evolution of supernovae originating from the first stellar generation. Our work in Ir I and Ir II [101] ions has led to the results summarized in Table 1.3.

Radiative lifetimes of eight odd-parity states of Pt II, in the energy range extending from 51 408 to 64 388 cm^{-1}, have been measured by means of the TR-LIF technique [84]. Free, singly ionized platinum ions were obtained in a laser-produced plasma and a tunable laser with 1.5 ns duration pulse was used to selectively excite the Pt^{+} ions. The comparison of the experimental results with HFR calculations emphasizes the importance of valence-valence correlation and of CPOL effects in this complex ion. A new set (164 transitions) of calculated transition probabilities has been reported.

In plasma physics, gold is used as an active medium in metal vapor lasers. Numerous laser transitions, in the spectral range 253–763 nm, have been observed when exciting a helium discharge in a gold-plated hollow cathode by Reid *et al.* [89]. Consequently, the determination of accurate radiative parameters of Au II excited states, including transition probabilities, is of great interest. In astrophysics, neutral and singly ionized gold (Au I, Au II) have been identified in Ap and Bp stars but, up to now, doubly ionized gold (Au^{++}) has been much less investigated than Au^{+} in both laboratory and stellar spectra. The only work concerning the observation of this ion in the CP HgMn-type stars κ Cancri and χ Lupi is due to Wahlgren *et al.* [98], and is related to the analysis of the spectra obtained with the Goddard High Resolution Spectrograph onboard the HST. Our contributions regarding gold ions concern the three species [18, 30, 33] and the results are summarized in Table 1.3.

The investigations of the heavy elements with $Z \geq 80$ include contributions to the determination of transition probabilities and lifetimes in Hg I [24], Tl I [16], Pb I and Pb II [11, 53, 83], Bi II [72] and Bi III [83] (see Table 1.3). More details can be found in the relevant publications.

1.4.3 The Lanthanide Ions

Extensive discussions about lifetime and transition probability determination in LI have been presented in our previous compilations on the subject [5, 7]. In particular, the difficulties associated with both the theoretical and the experimental investigations of the elements or ions of this group have been emphasized. We limit the present discussion to the results obtained since 2005.

Presently, a firm conclusion about the presence of radioactive elements, particularly Pm, in CP stars remains an open issue that can have important implications for our understanding of the composition of these stellar objects. In a recent paper, Goriely [41] has stressed the importance of spallation nucleosynthesis compared to diffusion processes as a possible explanation of the peculiar abundances spectroscopically determined at the surface of HD 101 065. Although it remains difficult to disentangle the effect of both processes theoretically, this conclusion does not necessarily reduce the role of the diffusion processes which have proven to be of first importance to understand the atmosphere of CP stars. In this context, the first theoretical transition probabilities have been obtained for a set of 46 Pm II transitions of astrophysical interest [34]. These data fill in a gap in astrophysics and should allow to establish, on a firmer basis, the presence of some lines of this radioactive element in the spectra of CP stars and, consequently, a quantitative investigation of the stellar Pm abundance. A search for Pm II lines in Przybylski's star (HD 101 065) and in HR 465 has been reported and discussed, supporting the detection of this ion.

The spectra of moderately ionized lanthanides are of considerable interest in several fields of physics including quantum information, lighting industry, laser materials, and stellar physics. Our knowledge of these spectra is still very fragmentary. Recently, there has been a revival in the interest in trebly ionized lanthanides, and several spectral analyses have been completed on the basis of VUV high resolution observations. For these reasons, HFR and MCDF calculations of atomic structure and transition rates have been carried out in trebly ionized lanthanum (La^{3+}, $Z = 57$) [21]. The calculations had to cope with CI effects but also with the very complex situation of the collapse of the 4f wave function. The results have been compared to experimental data obtained with BFS in the extreme ultraviolet, at ion energies that favor the production of the La IV spectrum. Besides transitions known from sliding spark discharges, many more lines were observed that have not yet been identified. TR measurements yielded 3 level lifetimes in La IV that agree roughly with the results of our own calculations. This work opens the way to additional investigations of the trebly ionized elements of the lanthanide group.

1.4.4 The DESIRE and DREAM Databases

About 700 lifetimes have been measured by time-resolved laser-induced fluorescence (TR-LIF) spectroscopy for the elements Rb to Xe, Cs to Rn and for the lanthanides and, in many cases, the corresponding BFs have been calculated using a HFR + CPOL approach [5, 7]. This combination of lifetime measurements with theoretical and, when possible, experimental e.g. obtained with the Fourier Transform Spectroscopy (FTS), BF determination has led to transition probabilities for about 64 000 (lanthanides) and about 15 000 transitions (fifth and sixth rows of the periodic table). All these results are stored in two databases i.e. the DREAM database: **D**atabase on **R**are **E**arths **A**t **M**ons University [10] and the **DE-SIRE database**: **D**atabas**E** for the **SI**xth **R**ow **E**lements [8, 35], respectively.

The main aim of these databases is to provide the scientific community with updated spectroscopic information and radiative parameters. The database contains information about the wavelengths, oscillator strengths, transition probabilities and radiative lifetimes of neutral, singly or doubly ionized elements.

In the database, for each line in a specific spectrum, the tables show, respectively:

- The wavelength (in Å) deduced from the experimental energy levels. These wavelengths are given in air above 2 000 Å and in vacuum below that limit.
- The lower level of the transition represented by its experimental value (in cm^{-1}), its parity [(e) for even and (o) for odd] and its J-value. Level energies are taken from the NIST compilations (http://www.nist.gov/physlab/data/asd.cfm) and, when needed, from subsequent publications.
- The upper level of the transition presented in the same way as the lower level.
- The calculated oscillator strength, $\log gf$, where $g = 2J + 1$ is the statistical weight of the lower level of the transition. More details about the computational procedure are given in the different publications listed in the database.
- The calculated transition probability, gA, in s^{-1}, where $g = 2J + 1$ is the statistical weight of the upper level of the transition.
- The cancellation factor, CF, as defined by Cowan [17]. Small values of this factor (typically CF < 0.01) indicate transitions possibly affected by severe cancellation effects in the calculation of the line strengths.

The tables are accessible directly on the web sites http://www.umh.ac.be/~astro/dream.shtml and http://www.umh.ac.be/~astro/desire.shtml. In our compilation, the spectra are classified in order of increasing Z-values and, for a given Z, according to the ionization degree. More details about the computational procedure are also given in the different publications listed in the compilation. Only transitions for which $\log gf > -4.0$ are reported in the tables. For some ions, experimental oscillator strengths or normalized f values obtained using measured lifetimes are given. In these cases, EXPT or NORM appears in the last column of the tables.

1.4.5 The Landé Factors

In astrophysics, strong magnetic fields have been detected in hot stars of the types O, B and A. Definite spectropolarimetric detections have been reported e.g. for Ap, Be or β Cephei stars, the field strength reaching in some cases several tens of kG [65].

Detailed investigations of these magnetic fields require the knowledge of accurate Landé g-factors. Many of these g-factors are unknown or poorly known, particularly for the heavy elements of the periodic table. In the past, some experimental data have been published in the successive NIST compilations (see e.g. [67]) but data for many levels, even in the case of low excitation energy, are still lacking or, when they exist, their accuracy is frequently suffering from the limitations inherent to "old" laboratory analyses.

For these reasons, Landé g-factors have been calculated, in intermediate coupling (HFR + CPOL approach), for 2 084 levels belonging to atoms or ions of the sixth row of the periodic table [22]. The results have been refined using least-squares fittings of the Hamiltonian eigenvalues to the observed energy levels (when available). The new results fill in some gaps in the existing data for a large number of levels belonging to ions of astrophysical interest and are expected to be useful for investigating magnetic fields in CP stars. This work extends the results reported for doubly ionized lanthanides [80]. In that paper, Landé g-factors have been calculated for over 1 500 energy levels ($57 \leq Z \leq 71$) using a similar approach.

Experimental methods have also been applied and Landé g-factors have been measured by TR-LIF and Zeeman quantum-beat techniques for the even-parity levels of the $J = 1$ $5pnp$ ($n = 11$–13, 15–19) and $J = 2$ $5pnp$ ($n = 11$–13, 15–19, 31, 32), $5pnf$ ($n = 4$, 5, 9–19, 22, 23) Rydberg series and for all the $5p7p$ and $5p8p$ perturbing levels of neutral tin [102, 106]. A two-color two-step excitation scheme was used in the experiment. The experimental results have been compared with theoretical g-values obtained by the multichannel quantum defect theory and the HFR model, respectively. In most cases, the theoretical values agree well with the experimental results.

Many additional Landé factors have been also obtained for a number of ions not mentioned in the present section. These values can be found in the relevant papers quoted in the last column of Tables 1.1 to 1.3.

Acknowledgements   The experimental part of the present work was realized in the framework of many collaborations involving the Lund Laser Centre (Prof. S. Svanberg and his collaborators), Stockholm University (Prof. S. Mannervik and his group), Changchun (Profs. Z.W. Dai, Z.K. Jiang and their group), the Bulgarian Academy of Sciences (K. Blagoev and collaborators), the University Complutense de Madrid and also the Liège University (Prof. H.-P. Garnir and collaborators). The theoretical part was carried out in collaboration with P. Quinet, P. Palmeri and V. Fivet from Mons University in Belgium. Financial support from the Belgian FRS-FNRS and from the Institut Interuniversitaire des Sciences Nucléaires is acknowledged. The experimental work carried out in Lund was supported by the EU-TMR Access-to-Large-Scale-Facility programme in the framework of LASERLAB-EUROPE. E.B. is Research Director of the FRS-FNRS.

References

1. S. Adelman, W.P. Bidelman, D. Pyper, Astrophys. J. Suppl. **40**, 371 (1979)
2. M. Asplund, N. Grevesse, A.J. Sauval, Nucl. Phys. A **777**, 1 (2006)
3. M. Asplund, N. Grevesse, A.J. Sauval, P. Scott, Annu. Rev. Astron. Astrophys. **47**, 481 (2009)

4. V.I. Azarov, W.-U.L. Tchang-Brillet, J.-F. Wyart, F.G. Meijer, Phys. Scr. **67**, 190 (2003)
5. É. Biémont, Phys. Scr. T **119**, 55 (2005)
6. É. Biémont, in *7th International Conference on Atomic and Molecular Data and Their Applications—ICAMDATA 2010*, Vilnius, Lithuania, ed. by A. Bernotas, R. Karazija, Z. Rudzikas. AIP Conf. Proceed., vol. 1344 (2011)
7. É. Biémont, P. Quinet, Phys. Scr. T **105**, 38 (2003)
8. É. Biémont, P. Quinet, J. Electron Spectrosc. Relat. Phenom. **144**, 27 (2005)
9. É. Biémont, N. Grevesse, M. Kwiatkovski, P. Zimmermann, Astron. Astrophys. **131**, 364 (1984)
10. É. Biémont, P. Palmeri, P. Quinet, Astrophys. Space Sci. **269**, 635 (1999). http://www.umh.ac.be/~astro/dream.shtml
11. É. Biémont, H.-P. Garnir, P. Palmeri, Z.S. Li, S. Svanberg, Mon. Not. R. Astron. Soc. **312**, 116 (2000)
12. É. Biémont, J. Lidberg, S. Mannervik, L.-O. Norlin, P. Royen, A. Schmitt, W. Shi, X. Tordoir, Eur. Phys. J. D **11**, 355 (2000)
13. É. Biémont, A. Derkatch, P. Lundin, S. Mannervik, L.-O. Norlin, D. Rostohar, P. Royen, P. Palmeri, P. Schef, Phys. Rev. Lett. **93**, 063003 (2004)
14. É. Biémont, K. Blagoev, J. Campos, R. Mayo, G. Malcheva, M. Ortiz, P. Quinet, J. Electron Spectrosc. Relat. Phenom. **144**, 23 (2005)
15. É. Biémont, V. Buchard, H.-P. Garnir, P.-H. Lefèbvre, P. Quinet, Eur. Phys. J. D **33**, 181 (2005)
16. É. Biémont, P. Palmeri, P. Quinet, Z. Dai, S. Svanberg, H.L. Xu, J. Phys. B, At. Mol. Opt. Phys. **38**, 3547 (2005)
17. É. Biémont, P. Quinet, C.J. Zeippen, Phys. Scr. **71**, 163 (2005)
18. É. Biémont, K. Blagoev, V. Fivet, G. Malcheva, R. Mayo, M. Ortiz, P. Quinet, Mon. Not. R. Astron. Soc. **380**, 1581 (2007)
19. É. Biémont, M. Clar, V. Fivet, H.-P. Garnir, P. Palmeri, P. Quinet, D. Rostohar, Eur. Phys. J. D **44**, 23 (2007)
20. É. Biémont, A. Ellmann, P. Lundin, S. Mannervik, L.-O. Norlin, P. Palmeri, P. Quinet, D. Rostohar, P. Royen, P. Schef, Eur. Phys. J. D **41**, 211 (2007)
21. É. Biémont, M. Clar, S. Enzonga Yoca, V. Fivet, P. Quinet, E. Träbert, H.-P. Garnir, Can. J. Phys. **87**, 1275 (2009)
22. É. Biémont, P. Palmeri, P. Quinet, J. Phys. B **43**, 074010 (2010). High Precision Atomic Physics, Special Issue
23. É. Biémont, K. Blagoev, L. Engström, H. Hartman, H. Lundberg, G. Malcheva, H. Nilsson, P. Palmeri, P. Quinet, Mon. Not. R. Astron. Soc. **414**, 3350 (2011)
24. K. Blagoev, V. Penchev, É. Biémont, Z.G. Zhang, C.-G. Wahlström, S. Svanberg, Phys. Rev. A **66**, 032509 (2002)
25. J. Campos, M. Ortiz, R. Mayo, É. Biémont, P. Quinet, K. Blagoev, G. Malcheva, Mon. Not. R. Astron. Soc. **363**, 905 (2005)
26. C.H. Corliss, W.R. Bozman, *Experimental Transition Probabilities for Spectral Lines of Seventy Elements*. NBS Monograph, vol. 53 (US Department of Commerce, Washington, 1962)
27. R.D. Cowan, *The Theory of Atomic Structures and Spectra* (University of California Press, Berkeley, 1981)
28. C.R. Cowley, http://www.astro.lsa.umich.edu/~cowley/ (2009)
29. A. Derkatch, L. Ilyinski, S. Mannervik, L.-O. Norlin, D. Rostohar, P. Royen, É. Biémont, Phys. Rev. A **65**, 062508 (2002)
30. S. Enzonga Yoca, É. Biémont, F. Delahaye, P. Quinet, C.J. Zeippen, Phys. Scr. **78**, 025303 (2008)
31. M. Eriksson, U. Litzén, G.M. Wahlgrén, D.S. Leckrone, Phys. Scr. **65**, 480 (2002)
32. V. Fivet, P. Palmeri, P. Quinet, É. Biémont, H.L. Xu, S. Svanberg, Eur. Phys. J. D **37**, 29 (2006)
33. V. Fivet, P. Quinet, É. Biémont, H.L. Xu, J. Phys. B, At. Mol. Opt. Phys. **39**, 3587 (2006)

34. V. Fivet, P. Quinet, É. Biémont, A. Jorissen, A.V. Yushchenko, S. Van Eck, Mon. Not. R. Astron. Soc. **380**, 771 (2007). Erratum: Mon. Not. R. Astron. Soc. **382**, 944 (2007)
35. V. Fivet, P. Quinet, P. Palmeri, É. Biémont, H.L. Xu, J. Electron Spectrosc. Relat. Phenom. **156**, 250 (2007). http://www.umh.ac.be/~astro/desire.shtml
36. V. Fivet, É. Biémont, L. Engström, H. Lundberg, H. Nilsson, P. Palmeri, P. Quinet, J. Phys. B, At. Mol. Opt. Phys. **41**, 015702 (2008)
37. V. Fivet, P. Quinet, P. Palmeri, É. Biémont, M. Asplund, N. Grevesse, A.J. Sauval, L. Engström, H. Lundberg, H. Hartman, H. Nilsson, Mon. Not. R. Astron. Soc. **396**(4), 2124 (2009)
38. S. Fraga, J. Karwowski, K.M.S. Saxena, *Handbook of Atomic Data* (Elsevier, Amsterdam, 1976)
39. M. Gallardo, M. Raineri, J.R. Almandos, É. Biémont, J. Phys. B **44**, 045001 (2011)
40. H.-P. Garnir, S. Enzonga Yoca, P. Quinet, É. Biémont, J. Quant. Spectrosc. Radiat. Transf. **110**, 284 (2009)
41. S. Goriely, Astron. Astrophys. **466**, 619 (2007)
42. I.P. Grant, Methods Comput. Chem. **2**, 1 (1988)
43. J. Gurell, É. Biémont, K. Blagoev, V. Fivet, P. Lundin, S. Mannervik, L.-O. Norlin, P. Quinet, D. Rostohar, P. Royen, P. Schef, Phys. Rev. A **75**, 052506 (2007)
44. S. Hameed, J. Phys. B **5**, 746 (1972)
45. S. Hameed, A. Herzenberg, M.G. James, J. Phys. B, At. Mol. Opt. Phys. **1**, 822 (1968)
46. P. Hannaford, R.M. Lowe, J. Phys. B, At. Mol. Phys. **16**, 4539 (1983)
47. H. Hartman, H. Nilsson, L. Engström, H. Lundberg, P. Palmeri, P. Quinet, É. Biémont, Phys. Rev. A **82**, 052512 (2010)
48. H. Holweger, E.A. Müller, Sol. Phys. **39**, 19 (1974)
49. S. Ivarsson, J. Andersen, B. Nordström, X. Dai, S. Johansson, H. Lundberg, H. Nilsson, V. Hill, M. Lundqvist, J.-F. Wyart, Astron. Astrophys. **409**, 1141 (2003)
50. S. Ivarsson, G.M. Wahlgren, Z. Dai, H. Lundberg, D.S. Leckrone, Astron. Astrophys. **425**, 353 (2004)
51. M. Jaschek, E. Brandi, Astron. Astrophys. **20**, 233 (1972)
52. D.L. Lambert, in *Cool Stars with Excesses of Heavy Elements*, ed. by M. Jaschek, P.C. Keenan (Reidel, Dordrecht, 1985)
53. Z.S. Li, S. Svanberg, É. Biémont, P. Palmeri, J. Zhangui, Phys. Rev. A **57**, 3443 (1998)
54. J. Lidberg, A. Al-Khalili, L.-O. Norlin, P. Royen, X. Tordoir, S. Mannervik, Nucl. Instrum. Methods Phys. Res. B **152**, 157 (1999)
55. M.D. Lukin, M. Fleischhauer, R. Cote, L.M. Duan, D. Jaksch, J.I. Cirac, P. Zoller, Phys. Rev. Lett. **87**, 037901 (2001)
56. H. Lundberg, S. Johansson, U. Litzén, G.M. Wahlgren, S. Leckrone, in *The Scientific Impact of the Goddard High Resolution Spectrograph*. ASP Conf. Ser., vol. 143 (1998), p. 343
57. H. Lundberg, L. Engström, H. Hartman, H. Nilsson, P. Palmeri, P. Quinet, É. Biémont, J. Phys. B, At. Mol. Opt. Phys. **43**, 085004 (2010)
58. P. Lundin, J. Gurell, L.-O. Norlin, P. Royen, S. Mannervik, P. Palmeri, P. Quinet, V. Fivet, É. Biémont, Phys. Rev. Lett. **99**, 213001 (2007)
59. G. Malcheva, K. Blagoev, R. Mayo, M. Ortiz, H.L. Xu, S. Svanberg, P. Quinet, É. Biémont, Mon. Not. R. Astron. Soc. **367**, 754 (2006)
60. G. Malcheva, S. Enzonga Yoca, R. Mayo, M. Ortiz, L. Engström, H. Lundberg, H. Nilsson, É. Biémont, K. Blagoev, Mon. Not. R. Astron. Soc. **396**, 2289 (2009)
61. G. Malcheva, R. Mayo, M. Ortiz, J. Ruiz, L. Engström, H. Lundberg, H. Nilsson, P. Quinet, É. Biémont, K. Blagoev, Mon. Not. R. Astron. Soc. **395**, 1523 (2009)
62. S. Mannervik, Phys. Scr. T **100**, 81 (2002)
63. S. Mannervik, Phys. Scr. T **105**, 67 (2003)
64. S. Mannervik, L. Broström, J. Lidberg, L.-O. Norlin, P. Royen, Hyperfine Interact. **108**, 291 (1997)
65. G. Mathys, in *Lectures Notes in Physics*, vol. 523 (Springer, Berlin, 1999), p. 95
66. J. Migdalek, W.F. Baylis, J. Phys. B, At. Mol. Opt. Phys. **12**, 1113 (1979)

67. C.E. Moore, *Atomic Energy Levels, vol. 3*. NSRDS-NBS, vol. 35 (US Govt Printing Office, Washington, 1958). Reissued 1971
68. C.E. Moore, M.G.J. Minnaert, J. Houtgast, *The Solar Spectrum 2935 Å to 8770 Å*. NBS Monograph, vol. 61 (US Department of Commerce, Washington, 1966)
69. D. Naujoks, K. Asmussen, M. Bessenrodt-Weberpals, S. Deschka, R. Dux, W. Engelhardt, A.R. Field, G. Fussmann, J.C. Fuchs, C. Garcia-Rosales, S. Hirsch, P. Ignacz, G. Lieder, K.F. Mast, R. Neu, R. Radtke, J. Roth, U. Wenzel, Nucl. Fusion **36**, 671 (1996)
70. H. Nilsson, L. Engström, H. Lundberg, P. Palmeri, V. Fivet, P. Quinet, É. Biémont, Eur. Phys. J. D **49**, 13 (2008)
71. H. Nilsson, H. Hartman, L. Engström, H. Lundberg, C. Sneden, V. Fivet, P. Palmeri, P. Quinet, É. Biémont, Astron. Astrophys. **511**, A16 (2010)
72. P. Palmeri, P. Quinet, É. Biémont, Phys. Scr. **63**, 468 (2001)
73. P. Palmeri, P. Quinet, É. Biémont, H.L. Xu, S. Svanberg, Mon. Not. R. Astron. Soc. **362**, 1348 (2005)
74. P. Palmeri, P. Quinet, É. Biémont, S. Svanberg, H.L. Xu, Phys. Scr. **74**, 297 (2006)
75. P. Palmeri, P. Quinet, É. Biémont, A.V. Yushchenko, A. Jorissen, S. Van Eck, Mon. Not. R. Astron. Soc. **374**, 63 (2007)
76. P. Palmeri, P. Quinet, V. Fivet, É. Biémont, H. Nilsson, L. Engström, H. Lundberg, Phys. Scr. **78**, 015304 (2008)
77. P. Palmeri, P. Quinet, V. Fivet, É. Biémont, C.R. Cowley, L. Engström, H. Lundberg, H. Hartman, H. Nilsson, J. Phys. B, At. Mol. Opt. Phys. **42**, 165005 (2009)
78. F.A. Parpia, C. Froese Fischer, I.P. Grant, Comput. Phys. Commun. **94**, 249 (1996)
79. L. Qin, N. Dauphas, M. Wadhwa, A. Markowski, R. Gallino, E.P. Janney, C. Bouman, Astrophys. J. **674**, 1234 (2008)
80. P. Quinet, É. Biémont, At. Data Nucl. Data Tables **87**, 207 (2004)
81. P. Quinet, P. Palmeri, É. Biémont, J. Quant. Spectrosc. Radiat. Transf. **62**, 625 (1999)
82. P. Quinet, P. Palmeri, É. Biémont, A. Jorissen, S. Van Eck, S. Svanberg, H.L. Xu, B. Plez, Astron. Astrophys. **448**, 1207 (2006)
83. P. Quinet, É. Biémont, P. Palmeri, H.L. Xu, J. Phys. B, At. Mol. Opt. Phys. **40**, 1705 (2007)
84. P. Quinet, P. Palmeri, V. Fivet, É. Biémont, H. Nilsson, L. Engström, H. Lundberg, Phys. Rev. A **77**, 022501 (2007)
85. P. Quinet, V. Fivet, P. Palmeri, É. Biémont, L. Engström, H. Lundberg, H. Nilsson, Astron. Astrophys. **493**, 711 (2009)
86. P. Quinet, V. Vinogradoff, P. Palmeri, É. Biémont, J. Phys. B, At. Mol. Opt. Phys. **43**, 144003 (2010)
87. P. Quinet, É. Biémont, P. Palmeri, L. Engström, H. Hartman, H. Lundberg, H. Nilsson, Astron. Astrophys. (2011, in press)
88. P. Quinet, P. Palmeri, É. Biémont, J. Phys. B **44**, 145005 (2011)
89. R.D. Reid, J.R. McNeil, G.J. Collins, Appl. Phys. Lett. **29**(10), 666 (1976)
90. K. Sadakane, Publ. Astron. Soc. Pac. **103**, 355 (1991)
91. M. Saffman, T.G. Walker, Phys. Rev. A **72**, 022347 (2005)
92. P. Schef, A. Derkatch, P. Lundin, S. Mannervik, L.-O. Norlin, D. Rostohar, P. Royen, É. Biémont, Eur. Phys. J. D **29**, 195 (2004)
93. P. Schef, P. Lundin, É. Biémont, A. Källberg, L.-O. Norlin, P. Palmeri, P. Royen, A. Simonsson, S. Mannervik, Phys. Rev. A **72**, 020501(R) (2005)
94. M. Schultz-Johanning, R. Kling, R. Schnabel, M. Kock, Z. Li, H. Lundberg, S. Johansson, Phys. Scr. **63**, 367 (2001)
95. V.V. Smith, Astron. Astrophys. **132**, 326 (1984)
96. C. Sneden, A. McWilliam, G.W. Preston, J.J. Cowan, D.L. Burris, B.J. Armosky, Astrophys. J. **467**, 819 (1996)
97. U.J. Sofia, D.M. Meyer, J.A. Cardelli, Astrophys. J. **522**, L137 (1999)
98. G.M. Wahlgren, D.S. Leckrone, S.G. Johansson, M. Rosberg, T. Brage, Astrophys. J. **444**, 438 (1995)

99. H.L. Xu, A. Persson, S. Svanberg, K. Blagoev, G. Malcheva, V. Pentchev, É. Biémont, J. Campos, M. Ortiz, R. Mayo, Phys. Rev. A **70**, 042508 (2004)
100. H.L. Xu, Z.W. Sun, Z.W. Dai, Z.K. Jiang, P. Palmeri, P. Quinet, É. Biémont, Astron. Astrophys. **452**, 357 (2006)
101. H.L. Xu, S. Svanberg, P. Quinet, P. Palmeri, É. Biémont, J. Quant. Spectrosc. Radiat. Transf. **104**, 52 (2007)
102. J. Xu, S. You, Y. Zhang, W. Zhang, Z. Ma, Y. Feng, Y. Xing, X. Deng, P. Quinet, É. Biémont, Z. Dai, J. Phys. B, At. Mol. Opt. Phys. **42**, 035001 (2009)
103. A.V. Yushchenko, Odessa Astron. Publ. **9**, 84 (1996)
104. A.V. Yushchenko, V.F. Gopka, C. Kim, F.A. Musaev, Y.W. Kang, in *The A-Star Puzzle*, ed. by J. Zverko, J. Zizovsky, S.J. Adelman, W.W. Weiss. Proc. IAU Symp., vol. 224 (Cambridge University Press, Cambridge, 2004), p. 224
105. A. Yushchenko, V. Gopka, S. Goriely, F. Musaev, A. Shavrina, C. Kim, Y. Woon Kang, J. Kuznietsova, V. Yuschenko, Astron. Astrophys. **430**, 255 (2005)
106. Y. Zhang, J. Xu, W. Zhang, S. You, Z. Ma, L. Han, P. Li, G. Sun, Z. Jiang, S. Enzonga-Yoca, P. Quinet, É. Biémont, Z. Dai, Phys. Rev. A **78**, 022505 (2008)
107. W. Zhang, S. You, C. Sun, Y. Zhang, J. Xu, Z. Ma, Y. Feng, H. Liu, P. Quinet, É. Biémont, Z. Dai, Eur. Phys. J. D **55**, 1 (2009)

Chapter 2
Atomic Structure Calculations Useful for Fusion and Astrophysics

Man Mohan, Jagjit Singh, Sunny Aggarwal, and Nupur Verma

Abstract In this work, we have reviewed the present status of atomic structure calculations for multi-electrons atoms and ions using different international computer codes. Direction is given for dealing with heavy atoms and ions where relativistic effects become as important as correlation effects. Separate paragraphs are devoted on the application of atomic data in the fields of Astrophysics & Fusion plasma including future International Thermonuclear Experimental Reactor (ITER) for harnessing fusion power.

2.1 Introduction

There is an unprecedented demand for accurate atomic data as it is needed urgently for the prediction and analysis of experimental data obtained by several space missions (e.g. AXAF, SOHO, Chandra, XMM, Hubble-Space Telescope etc.) [1]. Recently, greatly enhanced resolution of spectral lines have been achieved by space based over ground based instruments, in particular for ions with relatively high abundances, demanding a greater accuracy in available atomic data. Additionally, elements which had not been detected from ground based instruments have now been observed in stellar objects, particularly the heavy elements. Not only this, there is a great demand for accurate atomic data for finding the physical conditions like temperature, densities etc. in the present and future fusion research programmes like ITER, laser fusion etc. [2]. Apart from Astrophysics and fusion research, reliable atomic data are needed such as UV and X-ray lithography, production of X-ray lasers, Bio & Nanotechnology etc. [3]. Moreover, for understanding the fundamental interactions in nature and correlation in quantum many particle systems, one needs accurate atomic data.

In response to the growing demand of large and accurate atomic data for various applications, a number of powerful atomic codes have been developed during the last decade for light and heavy elements. However, every code has

M. Mohan (✉) · J. Singh · S. Aggarwal · N. Verma
Department of Physics and Astrophysics, University of Delhi, Delhi 110 007, India
e-mail: drmanmohan.05@gmail.com

M. Mohan (ed.), *New Trends in Atomic and Molecular Physics*,
Springer Series on Atomic, Optical, and Plasma Physics 76,
DOI 10.1007/978-3-642-38167-6_2, © Springer-Verlag Berlin Heidelberg 2013

its own capabilities and deficiencies. Some of the codes are CIV3 (Configuration Interaction Version3) [4], Superstructure (SS) [5], GRASP (General purpose Relativistic Atomic Structure Package) [6], AS (Autostructure) [7], MCDF (Multi-Configuration Hartree Fock) [8], Relativistic Many-Body Perturbation theory (RMBPT) [9], HFR (Hartree-Fock Relativistic HFR) [10], FAC (Flexible Atomic Code) [11], HULLAC (Hebrew University Lawrence Livermore Atomic Code) [12], RCI (Relativistic Configuration Interaction) [13], RATIP (Relativistic Atomic Transition and Ionization Properties) [14], HTAC (Hamansha and Tahat Atomic Code) [15], MR-MBPT (Multi-Reference Many-Body Perturbation Theory) [16] etc.

For light atoms with nuclear charge Z less than 20, electron correlation effects dominate. Relativistic correction can be added using a perturbation theory, such as Breit-Pauli approximation while Quantum electrodynamics (QED) corrections can be ignored. In this direction, Hibbert's CIV3 (Configuration Interaction) [4] and Werner Eisner's SS (Superstructure) [5] codes are very popular and have been widely used in various International research programmes like Opacity [17] and Ferrum [18] etc. For atoms with Z between 20 and 60, inner shell radial functions are altered sufficiently by relativistic effects and it is safer to use a relativistic atomic structure theory than a non-relativistic one. For heavy atoms with Z greater than 60, electron correlation effects and relativistic effects are strongly coupled making it necessary to use a relativistic theory which also incorporates electron correlations effects on the same footing. For treating heavy atoms there have been new developments and many codes in the relativistic domain have been developed by various authors. Among them, multi-configuration Hartree (Dirac) Fock (MCDF) [8] model based codes have been found very useful in ab-inito investigations. In the next paragraph we will discuss correlation effects important for a multi-electron system.

2.2 Correlation Effects

For including correlation effect, configuration interaction (CI) method uses a variational wavefunction which is a linear combination of configuration state functions (CSFs) built from spin orbitals. The full CI expansion includes all possible CSFs of the appropriate symmetry, which solves the Schrodinger equation within the space spanned by the one-particle basis set. The CI wavefunction is of the form

$$\Psi(LSJ) = \sum_{i=1}^{M} a_i \Phi_i(\alpha_i LSJ) \tag{2.1}$$

in LS coupling or

$$\Psi(J\Pi) = \sum_{i=1}^{M} a_i \Phi_i(\alpha_i L_i S_i J\Pi) \tag{2.2}$$

in LSJ coupling where Π is the parity of the state. For any particular set of configuration state function Φ_i, a_i is the set of components of one of the eigenvectors of Hamiltonian matrix element H_{ij}, where

$$H_{ij} = \langle \Phi_i | H | \Phi_j \rangle \tag{2.3}$$

We can write CI wavefunction for two electron system with ^{1}S symmetry as

$$\Psi\left(^1\text{S}\right) = a_1 \Phi\left(1\text{s}^2\,^1\text{S}\right) + a_2 \Phi\left(1\text{s}\,2\text{s}\,^1\text{S}\right) + a_3 \Phi\left(2\text{s}^2\,^1\text{S}\right)$$
$$+ a_4 \Phi\left(2\text{p}^2\,^1\text{S}\right) + a_6 \Phi\left(3\text{p}^2\,^1\text{S}\right) + \cdots \tag{2.4}$$

The CSFs are constructed from the combination of one electron orbitals of the form

$$\Phi(r, ms) = \frac{1}{r} P_{nl}(r) Y_{lm}(\theta, \varphi) \chi(ms) \tag{2.5}$$

For a particular orbital, the spherical harmonics Y_{lm} and spin eigenfunction $\chi(ms)$ are fixed and the radial function $P_{nl}(r)$ is determined from the inequality by Hylleraas-Undhein theorem

$$E_i \geq E_i^{exact}$$

The different ways of treating this variational approach are superposition of configurations (SOC) [19] and Multi-configuration Hartree-Fock (MCDF) [8] etc. The SOC method has been used in the Computer programme CIV3 [4] of Hibbert.

The absorption oscillator strength of a transition from initial state $|i\rangle$ to final state $|j\rangle$ in length form is given by

$$f_l = \frac{2}{3} \frac{\Delta E}{g_i} \sum_{m\text{-values}} \left| \langle \Psi_i | \sum_{k=1}^{N} r_k | \Psi_j \rangle \right|^2 \tag{2.6}$$

And in velocity form by

$$f_v = \frac{2}{3} \frac{1}{\Delta E g_i} \sum_{m\text{-values}} \left| \langle \Psi_i | \sum_{k=1}^{N} \nabla_k | \Psi_j \rangle \right|^2 \tag{2.7}$$

Where $\Delta E = E_j - E_i$ is the transition energy. The summation over m is the summation of m-degeneracy's of the two states and the inner summation is over the electrons of atom or ion. Here g_i is the statistical weight which represents the number of distinct m-values of the lower state. Comparison between length and velocity forms are usually done in LS coupling to assess the degree of electron correlation included in calculations.

For light to medium Z elements, one generally uses LSJ coupling in Breit-Pauli approximation. Breit-Pauli approximation arises from an expansion of the fully relativistic equation in powers of α (fine structure constant) while neglecting terms of order higher than α^2. Now the length form of oscillator strength is same as given by Eq. (2.6) but correction are made in velocity forms to get an agreement between f_L and f_v values. Usually, the corrections to velocity forms are not included in calculations and only length form is provided. The transition for $J_i \neq J_j = 0$ is forbidden in Breit-Pauli approximation. However, this transition becomes allowed if hyperfine structure operators are included in the Hamiltonian.

2.2.1 Type of Correlations

For accurate calculation of oscillator strength, transition probability, hyperfine structure effects etc., it is very important to include electron correlations. For the choice of configurations we generally follow the scheme of Sinanoglu and co-workers [20] as the number of electrons increases. In general, Hartree-Fock (HF) sea is defined as the set of orbitals occupied in the HF configuration plus all other orbitals that have the same or smaller n values. For examples HF sea for ground state of Fe^{20+} is 1s, 2s, and 2p. Correlations effects are of three types and are described below;

(a) *Internal Correlation*: Internal correlations are described by configurations built entirely from orbitals in the HF sea. e.g. Fe^{20+} (^3P): $1s^2 2s^2 2p^2$, $1s^2 2p^4$, $2s^2 2p^4$. The CI wavefunction for the ground state of Fe^{20+} in terms of the first two configurations is given by

$$\Psi\left(^3P\right) = a_1 \Phi_1\left(1s^2 2s^2 2p^2 \; ^3P\right) + a_2 \Phi_2\left(1s^2 2p^4 \; ^3P\right)$$

where the expansion coefficients a_1 and a_2 have values 0.930 and 0.367, respectively, with $a_1^2 + a_2^2 \approx 1$. These two configurations are very important to include as they are quite big in size and also because 2p function in the same region of space as 2s. The third configuration $2s^2 2p^4 \; ^3P$ is not as important as there is less chance of 6 electrons remaining in the same space.

(b) *Semi-internal Correlation*: Semi-internal Correlation is described by configurations built from $(N - 1)$ orbitals of the HF sea plus one other electron outside the HF sea. For example, in the case of Fe^{20+} (^3P), the semi-internal correlation is represented by $1s^2 2s^2 2p3p$ (^3P) and $1s^2 2p^3 4f$ (^3P) configurations. The expansion coefficients of the configuration belonging to semi-internal correlations are such that a_i lies between 0.05 and 0.01.

(c) *External Correlations*: For N-electrons, the external correlation is described by $(N - 2)$ electrons in the HF sea and two electrons outside the HF sea. For example in the case of Fe^{20+}, (^3P) $1s^2 2s^2 3p^2$ and $1s^2 (2s2p)$ (^3P) $3s3p$ (^3P) configurations represent the external correlations. The expansion coefficients for external correlation are very small and their values vary from 0.01 and 0.001. The internal and semi-internal correlations can be obtained from a finite CI calculation while the external correlations involve infinite configuration-interaction. Therefore, the HF, internal, semi-internal and some of external correlations should be used for practical calculations

2.3 Radial Function

The radial function is presented as a linear combination of Slater-type orbitals (STO),

$$P_{nl}(r) = \sum_{j=1}^{k} C_{jnl} \chi_{jnl}(r) \tag{2.8}$$

where

$$\chi_{jnl}(r) = \frac{(2\xi_{jnl})^{I_{jnl}+1/2}}{[(2I_{jnl})!]^{1/2}} r^{I_{jnl}} \exp(-\xi_{jnl}.r) \tag{2.9}$$

and STOs are chosen to satisfy the orthonormality condition

$$\int_0^\infty P_{nl}(r)P_{n'l}(r)dr = \delta_{nn'}, \quad l < n' \le n \tag{2.10}$$

If $k = n - l$, for a given value of I_{jnl} and ξ_{jnl}, the coefficients C_{jnl} are uniquely determined. If $k > n - l$, then some of the co-efficients can be treated as variational parameters and the others can be varied freely. Generally, I_{jnl} are chosen to be fixed integers. So the variational parameters are the exponent ξ_{jnl} and perhaps some of the C_{jnl}. Using these radial wavefunctions one can calculate the radial integrals in the Hamiltonian matrix.

2.4 Optimization Process

Since the jth eigenvalue E_j is an upper bound to the exact value of the jth energy level hence linear combination of E_i is an upper bound to the same linear combination, then E_i may be used as a variational functional, depending on the exponents ξ_{jnl} and the coefficients C_{jnl} of (2.10). Clearly the optimization process requires the functional values to be determined. When more configurations are added to HF, then one can either fix the HF radial functions as is done in SOC or can allow HF functions to vary as is done in MCHF [8]. An example of SOC based program is CIV3 code of Hibbert [4]. So, there are two ways of treating the variational principle.

In the first one the radial functions are chosen initially. The Hamiltonian matrix is set up and diagonalized to yield eigenvalue E_i and the coefficients a_i^j. Next, the radial functions are changed in a particular way and the process is repeated until there is no significant improvement in the final E_i. This scheme is the basis of SOC method and is particularly useful when P_{nl} depends on the variable parameter, and has been used in configuration-interaction code CIV3 of Hibbert [4]. The second case involves the use of eigenvalue E^j as a variational functional of the radial functional $P_{nl}(r)$ with some appropriate constraints. Now, the integro-differential HF type equation for P_{nl} can be derived. In this, P_{nl} and a_i are initially chosen to give a new set of radial functions after the HF type of equations. From this, the Hamiltonian matrix H can be set up and diagonalized to yield coefficients a_i. This process is continued till the self consistency is achieved. This process has been used in computer codes based on MCHF [8] scheme.

2.5 Hamiltonian Matrix Elements

Construction of Hamiltonian matrix is one of the important stages in configuration interaction wavefunctions. In general, we can write

$$H_{ij} = \langle \Phi_i | H | \Phi_j \rangle = \langle \Phi_i | H_0 + v | \Phi_j \rangle \tag{2.11}$$

where

$$H_0 = \sum_{i=1}^{N} -\frac{\hbar^2}{2m} \nabla_i^2 - \frac{Ze^2}{r_i}, \qquad v = \sum_{i<j=1}^{n} \sum \frac{e^2}{r_{ij}} \tag{2.12}$$

then the matrix element of one electron operator is written as

$$\langle \Phi_i | H | \Phi_j \rangle = \sum_{\sigma,\sigma'} x(\sigma, \sigma') \langle p_{n_\sigma l_\sigma} | -\frac{\hbar^2}{2m} \frac{d^2}{dr^2} - \frac{Ze^2}{r} + \frac{l_\sigma(l_\sigma + 1)}{2r^2} | p_{n_\sigma l_\sigma} \rangle \delta_{l_\sigma l_{\sigma'}}$$

$$\tag{2.13}$$

and the matrix element for two electron operator as:

$$\langle \Phi_i | v | \Phi_j \rangle = \sum_{\rho,\sigma,\rho',\sigma',k} y(\rho, \sigma, \rho', \sigma', k) R^k(n_\rho l_\rho, n_\sigma l_\sigma; n_\rho, l_{\rho'}, n_{\sigma'}, l_{\sigma'}) \tag{2.14}$$

where ρ, σ represents the subshells in the configuration Φ. Integrals in (2.14) and the two electron radial integral (R^k) in Eq. (2.14) are evaluated analytically in CIV3 [4]. The evaluation of coefficients $x(\sigma, \sigma')$ in Eq. (2.13) has been described and programmed by Hibbert [21]. The coefficients $y(\rho, \sigma, \rho', \sigma', k)$ has been described in terms of Racah Algebra by Fano [22].

2.6 Relativistic Atomic Structure Methods

There are several relativistic methods developed which are counterparts of non-relativistic Hartree-Fock, Configuration Interaction or many body perturbation methods. These are called the Dirac Fock (DF) method, relativistic configuration interaction method (RCI), and the relativistic many-body perturbation (RMBPT) respectively. All these methods depends upon the construction of N-electron atom or ion Dirac-Coulomb Hamiltonian given by

$$\hat{H}^{DC} = \sum_{i=1}^{N} \hat{H}_i + \sum_{i=1}^{N-1} \sum_{j=i+1}^{N} |\hat{r}_i - \hat{r}_j|^{-1} \tag{2.15}$$

where

$$\hat{H} = C \sum_{i=1}^{3} \vec{\alpha}_i \vec{p}_i + \beta c^2 + V_{nuc}(r) \tag{2.16}$$

is the single particle (one body) Hamiltonian consisting of kinetic energy and its interaction with the nucleus. c is the speed of light whereas, α and β are 4×4 Dirac matrices and are given by

$$\alpha_i = \begin{bmatrix} 0 & \sigma_i \\ \sigma_i & 0 \end{bmatrix}, \quad i = 1, 2, 3, \ldots, \quad \beta = \begin{bmatrix} 1 & 0 \\ 0 & -1 \end{bmatrix} \cdots \quad (2.17)$$

In the DF method, θ_{DF} is a linear combination of Slater determinants, which are antisymmetrized products of the relativistic one-electron orbitals θ_{nkm} given by

$$\phi_{nkm} = \frac{1}{r} \begin{pmatrix} P_{nk}(r) \chi_{km}(\theta, \varphi, \sigma) \\ -i Q_{nk}(r) \chi_{-km}(\theta, \varphi, \sigma) \end{pmatrix} \quad (2.18)$$

Where k is the Dirac angular quantum number, $k = \pm(j + 1/2)$ for $l = j \pm 1/2$, thus $j = k - 1/2$ ($k = -1$ for s electrons, $+1$ for $p_{1/2}$ electrons, -2 for $p_{3/2}$ electrons etc.), m is the projection of the angular momentum j, and P_{nk}, Q_{nk} are radial functions. The spin angular momentum $\chi_{km}(\theta, \Phi)$ is a 2 component function (spinor spherical harmonics) defined by

$$\chi_{km}(\theta, \phi) = \sum_{\sigma = \pm 1/2} \left\langle lm - \sigma \frac{1}{2}\sigma \middle| l \frac{1}{2} jm \right\rangle Y_l^{m-\sigma}(\theta, \phi) \phi^\sigma \quad (2.19)$$

With the Clebsch-Gorden Coefficient

$$\left\langle lm - \sigma \frac{1}{2}\sigma \middle| l \frac{1}{2} jm \right\rangle$$

and

$$\varphi_{1/2} - \begin{pmatrix} 1 \\ 0 \end{pmatrix}, \quad \varphi_{-1/2} = \begin{pmatrix} 0 \\ 1 \end{pmatrix}$$

are spinor basis functions. In the DF method, usually the wavefunction is associated with a specific total angular momentum J and the radial functions are determined by applying variational principle to the expectation value of H_{DC}

$$\delta \langle \Psi_{DF} | H_{DC} | \Psi_{DF} \rangle = 0 \quad (2.20)$$

In DF method the radial functions are represented in terms of Slater-type orbitals or numerical basis functions [23] or as a pure numerical functions. However, DF wavefunctions with the minimum number of radial functions representing only physically occupied orbitals are not flexible enough to account for the effects of full electron correlations.

To improve this in Relativistic Configuration Interaction method (RCI), the extra configurations are added to DF function for taking the correlation effects [13]. In the MCDF (Multi-Configuration Dirac Fock) method as used in the relativistic (GRASP) package developed by I.P. Grant and co-authors over the years [6] an atomic state function for an N-electron system with given total angular momentum J, M and parity P is approximated by a linear combination of n_c electronic configuration state functions (CSF)

$$|\psi_\alpha(PJM)\rangle = \sum_{i-1}^{n_c} C_i(\alpha) |\gamma_i(PJM)\rangle \quad (2.21)$$

Where $C_i(\alpha)$ are the expansion mixing co-efficients and α represents all information such as orbital occupation numbers, coupling, seniority numbers etc. The ASF's are usually chosen to the orthonormal, so that

$$\left(C_i(\alpha)\right)^+ C_j(\alpha) = \delta_{ij} \tag{2.22}$$

For expectation of Dirac-Hamiltonian we get the energy of the N-electron system as

$$\begin{aligned}
E_\alpha^{PJM} &= \left\langle \Psi_\alpha(PJM) \middle| H^{DC} \middle| \Psi_\alpha(PJM) \right\rangle \\
&= \sum_{ij} C_i^*(\alpha) C_i(\alpha) \left\langle \gamma_i^{PJM} \middle| H^{DC} \middle| \gamma_j^{PJM} \right\rangle \\
&= \left(C_\alpha^{DC}\right)^+ H^{DC} C_\alpha^{DC}
\end{aligned} \tag{2.23}$$

The Hamiltonian matrix, H^{DC} has elements

$$H_{rs}^{DC} = \left\langle \gamma_r^{PJM} \middle| H^{DC} \middle| \gamma_s^{PJM} \right\rangle \tag{2.24}$$

For the energy given by Eq. (2.24) to be stationary with respect to variations in mixing co-efficients $C_i(\alpha)$ subject to orthonormality condition given by Eq. (2.22) leads to eigenvalue problem for the mixing coefficients

$$\left(H^{DC} - E_\alpha^{DC} 1\right) C_\alpha^{DC} \tag{2.25}$$

where 1 is the $n_c \times n_c$ unit matrix. Thus the predicted atomic energy level E_α^{PJM} can be taken to be eigenvalues of H^{DC} and $C_{i\alpha}$ are the components of the associated eigenvectors.

In RCI method, one first obtains the radial functions for the basic orbital $^2S_{1/2}$, $^2P_{1/2}$ and $^2P_{3/2}$ e.g. for Be like ion by using hydrogenic radial functions with an effective charge. Next the radial functions are freezed and mixing coefficients are obtained by solving Eq. (2.25). However, in MCDF methods in general a set of trial radial functions are used to obtain the initial values of the mixing co-efficients (in Eq. (2.21)) while the radial functions are kept frozen, then the mixing co-efficients are frozen while the radial functions are varied, and so on until both the mixing coefficients and radial functions converge to the desired accuracy.

The wavefunctions obtained by MCDF method are more compact and better as compared to RCI method as it also takes account of correlation contributions from the continuum states. In this direction two codes are now used for studying atoms and ions. One is the Desclaux programme [24] and the other Grasp codes developed by I.P. Grant and colleagues [6]. In this series, the RATIP package is developed for studying the level structures, transition probabilities and Auger properties of atoms and ions with rather different shell structures.

There is another successful method called the Relativistic Many Body Perturbation Method (RMBPT) developed by Lindgren, Safronova et al. [9, 25] in which the coulomb repulsion term is replaced by a model one electron localized potential which results in the determination of a complete set of one electron wavefunctions using continuum, Next, using the difference between the coulomb repulsion term

Table 2.1 Comparison of our energy levels calculated with GRASP code with other results of O VII

Levels	HTAC-CIM[a]	HTAC-MBPT[b]	HTAC R-matrix[c]	NIST[d]	SS[e]	GRASP[f]
$1s^2$	0	0	0	0	0	0
1s2s	41.15145	41.15315	41.23979	41.23155	41.2438	41.0550
1s2p	41.73916	41.70903	41.80061	41.78724	41.7933	41.6125
1s2p	41.74001	41.71470	41.80129	41.78779	41.79422	41.6155
1s2p	41.74482	41.71581	41.80639	41.79280	41.7997	41.6216
1s2s	41.77862	41.72003	41.86696	41.81240	41.8074	41.6672
1s2p	42.17263	42.12810	42.23406	42.18438	42.2100	42.0238

a,b,c Calculations of Hamansha et al. [15]

d Values of data listed in NIST

e Calculations do Delahaye and Pradhan [29]

f Our calculated value using GRASP code

and model potential as a perturbation, perturbation series are developed for the total energy and wavefunction. Usually only the first and second order perturbations terms are included for finding energies. It is found that perturbation series converges rapidly for highly charged ions but for neutral atoms and lightly charge ions the convergence is very slow.

There are also other fully relativistic codes like HULLAC [12], SZ [26], ATOM [27], FAC [11] etc. based on solving Dirac equations. Among these, FAC code is quite popular and it is also available in public domain (http://sprg.ssl.berkley.edu). FAC is based on Relativistic Configuration Interaction method (RCIM) for expansion of the wavefunctions While the basis states are derived from the modified self-consistent Dirac-Fock slater interaction on a mean fictitious mean configuration with fractional occupation numbers representing the average electron cloud of the configurations included in the calculation. FAC is not only capable to do atomic calculations but is also capable to do electron-atom scattering using distorted wave approximation.

There have been recent developments where many powerful programmes are combined in one as a utility programme (HTAC) based on window for input for finding the atomic structure and transition rates for atoms and multi-electrons ions. In HTAC [15] three codes FAC (Flexible Atomic Code) [11], MR-MBPT (Multi-Reference Many Body Perturbation Theory) [16] and R-matrix method [28] are included. Hamansha et al. [15] using their utility programme have obtained atomic data for O VII ion and have compared it with NIST (National Institute of Standard and Technology) data and data obtained by Delahaye and Pradhan [29] using Superstructure (SS) code [5]. As can be seen from Table 2.1, HTAC results are in good agreement with NIST and SS data.

2.6.1 QED Corrections in Relativistic Codes

In heavy atoms there is another correction due to Quantum electrodynamics effect (QED) or Lamb shift which results in splitting of some of the degenerate levels obtained by solving the Dirac equation e.g. $^2S_{1/2}$ and $^2P_{1/2}$ levels of hydrogenic like ions. As it is difficult to express QED correction as an effective operator in the Hamiltonian [23], the QED corrections are simply added to the total energy while wavefunctions are determined by the usual method as described earlier. The first significant QED correction is a result of the finite size of the nucleus, while the second is due to a bound electron emitting a virtual photon and absorbing it again in the field of the nucleus. The third most significant QED correction is called the vacuum polarization and it is due to the creation and annihilation of virtual electron-positron pairs in the field of the nucleus. The fully relativistic GRASP code is capable of taking both Breit interaction and QED corrections [6].

Also sometimes the accuracy in the determination of atomic data like energy and transition probabilities by even simple experiments are quite high for forbidden transition as compared to the theoretical results obtained by any computer programme (codes) available. This is due to unavailability of perfect theory for many electron QED corrections such as mutual screening of bound electrons and as only the leading relativistic correction to the coulomb repulsion between the bound electrons is not known, unlike the relativistic corrections for one electron [23]. However for strong transitions in the dipole allowed, theoretical methods are capable to provide more accurate results as compared to experiment.

In Table 2.2, we have shown a few results obtained by GRASP and compared them with the other results.

2.7 Atomic Data Requirement in Sun and Other Astrophysical Objects

Atomic data, such as energy levels, transition probabilities, collisional excitation, photoionization and recombination rates are a necessary component in the modelling of astrophysical plasma. Accurate data is required for answering important astrophysical questions, including the synthesis of primordial He, stellar evolution models, yields for supernovae of various types etc. In astronomical observations, lines are detected in the wavelengths from infra-red to ultra-violet. Usually, neutral and singly ionized atomic species are prominent in the spectra of cool stars, the interstellar medium, active galactic nuclei, circumstellar nebulae, planetary nebulae, H II regions, and the supernovae. In e.g. neutral O I, Si II, Fe II etc. exist in low temperature plasmas and ionization bounded photoionized nebulae. Thus, the spectral features of these species e.g. emission or absorption lines gives the physical conditions of the plasma in the observed astrophysical object. However, there are frequent observations of spectra from highly ionized species from Sun and various

Table 2.2 Calculated energy value of Fe XV (in Ryd.) relative to ground state

Levels	CIV3[a]	GRASP1[b]	Vainshtein and Safronova[c]	Nahar and Pradhan[d]	Plante[e]	Our work
$1s2s\,^3S_1$	488.2951	487.5850	487.772	488.4178	487.7859	488.2644
$1s2p\,^3P_0$	490.5317	489.7075	489.897	490.3639	489.9135	490.2764
$1s2p\,^3P_1$	490.6819	489.8579	490.057	490.6359	490.0617	490.5336
$1s2s\,^1S_0$	490.6904	489.9113	490.082	490.6090	490.0948	490.5528
$1s2p\,^3P_2$	491.7732	490.9411	491.132	491.8239	491.1462	491.6592
$1s2p\,^1P_1$	493.2183	492.2885	492.454	493.1927	492.4769	493.0075
$1s3s\,^3S_1$	578.5578	577.7355	577.934	578.6069		578.4475
$1s3p\,^3P_0$	579.1994	578.3194	578.519	579.1439		579.0036
$1s3s\,^1S_0$	579.2159	578.3514	578.544	579.1815		579.0510
$1s3p\,^3P_1$	579.2416	578.3634	578.565	579.2063		579.0816
$1s3p\,^3P_2$	579.5643	578.6872	578.886	579.5752		579.4116
$1s3d\,^3D_2$	579.8917	578.9979	579.201	579.8828		579.7216
$1s3d\,^3D_1$	579.8977	579.0037	579.206	579.8817		579.7291
$1s3p\,^1P_1$	579.9612	579.0539	579.244	579.9602		579.7768
$1s3d\,^3D_3$	580.0152	579.1210	579.324	580.0178		579.8509
$1s3d\,^1D_2$	580.0363	579.1418	579.344	580.0370		579.8643

[a] Calculation of Kingston et al. [30]

[b] Calculation of Kingston et al. [30]

[c] Calculations of Vainshtein and Safronova [31]

[d] Calculations of Nahar and Pradhan [32]

[e] Calculations of Plante [33]

stellar objects (like late type stars) due to the various types of interaction of particles and photons with atoms and molecules [34].

In Astrophysics for explaining the theories of stellar nucleosynthesis and abundances it is imperative to obtain the atomic data for heavy elements e.g. lanthanide ions (LI) ($57 \leq Z \leq 71$). These elements are also usually found overabundant in chemical peculiar (CP) stars. In this direction, several groups in the world are engaged in obtaining the accurate atomic data [35]. Biemont and co-authors have used Cowan's suits of computer programmes, which are based on relativistic Hartree-Fock (HFR) approximation. [36] and include core-polarization (CPOL) effects. However, still a lot of work is required in the direction to explain the spectrum from various ion stages of these heavy ions which are continuously being observed [37]. More recently, the spacecraft observatories SOHO (solar and Heliospheric observatory) and CHANDRA X-ray (space astrophysical observatory) are spanning spectra from the solar upper atmosphere, through soft X-rays from comets to supernovae remnants. The interpretation of the spectral data collected by these laboratories require a very accurate modelling of astrophysical plasmas, which involve

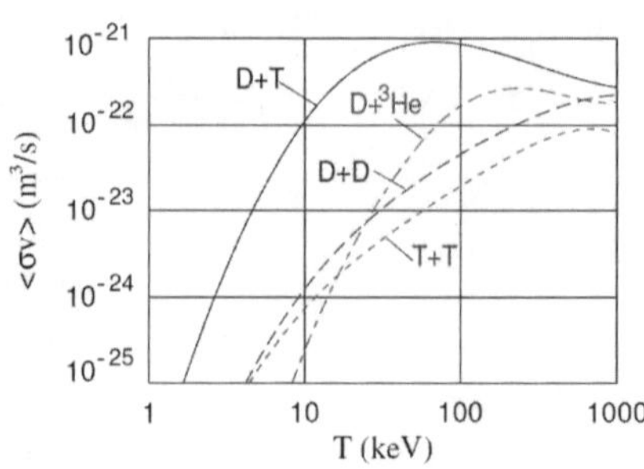

Fig. 2.1 Reaction parameter $\langle \sigma v \rangle$ of some fusion reactions in a thermal plasma of temperature T

lot of physical processes and need accurate atomic data. In this direction, there is a great initiative by Summars group in U.K. [38] by launching ADAS (Atomic data and analysis structure) project for compiling atomic data. The data is also useful for explaining the spectrum obtained from fusion experiments for energy production which is explained in the next paragraph.

2.7.1 Atomic Data for Fusion Plasma Diagnostics in ITER and Other Fusion Machines

Around the world, there is a quest to develop magnetic fusion machines for the production of energy which is without harmful CO_2 and without any possibility of accidental meltdown or long-lived actinide waste [39]. The energy is produced by fusion of two isotopes of hydrogen namely deuterium (D) and tritium (T). The cross-section for reaction

$$D + T \rightarrow He_2^4(3.5 \text{ MeV}) + n(14.1 \text{ MeV})$$

reaches a maximum value of $\sigma_{DT,max} = 4.9 \times 10^{-28}$ m^2 at a collision energy of 64 keV in the centre of mass frame. This fusion cross section is small against the relevant cross-sections for coulomb collisions between D^+ and T^+, and D^+ or T^+ with electrons which are 10^7 times higher. In these circumstances, only a very small fraction of accelerated D^+ ions would cause a fusion reaction, when injected to a tritium target, as there will be a great loss of energy due to coulomb-collisions with cold electrons. Thus for sustaining a fusion reaction, one has to confine hot DT plasma with almost equal temperature of ions and electrons. It is found that the DT reaction represented by the reaction parameter $\langle \sigma_D x \rangle$, where $\langle \rangle$ denotes the average over the Maxwellian distribution, has the highest value with a broad maximum for ion temperature T_i between 20 keV and 100 keV, while it decreases very fast below 10 keV ($T \sim 10^8$ K). The reaction parameter $\langle \sigma_{DT} \rangle$ of the DT reaction and a few other fusion reactions is shown in Fig. 2.1.

Presently, there are two fusion schemes which are going on for achieving plasma confinement. The first inertial confinement scheme uses a DT-pellet with a diameter of ~ 1 mm compressed by about a factor of 1000 and heated up very rapidly to $T \sim 10$ keV with a high power laser or ion beam source. Under these condition, due to inertia the plasma can be confined for a short duration of ~ 1 ns, due to these resulting fusion of DT mixtures in a controlled way and production of a large amount of energy.

The second confinement scheme uses magnetic field for confining the steady state plasma. The energy needed for sustaining the plasma heating is done through fusion reaction itself, i.e., from 3.5 MeV α-particles, which are stopped in plasma by coulomb-collisions. Around the world, the second scheme has become quite popular due to the successful demonstration of substantial energy production in the so called tokamak configuration devices: TFTR (Tokamak Fusion Test Reactor): 10.7 MW [40]; JET—Joint European Torus (16 MW) [41]. So far, JET is the largest and the most successful tokamak experiment which reached the peak fusion power of 16 MW for about 0.5 second.

Following the above success, a new plasma device ITER (International Thermonuclear Experimental Reactor) will be constructed in Cadarache, France with the collaboration of seven countries (EU, Japan, U.S.A., China, Republic of Korea, India and Russia). The goal of this machine is to produce ten times more power (~ 500 MW) than needed to heat the plasma [42]. The total budget for building ITER is around 4.7 billion Euros, and the first plasma is expected by 2016 (www.iter.org). The main parameters for ITER will be radius $R = 6.2$ m, toroidal magnetic field $B_E = 5.3$ Tesla (from superconducting coils), plasma current ≥ 15 mA, plasma volume $V = 840$ m^3 and surface area $A = 1000$ m^2. The ITER core plasma is expected to have an electron density of about $n_e \sim 10^{20}$ m^{-3} and electron temperature of about $T_e \sim 25$ keV ($\sim 10^8$ K). The interaction of burning plasma with the material wall will produce intense heat fluxes that challenges the material limit of the surrounding tokamak walls [43]. In the ITER design beryllium (Be) is planned for the main chamber wall while tungsten (W) is proposed for the ITER diverter quite away from the core plasma, where the main interaction with tokamak plasma is assumed to take place. The importance of tungsten (W) is due to its high thermal conductivity, its high melting point, and its resistance to sputtering and erosion. However, there is a possibility of high Z plasma impurity due to ions in various stages, although tungsten (W) does not strip fully so easily which makes it a tolerable fraction of impurity in plasma ($\sim 2 \times 10^{-5}$) [44]. Impurities such as Ar will be introduced to cool the plasma as it radiates around 75 % of heat flux. The diverter plasma will be having electron density range $N_e = 10^{20}$–10^{21} m^{-1} and electron temperature $T_e = 0.1$–100 eV (10^3–10^6 K). Other expected impurities in ITER are: He ash, carbon (C), beryllium (Be), tungsten (W), Copper (Cu), Iron (Fe), silicon (Si), Nickel (Ni), Titanium (Ti) as well as gases such as neon, argon, krypton and nitrogen that are introduced to dissipate the heat flowing to the diverter. The radiative power loss due to these impurities inhibits the plasma start-up and strongly influence the achievable plasma density limit.

2.7.2 Atomic Data for Nano-technology

The field of nanotechnology urgently requires the understanding the spectroscopy of heavy ion atoms useful in EUV lithography. In this direction, the EUV lithography at 135 Å is under great investigation with excited (in plasma) tin as the most probable fuel for the light sources [45]. The emission spectrum obtained experimentally shows a strong peak in the range 130–140 Å which appears due to $4p^6 4d^m$–$(4p^6 4d^{m-1} + 4p^5 4d^{m+1})$ transition in several ions, from Sn X ($m = 5$) to Sn XV ($m = 0$). Exact identification of this complex transition requires the study of similar transition in the iso-electronic ions of neighbouring lighter elements, which are still unknown. For example highly charged ions of In, Cd, Ag, Pd and Rh etc.

An accurate atomic data for modelling and diagnostic purpose is needed due to a great advancement in the spectroscopic technology requiring measurement of highly resolved spectrum of both Fusion and astrophysical plasma. The large-scale atomic data obtained so far by various atomic codes under different approximations and kept in various data-banks around the world revised for greater utility. There is also a growing need to calculate atomic data in the non-Maxwellian environments and a new data is required for heavy species [34].

2.8 Conclusions

We have discussed a number of recent developments in methodology resulting different computer codes for the calculation of atomic structure data needed for astrophysical, fusion plasmas & other fields: including Nano & Laser technologies. Lastly, we emphasize that; still there is great scope in future for the determination of accurate atomic data for heavy ions and atoms which are needed urgently in different fields as discussed in our work & which need fully relativistic treatment.

Acknowledgements M.M. is thankful to DST (India) and University of Delhi (India) under R & D research programme for funding this work. Sunny Aggarwal and Jagjit Singh are thankful to UGC (India) for junior research fellowship.

References

1. N.R. Badnell, Calculation of atomic data for astrophysics and fusion, in *Proceedings of the 7th Int. Conf. on Atomic and Mol. Data and Their Applications* Vilnius, Lithuania, 21–24 September 2010 ed. by A. Bernotas, R. Karazija, Z. Rudzikos. AIP Conf. Proceedings, vol. 1344 (AIP, New York, 2011), pp. 139–148
2. C.H. Skinner, Phys. Scr. T **134**, 014022 (2009)
3. M. Mohan (ed.), *Atomic Structure and Collision Processes* (Narosa, New Delhi, 2010)
4. A. Hibbert, Comput. Phys. Commun. **9**, 141 (1975)
5. W. Eissner, M. Jones, H. Nussbaumer, Comput. Phys. Commun. **8**, 270 (1974)
6. K.G. Dyall, I.P. Grant, C.T. Johnson, F.A. Parpia, E.P. Plummer, Comput. Phys. Commun. **55**, 425 (1989)
7. N.R. Badnell, J. Phys. B **19**, 3827 (1986)
8. I.P. Grant, B.J. Mckenzie, P.H. Norrington, D.F. Mayers, N.C. Pyper, Comput. Phys. Commun. **21**, 207 (1980)
9. I. Linderson, J. Morrison, *Atomic Many Body Theory* (Springer, Berlin, 1982)

10. R.D. Cowan, *The Theory of Atomic Structure and Spectra* (University of California Press, Berkeley, 1981)
11. M.F. Gu, Can. J. Phys. **86**, 675 (2008)
12. A. Bar Shalom, M. Klapisch, J. Oreg, J. Quant. Spectrosc. Radiat. Transf. **71**, 169 (2001)
13. K.T. Chen, M.H. Chen, Phys. Rev. A **53**, 2206 (1996)
14. S. Fritzsche, J. Electron Spectrosc. Relat. Phenom. **114**, 1155 (2001)
15. S. Hamansha, M. Abu-Allaban, A. Tahat, J. Appl. Sci. **11**, 2686 (2011)
16. M.J. Vilkas, Y. Ishikawa, K. Koc, J. Quantum Chem. **70**, 813 (1998)
17. The Opacity Project Team, *The Opacity Project*, vol. 1 (Institute of Physics Publications, Bristol, 1995)
18. J. Gurell, H. Nilsson, L. Engström, H. Lundberg, R. Blackwell-Whitehead, K.E. Nielsen, S. Mannervik, Astron. Astrophys. **511**, A68 (2010)
19. A. Hibbert, R. Glass, C.F. Fischer, Comput. Phys. Commun. **64**, 455 (1999)
20. I. Oksutzs, O. Sinanoglu, Phys. Rev. A **181**, 42 (1969)
21. A. Hibbert, Comput. Phys. Commun. **7**, 318 (1974)
22. U. Fano, Phys. Rev. A **67**, 140 (1965)
23. Y.K. Kim, Phys. Scr. T **73**, 19 (1997)
24. P. Indelicato, Private communication, 1999
25. M.I. Safronova, M.R. Johnson, V.I. Safronova, Phys. Rev. A **54**, 2850 (1996)
26. H.L. Zhang, D.H. Sampson, A.K. Mohanty, Phys. Rev. A **40**, 616 (1989)
27. M.Y. Amusia, L.V. Chernysheva, *Computation of Atomic Processes: A Handbook for ATOM Programs* (Institute of Physics Publishing, Bristol, 1997)
28. K.A. Berrington, W.B. Eissner, P.H. Norrington, Comput. Phys. Commun. **92**, 290 (1995)
29. F. Delahaye, A.K. Pradhan, J. Phys. B **35**, 3377 (2002)
30. P.H. Norrington, A.E. Kingston, A.W. Boone, J. Phys. B **33**, 1767 (2000)
31. S.N. Nahar, A.K. Pradhan, Astron. Astrophys. Suppl. Ser. **135**, 347 (1999)
32. L.A. Vainshtein, U.I. Safronova, Phys. Scr. **31**, 519 (1985)
33. D.R. Plante, W.R. Johnson, J. Sapirstein, Phys. Rev. A **49**, 3519 (1994)
34. M. Bergemann, J.C. Pickering, T. Gehren, Mon. Not. R. Astron. Soc. **401**, 1334 (2010)
35. E. Biemont et al., Can. J. Phys. **87**, 1275 (2009)
36. E. Biemont, Phys. Scr. T **119**, 55 (2005)
37. S. Goriely, Astron. Astrophys. **466**, 619 (2007)
38. H.P. Summers et al., Plasma Phys. **44**, B323 (2002)
39. C.H. Skinner et al., Phys. Scr. T **134**, 01402 (2009)
40. R.J. Hawryluk et al., Phys. Plasmas **5**, 1577 (1998)
41. A. Gibson, JET Team, Phys. Plasmas **5**, 1839 (1998)
42. M. Shimada et al., Nucl. Fusion **47**, 51 (2007)
43. G. Federic et al., Nucl. Fusion **41**, 1967 (2001)
44. B.J. Braams, H.K. Chung, Coordinated research projects of the IAEA atomic and molecular data unit, in *Proceedings of the 7th Int. Conf. on Atomic and Mol. Data and Their Applications* Vilnius, Lithuania, 21–24 September 2010 ed. by A. Bernotas, R. Karazija, Z. Rudzikos. AIP Conf. Proceedings, vol. 1344 (AIP, New York, 2011), 171–178, and reference therein
45. A.N. Ryabtsev, Spectroscopy of ionized atoms for nanotechnology, in *Proceedings of the 7th Int. Conf. on Atomic and Mol. Data and Their Applications* Vilnius, Lithuania, 21–24 September 2010 ed. by A. Bernotas, R. Karazija, Z. Rudzikos. AIP Conf. Proceedings, vol. 1344 (AIP, New York, 2011), 11–20

Chapter 3
Highly Charged Ions in Rare Earth Permanent Magnet Penning Traps

Nicholas D. Guise, Samuel M. Brewer, and Joseph N. Tan

Abstract A newly constructed apparatus at the United States National Institute of Standards and Technology (NIST) is designed for the isolation, manipulation, and study of highly charged ions. Highly charged ions are produced in the NIST electron-beam ion trap (EBIT), extracted through a beamline that selects a single mass/charge species, then captured in a compact Penning trap. The magnetic field of the trap is generated by cylindrical NdFeB permanent magnets integrated into its electrodes. In a room-temperature prototype trap with a single NdFeB magnet, species including Ne^{10+} and N^{7+} were confined with storage times of order 1 second, showing the potential of this setup for manipulation and spectroscopy of highly charged ions in a controlled environment. Ion capture has since been demonstrated with similar storage times in a more-elaborate Penning trap that integrates two coaxial NdFeB magnets for improved B-field homogeneity. Ongoing experiments utilize a second-generation apparatus that incorporates this two-magnet Penning trap along with a fast time-of-flight MCP detector capable of resolving the charge-state evolution of trapped ions. Holes in the two-magnet Penning trap ring electrode allow for optical and atomic beam access. Possible applications include spectroscopic studies of one-electron ions in Rydberg states, as well as highly charged ions of interest in atomic physics, metrology, astrophysics, and plasma diagnostics.

3.1 Background

A Penning trap combines a quadrupole electric potential with a strong magnetic field in order to trap a charged particle in three dimensions. The magnetic field provides radial confinement while the electric field provides confinement along the axis of the magnetic field. Penning traps see wide application in precision physics experiments

N.D. Guise (✉) · J.N. Tan
National Institute of Standards and Technology, Gaithersburg, MD 20899, USA
e-mail: nicholas.guise@nist.gov

S.M. Brewer
University of Maryland, College Park, MD 20742, USA

M. Mohan (ed.), *New Trends in Atomic and Molecular Physics*,
Springer Series on Atomic, Optical, and Plasma Physics 76,
DOI 10.1007/978-3-642-38167-6_3, © Springer-Verlag Berlin Heidelberg 2013

ranging from measurements of fundamental constants [10] to quantum information studies [3].

The use of Penning traps to recapture ions from accelerator or electron-beam ion trap (EBIT) sources has drawn longstanding interest at various facilities, as a means of extending the reach of spectroscopic studies with exotic ions/atoms and highly charged ions (HCI). Examples include antimatter studies [2, 6, 7], mass spectrometry and nuclear research at accelerator facilities [14], early experiments recapturing HCI from an EBIT source [25], and ongoing efforts in HCI laser spectroscopy [1] and mass measurement [24, 27]. The typical high-precision Penning trap utilized in these experiments requires a superconducting solenoid to generate its axial magnetic field of several Tesla. The associated cryostat makes a high-field Penning trap a large and expensive apparatus, similar in scale to an EBIT itself (see Sect. 3.2.2). A lower-field Penning trap, configured in similar geometry but using a room-temperature solenoid, may function as a pre-cooling stage; for example, evaporative cooling of Ar^{16+} ions was recently demonstrated in a 1.1 Tesla trap within the SMILETRAP II mass spectrometer [11].

An alternative Penning trap design relies upon rare-earth compounds such as $Nd_2Fe_{14}B$ (NdFeB) to generate the trapping magnetic field. While the magnetic field generated by rare earth magnets is typically less than 1 Tesla in the trapping region, the rare earth magnets are relatively inexpensive and allow increased flexibility in trap design. In earlier work, a compact Penning trap with its magnetic field generated by radially-oriented NdFeB wedges (assembled in two rings) was used to confine light ions loaded within a 77 K apparatus [9]; another trap that combined axially oriented magnets with epoxy-bonded arrays of radially-magnetized segments was used to store molecular anions at room temperature [29].

To facilitate operation at the high voltages necessary for catching highly charged ions from the NIST EBIT, we have developed two new compact Penning traps utilizing only cylindrical, axially-oriented NdFeB magnets in robust unitary architectures, with the trap electrodes and magnetic elements integrated as a unit. These traps are designed to provide easy access for laser and atomic beam manipulation of trapped ions, and to satisfy space constraints at the NIST EBIT facility. Here we report initial studies with highly charged ions produced in the EBIT, including bare nuclei, which are extracted and captured in these extremely compact Penning traps.

The primary goal of our experimental effort is to enable precise spectroscopic studies of one-electron (H-like) ions in circular states, synthesized from stored bare nuclei. Theoretical studies at NIST [12, 13] have shown that high angular momentum states of one-electron ions are better understood than low-lying states; in some cases, the energy level prediction for high-L states attains accuracy comparable to the precision of optical frequency combs. This offers a new avenue for the determination of fundamental constants, such as the Rydberg constant, if precise optical measurements can be realized to test theory. Such tests could potentially be useful in making improved measurements of the constants of nature that form the basis of modern metrology [17, 18]—efforts that are more compelling in the wake of the large discrepancy in proton radius determinations that resulted from recent muonic hydrogen Lamb-shift measurements [20, 22].

The ion-trapping techniques under development may also prove useful for other studies of highly charged ions. One example application (see Sect. 3.4) is measuring the lifetime of certain metastable states, where most previous studies have been performed inside an EBIT environment [15] or in Kingdon-type ion traps [5].

A brief outline of this paper follows:

- Section 3.2 reviews Penning trap basics, provides details of the compact NdFeB trap designs, and describes HCI extraction from the NIST EBIT.
- Section 3.3 presents initial results from HCI capture and storage.
- Section 3.4 discusses potential applications and future directions for this work.

3.2 Experimental Apparatus

3.2.1 Compact Penning Traps with Unitary Architecture

A basic cylindrical Penning trap (Fig. 3.1) consists of a ring electrode between two endcap electrodes. A near-quadrupole potential is created by applying voltage difference ΔV between the ring and the endcaps, forming a potential well along the z-axis but a potential hill in the radial dimension r. A magnetic field $\mathbf{B}$ applied along the z-axis provides radial confinement, completing the trap. In the canonical treatment of a single ion in a Penning trap [4], the trapped ion motion is described by a superposition of three oscillations: (1) a simple harmonic axial motion, at the "axial frequency" $2\pi\omega_z$; (2) a high-frequency cyclotron orbit around the axis of the magnetic field, at the trap-modified "cyclotron frequency" $2\pi\omega_c'$; and (3) a slower radial motion due to $\mathbf{E} \times \mathbf{B}$ drift, at the "magnetron frequency" $2\pi\omega_m$. These frequencies typically observe the hierarchy $\omega_c' \gg \omega_z \gg \omega_m$.

The axial frequency is defined by

$$\omega_z^2 = \lambda \frac{q\,\Delta V}{md^2},\tag{3.1}$$

where q is the ion charge, m is the ion mass, and parameters d and λ depend on trap geometry: d is an effective trap dimension given by

$$d^2 = \frac{1}{2}\left(z_0^2 + r_0^2/2\right),\tag{3.2}$$

while λ describes the degree to which the electrodes produce a quadrupole potential; $\lambda \approx 1$ for near-hyperbolic surfaces and small-amplitude oscillations. The distances z_0 and r_0 are shown in Fig. 3.1.

The free-space cyclotron frequency in a magnetic field of magnitude B is given by

$$\omega_c = \frac{q\,B}{m}.\tag{3.3}$$

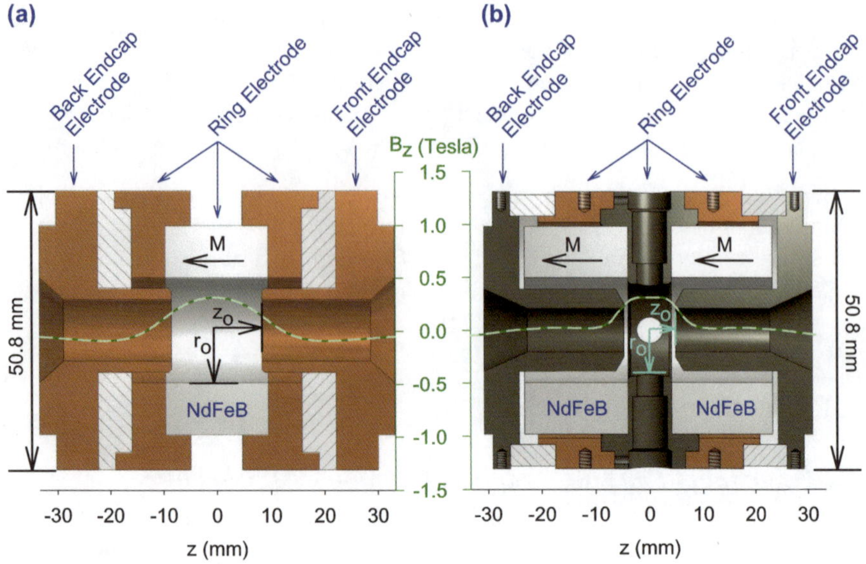

Fig. 3.1 Cross-sectional views of (**a**) one-magnet; and (**b**) two-magnet compact NdFeB Penning traps. The cylindrical symmetry axis is horizontal. Vector *M* indicates the direction of remnant magnetization for each NdFeB magnet. The on-axis magnetic field profile for each trap is plotted as an overlay (*green dashed line*). Detailed field profiles are provided in a separate publication [31]

The presence of the radial electrostatic potential in a Penning trap shifts this cyclotron frequency to its trap-modified value,

$$\omega_c' = \omega_c - \omega_m, \tag{3.4}$$

where ω_m is the frequency of the magnetron motion,

$$\omega_m = \frac{1}{2}\left(\omega_c - \sqrt{\omega_c^2 - 2\omega_z^2}\right). \tag{3.5}$$

To maintain localized orbits, the magnetron frequency must be real, hence

$$\omega_c^2 - 2\omega_z^2 > 0 \tag{3.6}$$

is a confinement criterion for Penning traps.

We present two designs for compact Penning traps utilizing NdFeB permanent magnets in a unitary architecture. Table 3.1 provides frequencies of motion for various HCI species of interest, in the one-magnet (Sect. 3.2.1.1) and two-magnet (Sect. 3.2.1.2) trap designs. Values of parameter λ (Eq. (3.1)) are determined by numerical simulation. Frequencies are calculated using Eqs. (3.1)–(3.5) for the simple case of a single ion oscillating near trap center. For ion clouds of significantly larger number or physical size, these frequencies are expected to shift due to trap anharmonicities and magnetic field gradients—precise trap frequencies in this regime may be obtained through numerical simulation.

Table 3.1 Approximate motional frequencies $2\pi\omega_z$, $2\pi\omega_c'$, and $2\pi\omega_m$ for various highly charged ions confined in the compact Penning trap designs of Fig. 3.1. The applied trapping well is $\Delta V = 10$ V, and the magnetic field has $B_z = 0.3$ T. Calculated values assume a single trapped ion in the limit of small-amplitude oscillations

Ion	One-magnet trap, $\lambda = 0.814$			Two-magnet trap, $\lambda = 0.854$		
	Axial freq. (kHz)	Cyclotron freq. (kHz)	Magnetron freq. (kHz)	Axial freq. (kHz)	Cyclotron freq. (kHz)	Magnetron freq. (kHz)
^{20}Ne^{10+}	415	2266	38.0	597	2224	80.1
^{20}Ne^{9+}	393	2036	38.0	566	1994	80.5
^{20}Ne^{8+}	371	1805	38.1	534	1763	80.9
^{40}Ar^{16+}	371	1806	38.1	534	1764	80.9
^{40}Ar^{15+}	359	1691	38.2	517	1648	81.2
^{40}Ar^{14+}	347	1576	38.2	500	1533	81.5
^{40}Ar^{13+}	334	1460	38.3	481	1417	81.8

3.2.1.1 One-Magnet Penning Trap

Our simplest trap design (Fig. 3.1a) uses a single NdFeB magnet as the ring electrode in an uncompensated open-endcap Penning trap. The NdFeB magnet is sandwiched between oxygen-free high-conductivity (OFHC) copper rings through which we apply the ring bias voltage. The trap endcap electrodes are machined from OFHC copper. MACOR ceramic glass spacers provide electrical isolation between neighboring electrodes. Characteristic trap dimensions for the one magnet trap are $r_o = 9.53$ mm and $z_o = 8.39$ mm. The value of parameter λ (Eq. (3.1)) is 0.814. Typical oscillation frequencies for representative ion species are provided in Table 3.1.

3.2.1.2 Two-Magnet Penning Trap

A design with two NdFeB magnets (Fig. 3.1b) increases the complexity of assembly but yields several advantages over the one-magnet design. Positioning the magnets behind electrical iron endcaps serves to minimize the effect of physical and magnetic imperfections, and the gap between the magnets now permits holes in the $z = 0$ plane, for laser or atomic beam access into the trapping region. Our two-magnet trap design also has improved B-field homogeneity in the range $z = 2$ mm to $z = -2$ mm, compared to the one-magnet design [31].

Characteristic trap dimensions for the two-magnet trap are $r_0 = 8.50$ mm and $z_0 = 4.74$ mm. The two NdFeB magnets dovetail onto an electrical iron yoke and two OFHC copper rings, together comprising the ring electrode. The iron yoke has four evenly spaced radial holes, one of which contains an aspheric lens for light collection. The endcap electrodes are machined from electrical iron. MACOR spacers are again used to provide electrical isolation between neighboring electrodes.

The value of parameter λ (Eq. (3.1)) is 0.854. Typical oscillation frequencies for representative ion species are provided in Table 3.1.

3.2.2 Production and Extraction of Highly Charged Ions

The electron-beam ion trap (EBIT) is a laboratory device for producing highly charged ions of various species. Invented in the late 1980s [16], EBITs now appear in multiple facilities worldwide. The NIST EBIT [8], in operation since 1993, incorporates an electron gun capable of generating a beam of current larger than 100 mA within an ultrahigh vacuum enclosure. The nearly mono-energetic electron beam is accelerated through high voltage, reaching kinetic energies up to 30 keV as it travels through the central axis of a stack of cylindrical drift tubes. Superconducting Helmholtz coils generate a strong axial magnetic field (≈ 3 Tesla) which compresses the radial extent of the electron beam, producing an extremely high current density. Highly charged ions are produced through repeated electron-impact ionization of an injected gas. These positively-charged ions are trapped in a configuration that resembles an open-endcap Penning trap, with the cylindrical drift tubes biased to form an electrostatic trapping potential along the axis, but with the highly compressed electron beam also providing radial confinement.

The EBIT produces a mixture of several charge states, determined by various tuning parameters and by the available ionizing energy of the electron beam. This HCI mixture is extracted by quickly ramping the EBIT middle drift tube up in voltage. As the pulse of ions accelerates out of the drift tube region, each ion acquires substantial kinetic energy, roughly equal to the ion charge times the effective EBIT operating voltage. Typical energies for the experiments described below are between 2 keV to 4 keV per unit charge. In the following sections we describe how this ion pulse is controlled and recaptured into a Penning trap made extremely compact with a unitary architecture.

3.2.2.1 EBIT Extraction Beamline

Figures 3.2 and 3.3 provide a schematic overview of the experiment. The approximate path length from the EBIT to the Penning trap capture/detection region is 7.2 meters. Specifics of the NIST EBIT extraction beamline are provided elsewhere [23].

The various tuning and focusing elements in the extraction beamline nominally maintain ion kinetic energy, and all but the analyzing magnet act equally on ions of different mass/charge. The analyzing magnet allows selection of a particular ion charge state. At the proper analyzing magnetic field value for a given mass/charge ratio, the ion is deflected by the necessary amount to continue through the beamline; it then passes through the Penning trap and strikes a Faraday cup or microchannel plate (MCP) for detection.

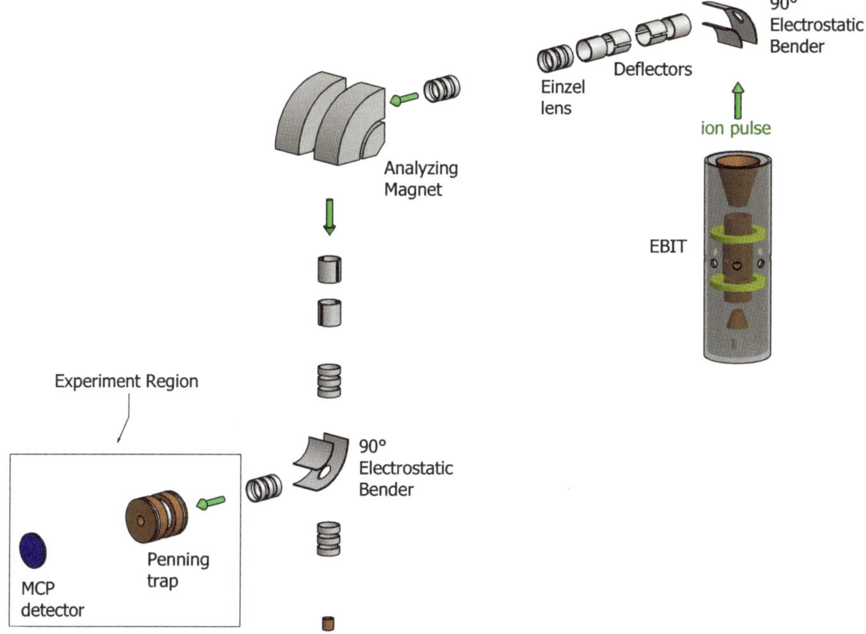

Fig. 3.2 Cartoon of extracted ion path from EBIT to the compact Penning trap. Details of the experiment region are shown in Fig. 3.3. Details of EBIT beamline elements are as previously published [23]

Figure 3.4 shows the result of a "mass scan" performed by sweeping the analyzing magnetic field and tracking the detected ion signal on the time-of-flight MCP. Charge states are identified from the spacing between peaks and the known ion kinetic energy. MCP signals may then be adjusted for the charge state of each ion to extract the actual number of each charge state striking the detector (Fig. 3.4a). The assignment of charge states is confirmed by plotting the arrival time of each ion at the TOF detector, measured as a delay from the EBIT dump trigger (Fig. 3.4b). This arrival time, dominated by the transit time through the EBIT extraction beamline, shows the expected linear dependence on the square root of ion mass/charge ratio, for lossless motion in conservative electric potentials. In particular, the time t required for an ion to traverse the extraction beamline of length L is given by

$$t = \int_0^L \frac{dx}{v(x)} = \sqrt{\frac{m}{q}} \int_0^L \frac{dx}{\sqrt{2(V_0 - V(x))}}, \qquad (3.7)$$

where the x-coordinate is directed along the beamline, $v(x)$ is the ion velocity, $V(x)$ is the electric potential at position x (except near deflectors or focusing elements, $V(x) \approx 0$), and V_0 is the effective operating voltage of the EBIT, i.e. the initial kinetic energy of an extracted ion is $q V_0$.

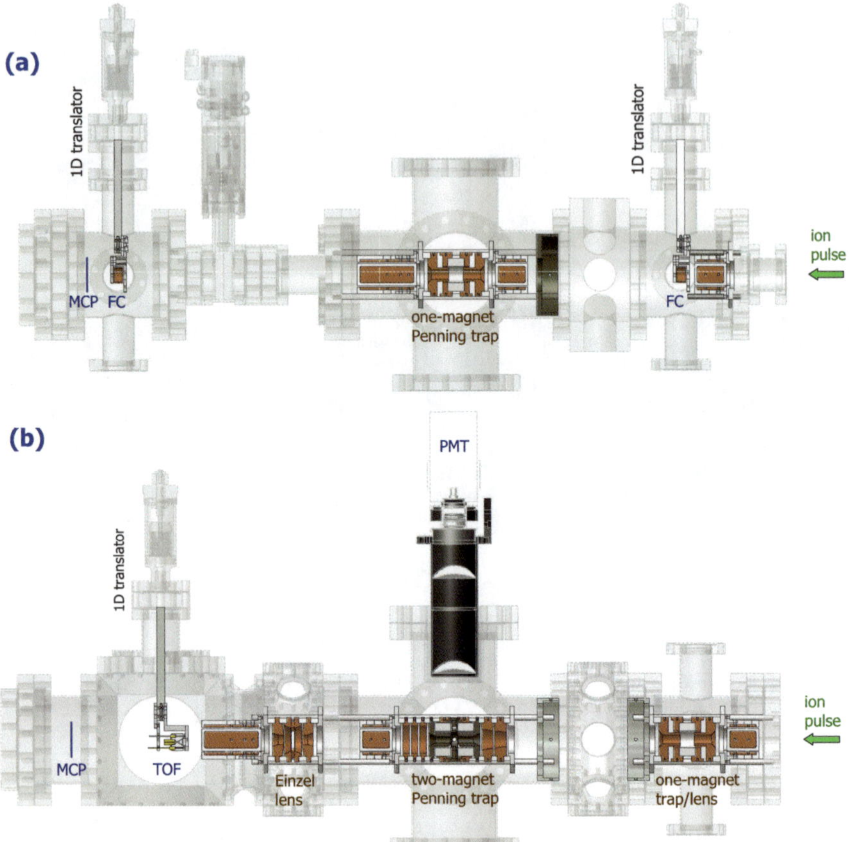

Fig. 3.3 Detailed view of the experiment region for two different Penning trap configurations: (**a**) Beamline incorporating the one-magnet Penning trap (Fig. 3.1a) and three sets of steering plates; detection elements include retractable Faraday cups (FC) and a position-sensitive microchannel plate (MCP). (**b**) Beamline incorporating the two-magnet Penning trap (Fig. 3.1b), with additional focusing elements, a retractable time-of-flight MCP detector (TOF), and visual access to a photomultiplier tube (PMT)

After setting the analyzing magnetic field to isolate the charge state of interest, we can use steering and focusing elements to optimize the beam image on the position-sensitive MCP detector. Figure 3.5 shows representative beam spot images taken while operating the one-magnet Penning trap as an Einzel lens for focusing—this is the primary function of the one-magnet trap in our two-magnet experiment configuration (Fig. 3.3b). The ion beam is centered by applying correction voltages to three sets of orthogonal steering plates, apertured through the trap electrodes, and focused to a spot as small as ≈ 3 mm diameter on the detector.

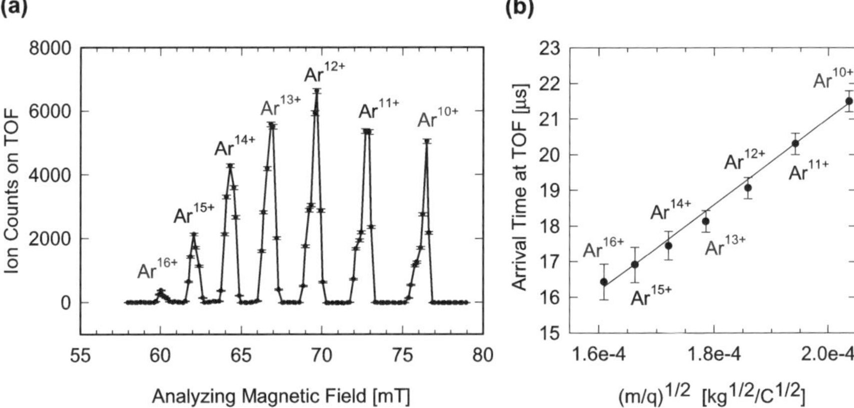

Fig. 3.4 (a) Typical "mass scan" of argon ions extracted from the EBIT with $V_0 \approx 2.85$ kV (see Eq. (3.7)). The number of ions is counted by measuring the signal on a time-of-flight MCP detector as the analyzing magnetic field is varied to select out different mass/charge states. (b) Arrival times of various argon ions at the detector, relative to the EBIT dump trigger. *Error bars* represent 1σ uncertainty

3.2.2.2 Capture and Detection of Extracted Ions

To capture ions extracted from the EBIT, the Penning trap electrodes are biased in a step configuration, with the front endcap "low," the back endcap "high," and the ring electrode at an intermediate voltage. Each electrode then receives an additional constant high voltage bias, roughly equal to the effective EBIT operating voltage, in order to match the kinetic energy of incoming ions. As the ions approach the Penning trap, they climb the potential hill and slow down; once inside the Penning trap region, the ions are then captured by rapidly increasing the front endcap voltage to match the back endcap voltage, closing the trap and creating a local potential well centered on the ring electrode. Once captured, ions may be held for several seconds with the trap voltages in this storage configuration, then dumped to a detector by lowering the back endcap voltage. High voltage pulsers, capable of rise times <100 ns, are used to accomplish the fast switching between voltage settings for ion capture, storage, and dumping.

A simplified timing diagram for the experiment is shown in Fig. 3.6. Ions are extracted from the EBIT by pulsing the middle drift tube up in voltage. After waiting some amount of time (t_{capture}) for the extracted ions to transit from the EBIT, the front endcap then pulses up in voltage to capture ions just after they arrive in the Penning trap region. After an additional storage time (t_{storage}), the Penning trap back endcap pulses down in voltage to dump the ions towards the MCP detector. A trigger pulse simultaneous with the back endcap pulser allows synchronization of the detection electronics.

The Penning trap front endcap must pulse closed at precisely the right time in order to catch any ions. If the trap closes too soon, the ions will not yet have reached

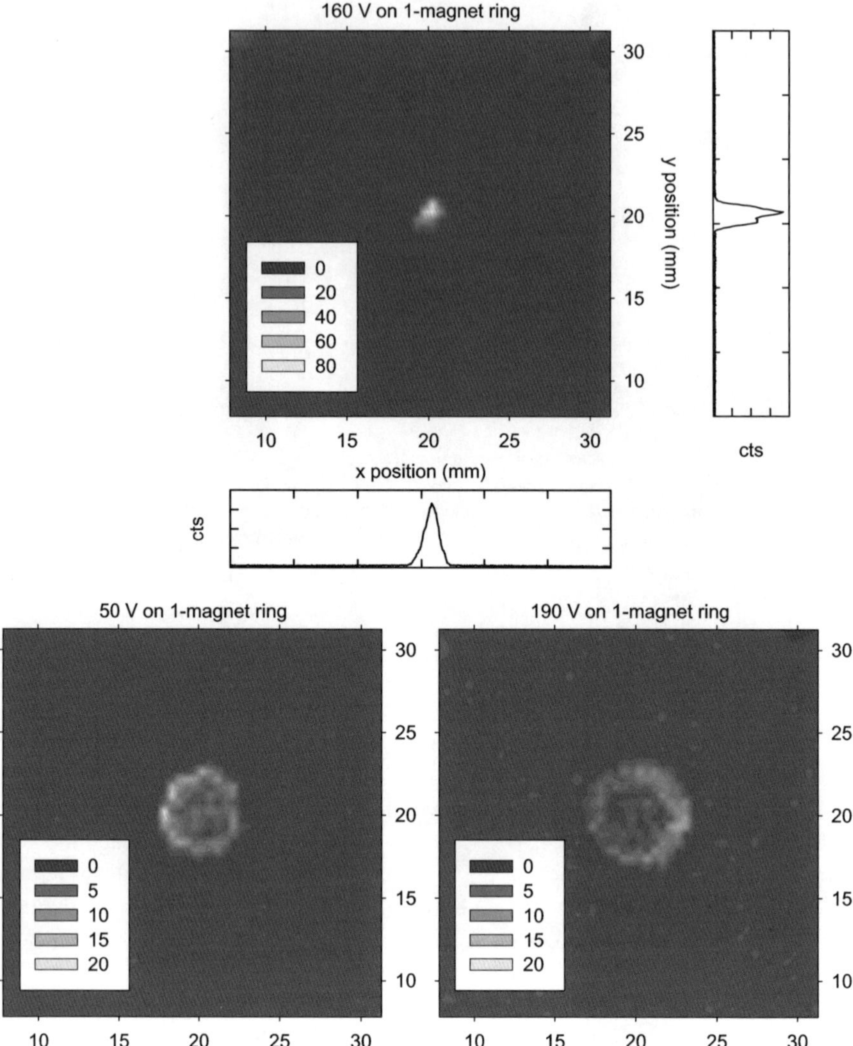

Fig. 3.5 Beam spots observed on the XY position-sensitive MCP detector. Ne^{10+} ions passing through the one-magnet experiment beamline (Fig. 3.3a) are focused by applying different tuning potentials to the one-magnet Penning trap, which functions as an Einzel lens. *Shading* indicates number of ions

the trapping region; if too late, they will have already exited, having been turned around by the high back endcap potential. Experimentally we find that timing resolution of order 10 ns is sufficient to optimize the "capture time" at which the trap is closed. To determine the best capture settings for a given ion species, we set the storage time to a short and fixed value (typically 1 ms), then sweep the capture time to optimize the number of detected ions. Results of this optimization are shown

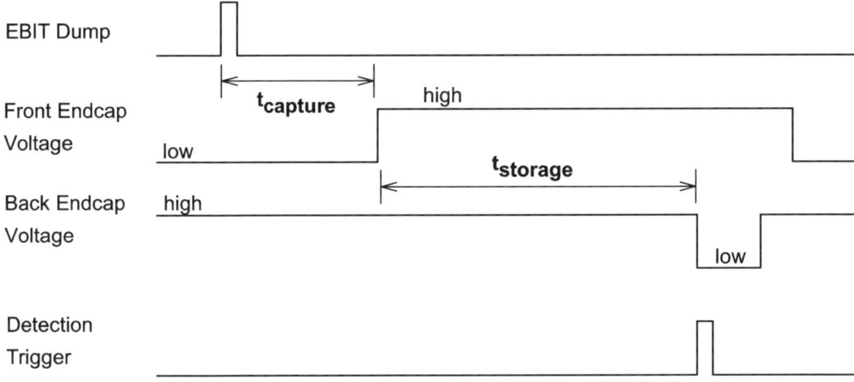

Fig. 3.6 Simplified timing diagram for ion capture, storage, and detection

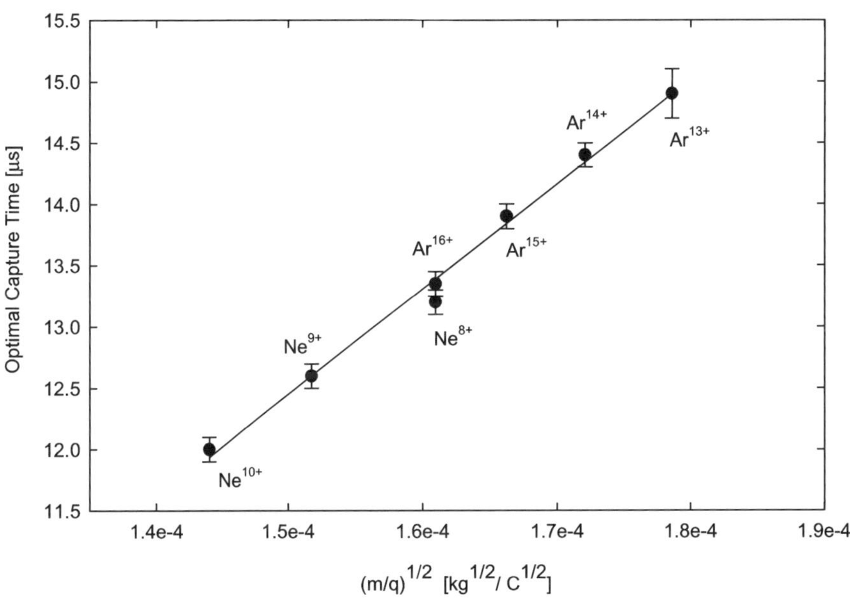

Fig. 3.7 Optimal capture time ($t_{capture}$ in Fig. 3.6) for various ion species in the two-magnet Penning trap. The effective EBIT extraction voltage (V_0 in Eq. (3.7)) is ≈ 4.3 kV. *Error bars* represent 1σ uncertainty

in Fig. 3.7 for various ion species captured in the two-magnet Penning trap. As in Fig. 3.4b, this is essentially a measure of transit time from the EBIT to the Penning trap, and we observe the expected linear dependence on the square root of ion mass/charge ratio (Eq. (3.7)).

3.3 Experimental Results

We have demonstrated the effectiveness of compact Penning traps with unitary architecture for capture and storage of highly charged ions, using both trap designs presented in Fig. 3.1. Initial experiments with highly charged ions extracted from the NIST EBIT utilized the one-magnet trap and the experiment beamline shown in Fig. 3.3a. Recent experiments utilize the two-magnet trap and the more complex setup of Fig. 3.3b. Capture and detection follows the scheme described in Fig. 3.6. Sample results are presented below.

3.3.1 One-Magnet Penning Trap with Position-Sensitive MCP Detector

Ion species captured in the one-magnet Penning trap include Ne^{10+}, Ne^{9+}, Ne^{8+}, Ar^{16+}, Ar^{15+}, Ar^{14+}, Ar^{13+}, and N^{7+}. The only means of ion detection in the one-magnet experiment configuration (Fig. 3.3a) is a position-sensitive MCP detector. This detector has an active area of diameter 40 mm, with a resistive anode encoder that correlates each ion count event with a position of impact on the MCP. While useful for beam tuning (e.g. Fig. 3.5), this encoding has the unwanted side-effect of relatively slow response time, of order 1 μs. To avoid dead-time errors generated by large ion count rate, we use a purposefully inefficient dump scheme. When dumping ions from the Penning trap, the back endcap voltage is set equal to or slightly higher than the ring voltage. Without a strong potential gradient across the trap, many ions escape to the electrode walls, but a fraction of the ions spill out towards the detector. For trap storage times longer than 0.5 seconds, this scheme results in ion numbers low enough to be individually counted without evidence of pileup on the position-sensitive detector.

Figure 3.8 shows the result of storage time measurements in the one-magnet Penning trap. The "low" voltage applied to the back endcap electrode while dumping the trap is set nominally equal to the ring voltage, intentionally reducing detected ion counts as described above. Storage times are measured by fitting a single exponential decay curve to the data. Residual gas pressure in the trapping chamber was roughly 2.4×10^{-7} Pa (1.8×10^{-9} Torr).

The one-magnet trap beamline (Fig. 3.3a) is intended primarily as a proof-of-concept prototype; the position-sensitive detector confirms the capture of HCI but cannot count large ion fluxes nor distinguish between different charge states. However, the storage times observed in Fig. 3.8, accounting for higher background gas pressure, are comparable to those observed in the two-magnet trap for the sum of all charge states (Fig. 3.9c, Sect. 3.3.2). The simplicity of the one-magnet design makes it an attractive option as a basic ion trap or focusing element; addition of a TOF detector (Sect. 3.3.2), for example, would enable further studies of charge-exchange and trap dynamics.

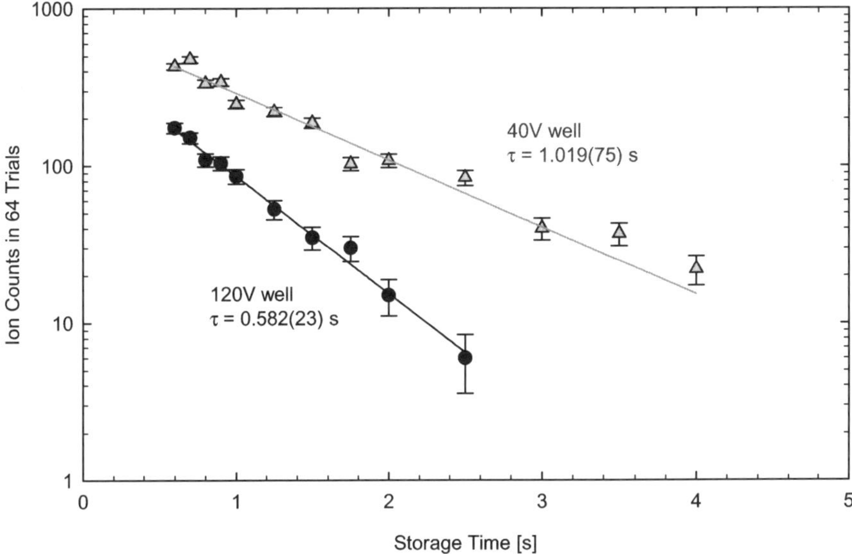

Fig. 3.8 Storage of Ne^{10+} ions in the one-magnet Penning trap. The number of ions detected on the position-sensitive MCP detector is plotted as a function of storage time, with applied potential differences of $\Delta V = 40$ V (*triangles*) and $\Delta V = 120$ V (*circles*) between ring and endcap electrodes. *Error bars* represent 1σ uncertainty

3.3.2 Two-Magnet Penning Trap with Time-of-Flight (TOF) MCP Detector

Ion species captured in the two-magnet Penning trap include Ne^{10+}, Ne^{9+}, Ne^{8+}, Ar^{16+}, Ar^{15+}, Ar^{14+}, Ar^{13+}, and Kr^{17+}. The two-magnet experiment beamline (Fig. 3.3b) incorporates the one-magnet trap as a focusing element and contains two additional detection options compared to the one-magnet trap beamline. First, trapped ions may be detected optically via a photomultiplier tube (Sect. 3.4). Second, a time-of-flight (TOF) MCP detector, mounted on a one-dimensional translator, may be inserted into the beamline in front of the position-sensitive detector. The TOF detector has active diameter of 8 mm and response time of order 1 ns. High voltage bias across the TOF MCP selects one of two operating modes: pulse-counting or analog amplification. Our experiments generally utilize the analog mode, in which the TOF detector functions as a fast current amplifier with gain of 10^5 to 10^6. Ion counts are extracted from the TOF signal strength by accounting for the known charge state of the incident ion.

Figure 3.9 shows ion storage measurements in the two-magnet Penning trap. The back endcap dump voltage is set to 400 V below the ring voltage, ramping the ions out in a narrow pulse; since the TOF detector is capable of much higher ion count rates than the position-sensitive detector, pileup errors (Sect. 3.3.1) are not observed even for storage times as short as ≈ 1 ms. Time resolution of the TOF detector is

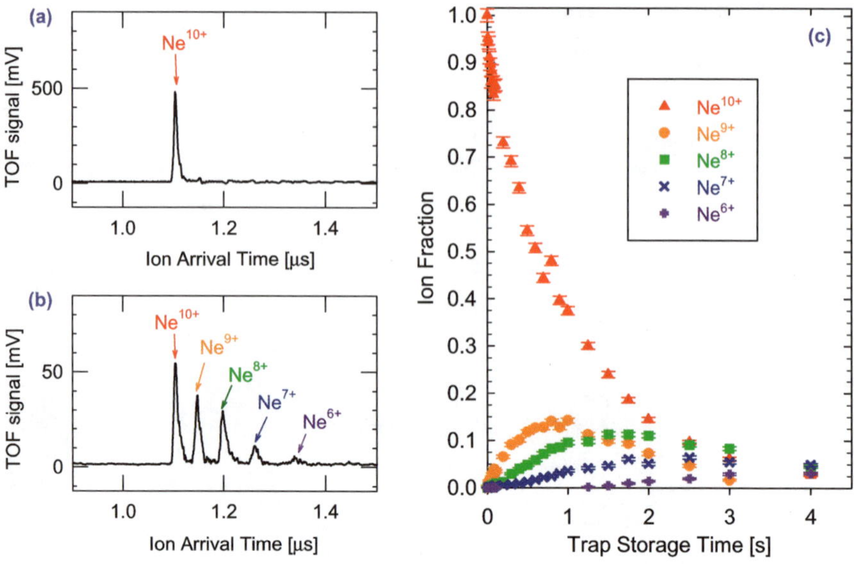

Fig. 3.9 Detection of highly charged ions in the two-magnet Penning trap, demonstrating charge-state resolution with the time-of-flight MCP detector (TOF). Neon ions are confined in the trap for storage times ranging from 1 ms to 4 s, with applied potential difference $\Delta V = 10$ V between the ring and endcap electrodes. Output of the TOF detector versus arrival time relative to the Penning trap dump trigger is shown for storage times of (**a**) 1 ms and (**b**) 2 s. The detector signal scale is magnified by $10\times$ from (**a**) to (**b**). The TOF signal peak for each charge state is converted to ion counts, and the evolution of charge states via charge exchange is shown in (**c**), normalized to the initial population of Ne^{10+} ions observed at the 1 ms storage time. *Error bars represent 1σ uncertainty*

sufficiently fast to permit the observation of ion charge-exchange processes. The ion species initially loaded in the two-magnet trap undergoes charge-exchange collisions with residual background gas. The product ion species will in general still oscillate with frequencies that satisfy the Penning trap confinement criterion (Eq. (3.6), Table 3.1), hence it will remain trapped and can continue to charge-exchange down to lower charge states. Figure 3.9c shows the evolution of lower charge states as a function of storage time in the two-magnet trap. The charge-exchange products are detected as separate peaks in the TOF detector signal (Fig. 3.9a-b), delayed from the initial ion peak due to higher mass/charge.

Residual background gas pressure for these measurements is 1.7×10^{-7} Pa (1.3×10^{-9} Torr). We observe an exponential decay in the number of ions detected on the TOF MCP detector as a function of trap storage time: the initial charge state (Ne^{10+}; red triangles in Fig. 3.9c) decays with time constant 1.09(2) s, and the sum of all charge states decays with time constant 2.41(6) s. Decay rates of both the initial charge state (dominated by charge-exchange), and of the sum of all charge states (dominated by ion cloud expansion/loss), are observed to increase linearly with background gas pressure [31].

3.4 Summary and Outlook

We have developed compact Penning traps that integrate cylindrical NdFeB permanent magnets and trap electrodes in unitary architectures. These traps are used to capture and store highly charged ions extracted from the NIST EBIT source, in particular charge states of neon, argon, krypton, and nitrogen. Ion storage times of order 1 second, limited primarily by background gas pressure, are demonstrated in both a one-magnet trap designed for simplicity of construction and in a two-magnet trap designed for field homogeneity and optical access to the trapping region. The unitary architecture could find various applications in compact instrument development, for example as portable mass analyzers.

Our ongoing work at NIST focuses on ion studies in the two-magnet Penning trap. Measurements of collisions, charge exchange rates, effective ion temperatures, and cooling rates are possible in the existing apparatus. Base pressure as low as 1.0×10^{-7} Pa (7.8×10^{-10} Torr) has been obtained at room temperature, yielding ion storage times up to 3.8 seconds. A precision leak valve and gas injector nozzle allow for controlled modification of the background gas pressure in the trapping chamber. An upcoming publication will provide details on the capture process and energy distribution of ions extracted from the EBIT.

The demonstrated ion storage times of >1 second are sufficient for various spectroscopic measurements of interest to astrophysics and plasma diagnostics. One application particularly suited to this apparatus is measurement of metastable lifetimes. Well-established efforts at NIST and other EBIT facilities have used interference filters or monochromators to detect fluorescence emitted by ions decaying from metastable levels [32], e.g. observing the magnetic dipole decay from Ar XIV $2p\ ^2P_{3/2}$ to $2p\ ^2P_{1/2}$, where the metastable $2p\ ^2P_{3/2}$ level is populated during ion production in the hot EBIT plasma [15, 26]. Most such measurements in the past have been performed inside an EBIT, producing highly charged ions and then turning off the electron beam long enough to measure fluorescence. However, since the extraction time of ions from the NIST EBIT is of order 10 μs, lifetimes of milliseconds or longer may also be observed by detecting fluorescence from ions recaptured in our compact Penning trap. The two-magnet trap beamline (Fig. 3.3b) includes a lens system with filter and photomultiplier tube for this purpose. Working with extracted ions in the controlled Penning trap environment allows direct study of certain systematics; for example, pressure dependence of the decay lifetime may be explicitly measured. Potential disadvantages include a limited detection solid angle and the possibility of new systematic effects due to trapped ion cloud dynamics. To increase detection solid angle and enable a second systematic check, we are investigating an alternative radiofrequency (RF) ion trap with electrodes that contain large slits for optical access. The RF trap (currently functioning as an Einzel lens) is located in the experiment beamline (Fig. 3.3b) between the two-magnet Penning trap and TOF detector; it could be loaded either directly from the EBIT or following an initial ion capture stage in the Penning trap.

Longer term applications are motivated by efforts at NIST to "engineer" ion states that are not readily produced in the EBIT. Recent proposals involve producing hydrogenlike ions in circular Rydberg states, for precision spectroscopy to test theory and measure fundamental constants [12, 13, 30]. We have now demonstrated an initial step in realizing this proposal—the capture of bare nuclei (Ne^{10+}) in a suitable ion trap. One possibility for attaching electrons with high angular momentum is charge-exchange with Rydberg atoms [28]; an atomic beam could be introduced through one of the remaining holes in the two-magnet Penning trap.

For related work with low-Z ions, for which the full energy of the NIST EBIT is unnecessary, the NdFeB Penning traps may also be integrated into a new apparatus with a gas injector and small electron gun, enabling similar experiments at greatly reduced operating cost. Compact HCI sources based on permanent magnets have been demonstrated at several other facilities; designs range from early "warm" EBIT designs utilizing two rings of $SmCo_5$ [21] or NdFeB [19], to a recent low-energy EBIT featuring 40 NdFeB and soft iron elements [33].

Acknowledgements This research was performed while one author (NDG) held a National Research Council Research Associateship Award at NIST.

References

1. Z. Andjelkovic, S. Bharadia, B. Sommer, M. Vogel, W. Nörtershäuser, Towards high precision in-trap laser spectroscopy of highly charged ions. Hyperfine Interact. **196**, 81–91 (2010). doi:10.1007/s10751-009-0155-x
2. G.B. Andresen, M.D. Ashkezari, M. Baquero-Ruiz, W. Bertsche, P.D. Bowe, E. Butler, C.L. Cesar, M. Charlton, A. Deller, S. Eriksson, J. Fajans, T. Friesen, M.C. Fujiwara, D.R. Gill, A. Gutierrez, J.S. Hangst, W.N. Hardy, R.S. Hayano, M.E. Hayden, A.J. Humphries, R. Hydomako, S. Jonsell, S.L. Kemp, L. Kurchaninov, N. Madsen, S. Menary, P. Nolan, K. Olchanski, A. Olin, P. Pusa, C.O. Rasmussen, F. Robicheaux, E. Sarid, D.M. Silveira, C. So, J.W. Storey, R.I. Thompson, D.P. van der Werf, J.S. Wurtele, Y. Yamazaki, Confinement of antihydrogen for 1,000 seconds. Nat. Phys. **7**, 558–564 (2011). doi:10.1038/nphys2025
3. M.J. Biercuk, H. Uys, A.P. Vandevender, N. Shiga, W.M. Itano, J.J. Bollinger, High-fidelity quantum control using ion crystals in a Penning trap. Quantum Inf. Comput. **9**(11), 920–949 (2009). http://dl.acm.org/citation.cfm?id=2012098.2012100
4. L.S. Brown, G. Gabrielse, Geonium theory: Physics of a single electron or ion in a Penning trap. Rev. Mod. Phys. **58**, 233–311 (1986). doi:10.1103/RevModPhys.58.233
5. D.A. Church, D.P. Moehs, M.I. Bhatti, Precision and accuracy of lifetimes of metastable levels using the Kingdon trap technique. Int. J. Mass Spectrom. **192**(1–3), 149–155 (1999). doi:10.1016/S1387-3806(99)00118-9
6. G. Gabrielse, X. Fei, K. Helmerson, S.L. Rolston, R. Tjoelker, T.A. Trainor, H. Kalinowsky, J. Haas, W. Kells, First capture of antiprotons in a penning trap: a kiloelectronvolt source. Phys. Rev. Lett. **57**, 2504–2507 (1986). doi:10.1103/PhysRevLett.57.2504
7. G. Gabrielse, P. Larochelle, D. Le Sage, B. Levitt, W.S. Kolthammer, R. McConnell, P. Richerme, J. Wrubel, A. Speck, M.C. George, D. Grzonka, W. Oelert, T. Sefzick, Z. Zhang, A. Carew, D. Comeau, E.A. Hessels, C.H. Storry, M. Weel, J. Walz, Antihydrogen production within a Penning-Ioffe trap. Phys. Rev. Lett. **100**, 113001 (2008). doi:10.1103/PhysRevLett.100.113001
8. J.D. Gillaspy, First results from the EBIT at NIST. Phys. Scr. **T71**, 99–103 (1997). http://stacks.iop.org/1402-4896/1997/i=T71/a=017
9. V. Gomer, H. Strauss, D. Meschede, A compact Penning trap for light ions. Appl. Phys. B, Lasers Opt. **60**, 89–94 (1995). doi:10.1007/BF01135848

10. D. Hanneke, S. Fogwell, G. Gabrielse, New measurement of the electron magnetic moment and the fine structure constant. Phys. Rev. Lett. **100**, 120801 (2008). doi:10.1103/PhysRevLett.100.120801

11. M. Hobein, A. Solders, M. Suhonen, Y. Liu, R. Schuch, Evaporative cooling and coherent axial oscillations of highly charged ions in a Penning trap. Phys. Rev. Lett. **106**, 013002 (2011). doi:10.1103/PhysRevLett.106.013002

12. U.D. Jentschura, P.J. Mohr, J.N. Tan, B.J. Wundt, Fundamental constants and tests of theory in Rydberg states of hydrogenlike ions. Phys. Rev. Lett. **100**, 160404 (2008). doi:10.1103/PhysRevLett.100.160404

13. U.D. Jentschura, P.J. Mohr, J.N. Tan, Fundamental constants and tests of theory in Rydberg states of one-electron ions. J. Phys. B, At. Mol. Opt. Phys. **43**(7), 074002 (2010). http://stacks.iop.org/0953-4075/43/i=7/a=074002

14. H.-J. Kluge, K. Blaum, F. Herfurth, W. Quint, Atomic and nuclear physics with stored particles in ion traps. Phys. Scr. **T104**, 167–177 (2003). http://stacks.iop.org/1402-4896/2003/i=T104/a=034

15. A. Lapierre, J.R. Crespo López-Urrutia, J. Braun, G. Brenner, H. Bruhns, D. Fischer, A.J. González Martínez, V. Mironov, C. Osborne, G. Sikler, R. Soria Orts, H. Tawara, J. Ullrich, V.M. Shabaev, I.I. Tupitsyn, A. Volotka, Lifetime measurement of the Ar XIV $1s^2 2s^2 2p^2 P^o_{3/2}$ metastable level at the Heidelberg electron-beam ion trap. Phys. Rev. A **73**, 052507 (2006). doi:10.1103/PhysRevA.73.052507

16. M.A. Levine, R.E. Marrs, J.R. Henderson, D.A. Knapp, M.B. Schneider, The electron beam ion trap: A new instrument for atomic physics measurements. Phys. Scr. **1988**(T22), 157 (1988). http://stacks.iop.org/1402-4896/1988/i=T22/a=024

17. P.J. Mohr, Defining units in the quantum based SI. Metrologia **45**(2), 129–133 (2008). http://stacks.iop.org/0026-1394/45/i=2/a=001

18. P.J. Mohr, The quantum SI: a possible new international system of units, in *Current Trends in Atomic Physics*, ed. by S. Salomonson, E. Lindroth. Advances in Quantum Chemistry, vol. 53 (Academic Press, San Diego, 2008), pp. 27–36. doi:10.1016/S0065-3276(07)53003-0. Chapter 3

19. K. Motohashi, A. Moriya, H. Yamada, S. Tsurubuchi, Compact electron-beam ion trap using NdFeB permanent magnets. Rev. Sci. Instrum. **71**(2), 890–892 (2000). doi:10.1063/1.1150323

20. F. Nez, A. Antognini, F.D. Amaro, F. Biraben, J.M.R. Cardoso, D. Covita, A. Dax, S. Dhawan, L. Fernandes, A. Giesen, T. Graf, T.W. Hänsch, P. Indelicato, L. Julien, C.-Y. Kao, P.E. Knowles, E. Le Bigot, Y.-W. Liu, J.A.M. Lopes, L. Ludhova, C.M.B. Monteiro, F. Mulhauser, T. Nebel, P. Rabinowitz, J.M.F. Dos Santos, L. Schaller, K. Schuhmann, C. Schwob, D. Taqqu, J.F.C.A. Veloso, F. Kottmann, R. Pohl, Is the proton radius a player in the redefinition of the International System of Units? Philos. Trans. R. Soc. A, Math. Phys. Eng. Sci. **369**(1953), 4064–4077 (2011). doi:10.1098/rsta.2011.0233

21. V.P. Ovsyannikov, G. Zschornack, First investigations of a warm electron beam ion trap for the production of highly charged ions. Rev. Sci. Instrum. **70**(6), 2646–2651 (1999). doi:10.1063/1.1149822

22. R. Pohl, A. Antognini, F. Nez, F.D. Amaro, F. Biraben, J.M.R. Cardoso, D.S. Covita, A. Dax, S. Dhawan, L.M.P. Fernandes, A. Giesen, T. Graf, T.W. Hänsch, P. Indelicato, L. Julien, C.-Y. Kao, P. Knowles, E.-O.L. Bigot, Y.-W. Liu, J.A.M. Lopes, L. Ludhova, C.M.B. Monteiro, F. Mulhauser, T. Nebel, P. Rabinowitz, J.M.F. dos Santos, L.A. Schaller, K. Schuhmann, C. Schwob, D. Taqqu, J.F.C.A. Veloso, F. Kottmann, The size of the proton. Nature **466**(7303), 213–218 (2010). doi:10.1038/nature09250

23. L.P. Ratliff, E.W. Bell, D.C. Parks, A.I. Pikin, J.D. Gillaspy, Continuous highly charged ion beams from the National Institute of Standards and Technology electron-beam ion trap. Rev. Sci. Instrum. **68**, 1998 (1997). doi:10.1063/1.1148087

24. J. Repp, C. Böhm, J. Crespo López-Urrutia, A. Dörr, S. Eliseev, S. George, M. Goncharov, Y. Novikov, C. Roux, S. Sturm, S. Ulmer, K. Blaum, PENTATRAP: a novel cryogenic multi-

Penning-trap experiment for high-precision mass measurements on highly charged ions. Appl. Phys. B, Lasers Opt. **107**(4), 983–996 (2012). doi:10.1007/s00340-011-4823-6

25. D. Schneider, D.A. Church, G. Weinberg, J. Steiger, B. Beck, J. McDonald, E. Magee, D. Knapp, Confinement in a cryogenic Penning trap of highest charge state ions from EBIT. Rev. Sci. Instrum. **65**(11), 3472–3478 (1994). doi:10.1063/1.1144525
26. F.G. Serpa, J.D. Gillaspy, E. Träbert, Lifetime measurements in the ground configuration of Ar^{13+} and Kr^{22+} using an electron beam ion trap. J. Phys. B, At. Mol. Opt. Phys. **31**(15), 3345 (1998). http://stacks.iop.org/0953-4075/31/i=15/a=008
27. V. Simon, P. Delheij, J. Dilling, Z. Ke, W. Shi, G. Gwinner, Cooling of short-lived, radioactive, highly charged ions with the TITAN cooler Penning trap. Hyperfine Interact. **199**, 151–159 (2011). doi10.1007/s10751-011-0309-5
28. C.H. Storry, A. Speck, D.L. Sage, N. Guise, G. Gabrielse, D. Grzonka, W. Oelert, G. Schepers, T. Sefzick, H. Pittner, M. Herrmann, J. Walz, T.W. Hänsch, D. Comeau, E.A. Hessels, First laser-controlled antihydrogen production. Phys. Rev. Lett. **93**, 263401 (2004). doi:10.1103/PhysRevLett.93.263401
29. L. Suess, C.D. Finch, R. Parthasarathy, S.B. Hill, F.B. Dunning, Permanent magnet Penning trap for heavy ion storage. Rev. Sci. Instrum. **73**(8), 2861–2866 (2002). doi:10.1063/1.1490411
30. J.N. Tan, S.M. Brewer, N.D. Guise, Experimental efforts at NIST towards one-electron ions in circular Rydberg states. Phys. Scr. **2011**(T144), 014009 (2011). http://stacks.iop.org/1402-4896/2011/i=T144/a=014009
31. J.N. Tan, S.M. Brewer, N.D. Guise, Penning traps with unitary architecture for storage of highly charged ions. Rev. Sci. Instrum. **83**(2), 023103 (2012). doi:10.1063/1.3685246
32. E. Träbert, Atomic lifetime measurements employing an electron beam ion trap. Can. J. Phys. **86**(1), 73–97 (2008). doi:10.1139/p07-099
33. J. Xiao, Z. Fei, Y. Yang, X. Jin, D. Lu, Y. Shen, L. Liljeby, R. Hutton, Y. Zou, A very low energy compact electron beam ion trap for spectroscopic research in Shanghai. Rev. Sci. Instrum. **83**(1), 013303 (2012). doi:10.1063/1.3675575

Chapter 4
Dominance of Higher-Order Contributions to Electronic Recombination

P.H. Mokler, C. Beilmann, Z. Harman, C.H. Keitel, S. Bernitt, J. Ullrich, and J.R. Crespo López-Urrutia

Abstract Inter-shell higher-order electronic recombination is reported for highly charged ions of Ar, Fe and Kr, where simultaneously to the K-shell excitation one or two additional L electrons are excited at a resonant recombination of a free electron. The strengths of resonant higher-order recombination processes increase dramatically with decreasing atomic number Z. Already for highly charged Ar ions the 2nd-order can overwhelm that of the 1st-order dielectronic recombination considerably. The experimental findings are confirmed by relativistic distorted wave calculations employing multi-configuration Dirac-Fock bound-state wave functions with configuration mixing. The found effect points to the eminent importance of correlation effects for highly charged, low-Z ions.

4.1 Motivation and Introduction

Photon-electron and electron-electron interactions are the most fundamental processes governing the world of matter in general and in particular that of plasmas. In calculating these interactions usually independent-particle models are used where normally only the most relevant two partners speak with each other and, for instance, in an atomic system the rest of electrons is considered as pure spectators providing some screening. With the advances both, in experimental techniques and in theory higher-order (HO) processes came more into focus in recent years (see e.g. [2, 13, 14] and references cited there). Nevertheless, still nowadays higher-order effects are normally considered of minor importance. However, this is in no way true

P.H. Mokler (✉) · C. Beilmann · Z. Harman · C.H. Keitel · S. Bernitt · J. Ullrich ·
J.R. Crespo López-Urrutia
Max-Planck Institute for Nuclear Physics, Saupfercheckweg 1, 69117 Heidelberg, Germany
e-mail: mokler@mpi-hd.mpg.de

Z. Harman
ExtreMe Matter Institute (EMMI), Planckstrasse 1, 64291 Darmstadt, Germany

J. Ullrich
Physikalisch Technische Bundesanstalt (PTB), Bundesallee 100, 38116 Braunschweig, Germany

M. Mohan (ed.), *New Trends in Atomic and Molecular Physics*,
Springer Series on Atomic, Optical, and Plasma Physics 76,
DOI 10.1007/978-3-642-38167-6_4, © Springer-Verlag Berlin Heidelberg 2013

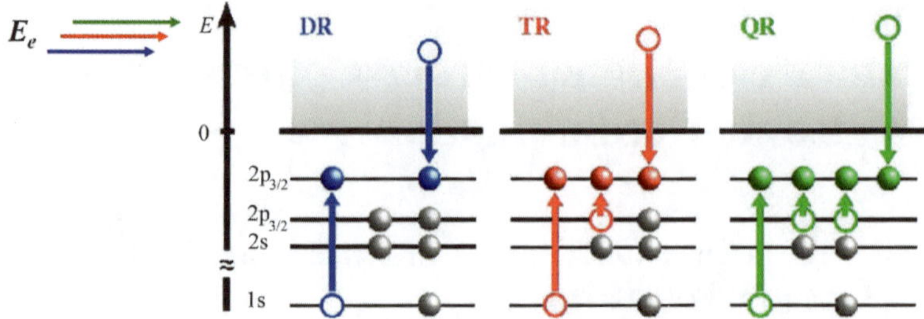

Fig. 4.1 Diagrams for K-L inter-shell resonant recombination of 1st, 2nd and 3rd order—dielectronic (DR), trielectronic (TR) and quadruelectronic (TR) recombination (*blue*, *red* and *green/left*, *middle* and *right*), respectively

in all cases. Here we proof the dominance of higher-order electron-electron interaction for the case of electron recombination of highly-charged, medium-heavy ions. The higher-order electron recombination can even overwhelm the first-order dielectronic recombination (DR) [4]. Hence, higher-order effects can play an essential role in determining the parameters of both, dynamic and equilibrium of plasmas. Consequently, higher-order effects may determine relevant plasma parameters and have to be included in the corresponding simulation codes. In particular, these processes may modify the opacities of plasmas and by this the transport of energy through them.

Autoionization of an excited atomic system and electronic recombination are the prominent show cases for electron-electron interaction. Both cases can be considered essentially as time reversed processes (see e.g. [10, 12]). For dielectronic recombination a free electron interacts with a bound one where the free electron is captured to a bound state transferring its energy to a bound one, see left-side diagram in Fig. 4.1. DR is a resonant process normally resulting in a doubly-excited state. This auto-ionizing state may decay radiatively toward a stable ground state thus completing the recombination:

$$A^{q+} + e \rightarrow \left[A^{(q-1)+}\right]^{(n+1)*} \rightarrow A^{(q-1)+} + \sum_i h\nu_i \tag{4.1}$$

Here A^{q+} is the highly-charged ion with the initial charge state $q+$, n the order of the process being 1 for DR (with the doubly excited intermediate state, $[A^{(q-1)+}]^{**}$) and $h\nu_i$ is (are) the emitted photon(s). For higher-order electronic recombination (with order n) more than one initially bound electron are excited simultaneously. In Fig. 4.1 aside the first-order DR process, the second-order trielectronic recombination (TR, with $n = 2$) and third-order quadruelectronic recombination (QR, with $n = 3$) processes are sketched for one K electron excitation with additional one or two L electrons being excited (TR and QR, respectively).

Using the principle of detailed balance the electronic recombination rates, i.e. the resonance strengths for all cases are given by:

$$S_R^n = f_R^n \cdot E_e^{-1} \cdot A_a \cdot \left[A_r \Big/ \left(\sum A_r + \sum A_{a(nR)}\right)\right] \tag{4.2}$$

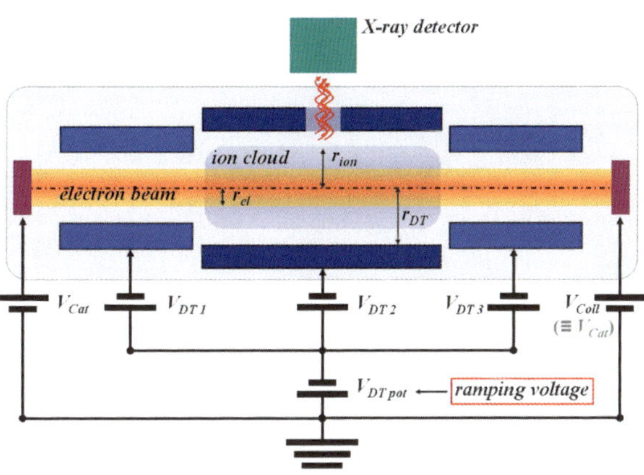

Fig. 4.2 Experimental arrangement at the EBIT: Symmetric potentials have been applied (both to the cathode, and the collector, $V_{Cat} = V_{Coll}$, and to the non-central drift tubes, $V_{DT1} = V_{DT3}$). The whole drift tube assembly (including the central one, V_{DT2}) is ramped by the common platform whereby the emitted X-rays are monitored

where n again is the order of the process, f_R^n is a statistical factor accounting for the weights of the levels involved, E_e the energy of the initially free electron, A_a the responsible auto-ionizing rate (Auger rate), $A_r/(\sum A_r + \sum A_{a(nR)})$ the fluorescence yield for the considered radiative decay branch (A_r) with respect to all possible decay possibilities (cf. [11]). HO recombination processes, nR with $n > 1$, can only happen by interaction mixing of the bound electrons, that means, via correlations between the electrons in the intermediate state

$$|nR\rangle = \sum_n c_n |config_n\rangle \tag{4.3}$$

where c_n are the mixing coefficients for the different contributing configurations $config_n$. In order to account theoretically for correlations in the HO processes, multi-configuration Dirac-Fock bound-state wave functions (i.e. including configuration mixing) have been performed [2, 8].

Intra-shell HO recombination processes—employing electrons of one common shell, the L shell—have already been communicated for highly charged ions earlier [13, 14]. Here we report on *inter-shell HO* processes where electrons of different electronic shells are involved [2, 4]. Especially we deal with simultaneous excitation of one K-shell electron together with one or two L electrons under the recombining capture of a free electron also to the L shell, resulting in a multiply excited intermediate state in the L shell, see Fig. 4.1. In analogy to the Auger nomenclature we assign the K-L inter-shell recombination as dielectronic—K-LL DR, as trielectronic—KL-LLL TR and as quadruelectronic—KLL-LLLL QR (1st, 2nd and 3rd order process, respectively).

4.2 Experimental Technique and Overview Spectrum

The inter-shell resonant recombination experiments have been performed at the electron beam ion traps (EBIT) available in Heidelberg [5–7]. For the different ions investigated (Ar, Fe and Kr) different EBITs have been used. In Fig. 4.2 the

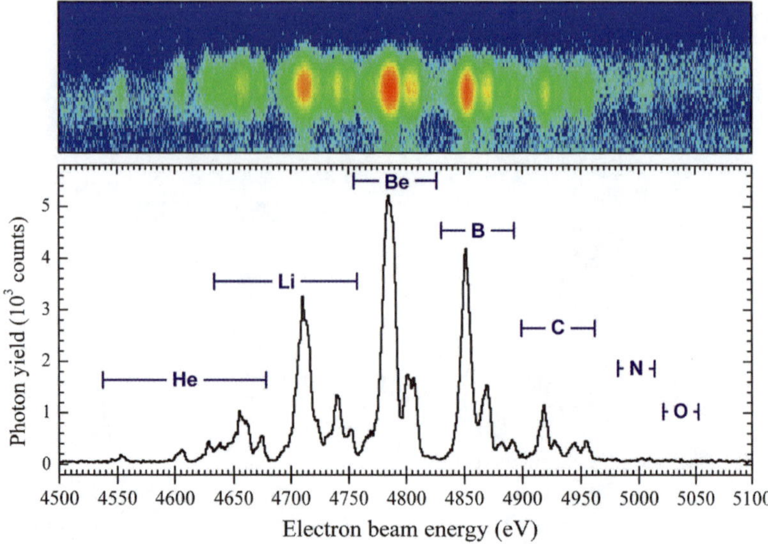

Fig. 4.3 Recombination associated with Fe-Kα emission for a trapped ensemble of highly charged Fe ions. At the *top* the two-dimensional event plot of emitted photons as a function of electron beam energy and X-ray energy around the Fe-Kα emission is displayed. At the *bottom* the projection along the Fe-Kα radiation on the electron beam axis is given, showing the electronic recombination spectrum for the different charged Fe ions indicated by bars

principle of the experimental arrangement with EBIT is shown. In short: An energetic electron beam compressed by a strong magnetic field ionizes atomic species to high charge states in the trap center. Here the ion cloud is confined radially by both the axial magnetic field and, mainly, the space charge potential of the compressed electron beam; in axial direction the ion cloud is confined by properly selected potentials at the non-central drift tubes. After cooking the ions by continuous electron bombardment to high charge states the ions are cooled by lowering the potential at the off-center drift tubes. Thus hot ions will escape axially the active trap volume— dubbed forced axial cooling [3]. Using the remaining cooled ions alone, the measuring cycle is started. By tuning the electron beam energy over the resonance region, electronic recombination may occur into the trapped and cold highly charged ions. These recombined ions are in highly excited states (cf. Fig. 4.1) and stabilize by X-ray emission which is monitored through an X-ray detector. The characteristic X-ray intensity as function of the electron beam energy reveals the resonances of the electronic recombination processes.

In Fig. 4.3 an overview on highly charged Fe ions is given as an example. The two-dimensional event plot at the top shows the emitted X-ray intensity as a function of the ramped electron beam energy and the X-ray energy of the emitted photons around the Fe-Kα emission. The bright spots manifest the electronic recombination resonances with intermediate K-shell excitation and the emission of a stabilizing Kα emission. A projection of the photon intensity for a cut around the shown Fe-Kα

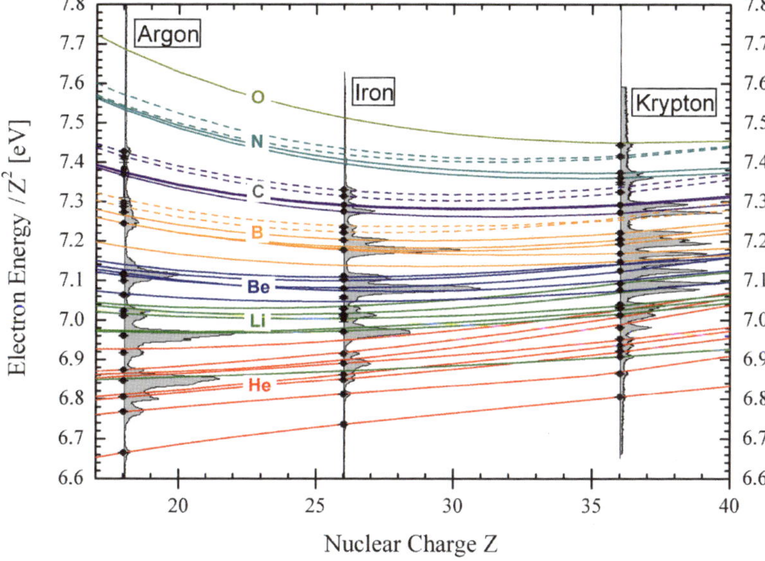

Fig. 4.4 Electronic recombination resonances normalized by Z^2 as a function of the atomic number Z. The curves show the results of MCDF calculations (this work) color coded on the ionic sequence. The *full lines* give the resonance energies for DR and the *dashed* ones those for the higher-order TR process. The measured resonance spectra are overlaid for the case of Ar, Fe and Kr

X-ray energies onto the electron beam axis yields the K-LL(+HO) electron recombination spectrum displayed at the bottom. As differently charged ions are present in the ion cloud at the trap the spectrum comprises recombination resonances from initially He- to O-like species: The resonance regions for the differently charged species are marked. Similar spectra have been found for the case of Kr and Ar ions. In the following we will concentrate mainly on the C-like species. In order to compare the recombination spectra from ions with different atomic numbers Z we refer to H-like ions and use a Rydberg scaling, i.e. we normalize the electron energy by the K-electron binding energy increasing in the H-like case with Z^2.

4.3 Results and Discussion

For an overview the results for Ar, Fe and Kr ions are compared to our multi-configuration Dirac Fock calculations (MCDF). In Fig. 4.4 the resonance energies normalized by Z^2 are given as function of the atomic number Z. The lines are color coded according to the ionic sequences. Obviously, the curves deviate from horizontal lines expected for a pure Rydberg scaling. As we are in the transition region—we find at low Z the influence of screening increasing toward lower Z values and, increasing independently on that, with the number of available electrons. At the higher Z region we realize already the relativistic influence increasing again with Z. On

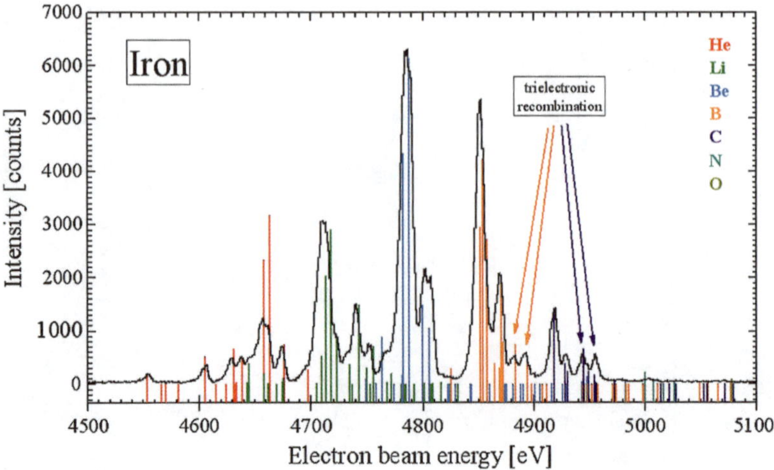

Fig. 4.5 Comparison of the measured electronic recombination spectrum for Fe with the MCDF calculations. The length of the lines color coded on the ionic sequence give the calculated resonance strengths. The TR resonances for B-like and C-like species are marked

this graph our measured spectra are overlaid showing a very good agreement in energy positions with theory. Here we point to the difference in full and dashed lines. Whereas the full lines give the calculations for the DR resonances, the dashed lines represent the results for higher-order TR processes. From the separation of lines it is clear that TR is most likely to be found for N-, C- and B-like sequences. Indeed, an inspection of the experimental spectra shows the corresponding TR lines are best visible for the Fe case.

Figure 4.5 shows the Fe results in more detail. The calculated energy positions are given by the lines. Additionally, the calculated line strengths are indicated by the length of the lines. However, only relative line intensities within one ionic sequence can be compared directly. The abundances for the different ion species have been estimated from the actual running parameters for the EBIT. Within this slight restriction a very good agreement between calculations and measurements has to be stated. This excellent agreement is also found for HO electronic recombination, in particular for TR, see the marked double lines for the B- and C-like sequence. The TR resonances are found as prominent structures on the high energy side of the corresponding DR resonances with an intensity of a few tenth of the DR strength.

Blowing up the Fe spectrum in Fig. 4.5 we find also for the N-like sequence intershell TR resonances with comparable relative intensity also in consonance with theory. Moreover, similar observations have been made for neighboring atomic species of Ar and Kr. In Fig. 4.6 we compare for the different atomic numbers the normalized spectra (Rydberg scaling) for the C-like sequence. The TR resonances are best visible for the case of Fe; for Kr ions the TR lines are already weak, whereas for Ar ions the TR intensity overwhelms the DR one. Despite the absolute experimental

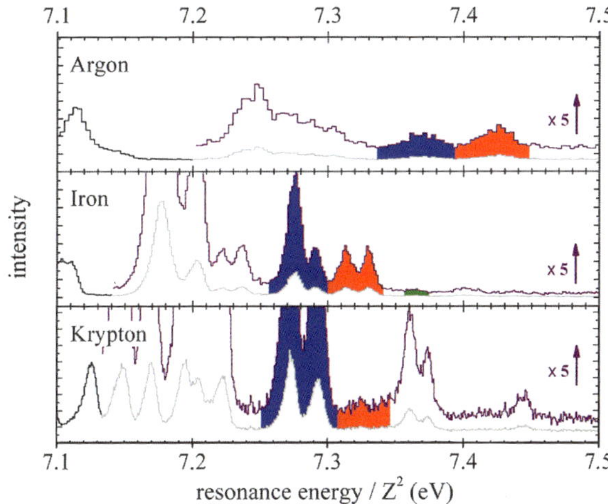

Fig. 4.6 Electronic recombination spectra normalized in energy by Z^2 for Ar, Fe and Kr ions for the region of C-like species. The DR and TR resonances for the C sequence are marked by color (*blue* and *red*, respectively). The TR/DR intensity ratio varies strongly with the atomic number and increases dramatically for low Z values

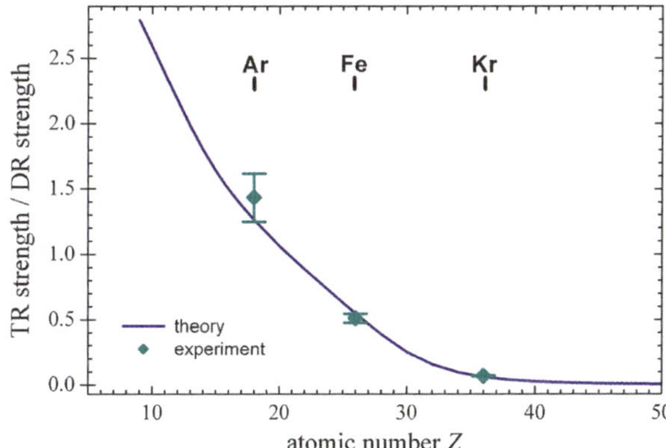

Fig. 4.7 TR/DR resonance strengths ratio as function of atomic number Z for the C-like ion sequence: experimental values in comparison with theory

resolution is also excellent in the case of Ar the double lines are merged together since the line separation gets so small.

On a first glance, the dramatic increase in the relative TR/DR intensity ratio with decreasing atomic number is striking indeed. For low-Z species the HO electronic recombination strength overwhelms that of the first order. In Fig. 4.7 the found experimental TR/DR intensity ratio plotted as a function of the atomic number Z is compared to the corresponding ratio calculated by theory. Also in this case a perfect agreement with theory can be stated, confirming the dramatic increase in HO electronic recombination at low Z relative to the corresponding first-order DR. Similar results have been established for the neighboring sequences of B- and N-like ions. In addition, here a perfect agreement between experiment and theory is observed too. All these observations confirm the eminent importance of HO recombination and electron correlation effects for lower Z species.

In order to understand this unexpected strong Z dependence of the relative TR strengths one has to consider the general dependences of the Auger rates, the radiative rates and the corresponding fluorescence yields. In a simple approximation not considering correlation effects the Auger rates are Z independent (Z^0), whereas the radiative rates increase with the forth power of the atomic number (Z^4). According to Eq. (4.2) we can parameterize the strength for DR by [4, 9]

$$S_{DR} \approx \left[b \cdot Z^{-2} + c \cdot Z^2 \right]^{-1} \qquad (4.4)$$

with b and c constants referring to the Auger and radiative decay channels. Hence, for lower Z values the DR strengths decrease due to the small fluorescence yield, i.e. despite the Auger rates stay about constant there the radiative rates decrease so strongly that the formed intermediate systems decay back to the continuum with high probability—also called resonance scattering. So for the K-LL DR strengths we expect a Z^2 increase for low Z values and a Z^{-2} decrease for high Z values with a smooth maximum around Fe.

For the higher-order TR the Auger rates have to be re-considered, here the electron correlations modify the production rates [4]. As an additional electron is involved, these rates depend on Z. Compared with the central Coulomb force the electron-electron interaction will decrease with Z^{-1} and consequently the TR capture rates will show a general Z^{-2} dependence. In summary, the TR strengths can be parameterized by

$$S_{TR} \approx Z^{-2} \cdot \left[a + bZ^{-2} + c \cdot Z^2 \right]^{-1} \qquad (4.5)$$

with the constant a referring to the higher-order Auger process. Also in the HO case the strength increases at low Z values and decreases again for higher Z values. However, for KL-LLL TR the maximum shifts to considerably lower Z values around Si. In consequence, the relative strength ratio TR/DR for the C-like sequence increases down to Ne to a value of about 2.5 [4]. That means the second-order TR strength overwhelms that of the first-order DR by a factor of 2.5.

Now we consider the next higher order, the QR. For the Fe spectrum shown in Fig. 4.6 we find also a clear indication on QR beyond the TR resonances (marked by green color). Unfortunately, background contributions can not be excluded completely. Nevertheless, the QR contribution for C-like Fe is in the order of 1 % of the DR contribution. This order of magnitude is supported by the calculations too. Moreover, in other ionic species indications for QR resonances have also been observed and fit to the theoretical results.

For completeness we compare now for Fe the results for the different ionic sequences, see Fig. 4.8. The experimental values for the TR/DR strength ratio are compared to the theoretical values. Good agreement is observed for Be- to N-like ions. For Be-like ions no TR is expected due to parity reasons [2]. On the other hand here the QR contribution is predicted to be somewhat strong. We emphasize that due to open and closed channels KLL-LLLL QR is only possible for Be- B- and C-like species, whereas KL-LLL TR is possible from Li-like to N-like ions.

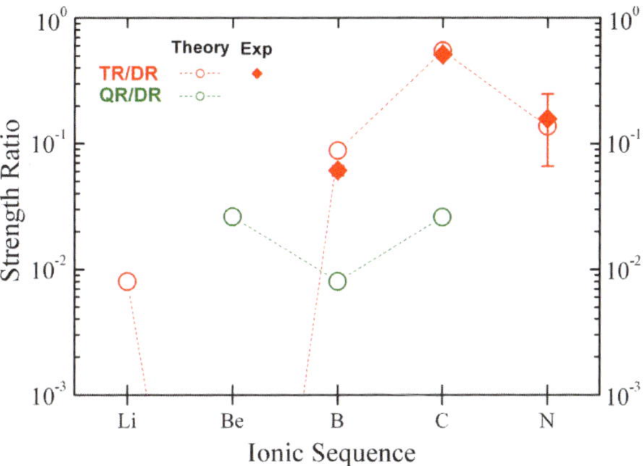

Fig. 4.8 Ratio of the higher-order electronic recombination strengths to the first-order DR strengths for Fe ions (TR/DR *red color*, QR/DR *green color*). In order to guide the eye the theoretical points (*open circles*) are connected by *dotted lines*

4.4 Summary and Outlook

Unexpected strong contributions of higher-order electronic recombination have been found for inter-shell processes where high momentum transfers between electrons of different shells are involved. Those prominent contributions of these higher-order processes are not really expected, even not for medium-low Z ions. However, in some cases the 2nd-order inter-shell recombination process, i.e. the KL-LLL TR strength, even overwhelms considerably the 1st-order K-LL DR contribution. Even the 3rd-order electronic recombination process, KLL-LLLL QR, seems not to be negligible. All the observations, which agree nicely with MCDF calculations including configuration mixing, point to the eminent importance of correlation effects, especially for lower Z ions. This finding urges to reconsider plasma models predicting the state of matter in fusion devices or in astrophysical matter. Hence, it seems compulsory that these HO effects have to be included in plasma codes in future.

These findings were only possible by the achievements made in experimental techniques. With the forced cooling technique applied in the EBIT, closely spaced recombination resonances could be resolved and thus the higher-order recombination processes have been established. Even small contributions from the third order recombination, the inter-shell QR processes could be seen. This process is especially interesting as for special ionic sequences the second order TR process is forbidden due to parity reasons and the QR is more prominent there. This is an additional and important case for future research. Moreover, the knowledge on higher-order recombination processes provides a novel bridge to understand more complex multi-electron processes like shake-up/off in outer shells especially for heavier species (cf. [1]). Furtheron, we expect that by the correlated excitation process all the involved electrons are entangled. Through the higher-order processes discussed here the door to higher-order entanglement may be opened in future.

References

1. H. Aksela, A. Aksela, N. Kabachnik, VUV and soft X-ray photoionization, in *Physics of Atoms and Molecules*, ed. by U. Becker, D. Sirley (Plenum, New York, 1996), p. 401
2. C. Beilmann et al., Phys. Rev. A **80**, 050702(R) (2009)
3. C. Beilmann et al., J. Instrum. **5**, C09002 (2010)
4. C. Beilmann et al., Phys. Rev. Lett. **107**, 143201 (2011)
5. C. Beilmann et al., Phys. Rev. A (2013, submitted). arXiv:1306.1029
6. J.R. Crespo Lopez-Urrutia et al., Phys. Scr. T **80**, 502 (1999)
7. S.W. Epp et al., Phys. Rev. Lett. **98**, 183001 (2007)
8. Z. Harman et al., Phys. Rev. A **73**, 052711 (2006)
9. A.P. Kavanagh et al., Phys. Rev. A **81**, 022712 (2010)
10. D.A. Knapp et al., Phys. Rev. A **47**, 2039 (1993)
11. P.H. Mokler, F. Folkmann, Top. Curr. Phys. **5**, 201 (1978)
12. C. Schippers, J. Phys. Conf. Ser. **163**, 012001 (2009)
13. S. Schippers et al., Int. Rev. At. Mol. Phys. **1**, 109 (2010)
14. M. Schnell et al., Phys. Rev. Lett. **91**, 043001 (2003)

Chapter 5
High Accuracy Non-LTE Modeling of X-Ray Radiation in Dense Matter

Arati Dasgupta, Robert W. Clark, John L. Giuliani, Ward J. Thornhill, John P. Apruzese, Brent Jones, and Dave J. Ampleford

Abstract X-rays are emitted from a variety of astrophysical objects in the universe. With the advancement of experimental technologies, intense and very bright X-ray sources are also being produced in the laboratory. Similar progress in theoretical investigations has made it possible to accurately model the radiation and spectroscopy of X-rays from both laboratory and astrophysical sources. Present-day Z-pinch experiments generate 200 TW peak power, 5–10 ns duration X-ray bursts that provide new opportunities to advance radiation science. The experiments spotlight the underlying atomic and plasma physics and offer inertial confinement fusion and astrophysics applications. Spectroscopy is a key diagnostic tool and its reliability depends on the accuracy and reliability of the atomic and plasma physics models used to interpret the data. We report the current status of our theoretical investigations of X-ray spectroscopy using state-of-the-art atomic and plasma modeling to analyze the data obtained from Z machine at the US Sandia National Laboratories. Analysis used for Z-pinches can also be used to study ICF and astrophysical plasmas where laboratory measurements and simulations are the only means to interpret observed data.

5.1 Introduction

Many astrophysical objects emit, fluoresce, or reflect X-rays. These X-rays are found in very low-density galactic objects such as supernova remnants, interstellar media, stellar coronae and in extremely dense accreting black holes or neutron stars. Some solar system bodies also emit X-rays. A combination of many unresolved

A. Dasgupta (✉) · J.L. Giuliani · W.J. Thornhill · J.P. Apruzese
Plasma Physics Division, Naval Research Laboratory, Washington, DC, USA
e-mail: arati.dasgupta@nrl.navy.mil

R.W. Clark
Berkeley Scholars Inc., Springfield, VA, USA

B. Jones · D.J. Ampleford
Sandia National Laboratories, Albuquerque, NM, USA

M. Mohan (ed.), *New Trends in Atomic and Molecular Physics*,
Springer Series on Atomic, Optical, and Plasma Physics 76,
DOI 10.1007/978-3-642-38167-6_5, © Springer-Verlag Berlin Heidelberg 2013

X-ray sources is thought to produce the observed X-ray background. The X-ray continuum can arise from bremsstrahlung, blackbody radiation, synchrotron radiation, collisions of fast protons with atomic electrons, and atomic recombination. Recent high resolution spectral observations of astrophysical objects provide important information on conditions and locations of X-ray-emitting plasmas. X-ray images from the Chandra X-ray Observatory, which was launched by Space Shuttle Columbia in 1999, can better define the hot, turbulent regions of space. This increased clarity can help scientists answer fundamental questions about the origin, evolution, and density of the universe. On June 14, 2012, the National Aeronautics and Space Administration (NASA) in the US launched an orbiting telescope, the Nuclear Spectroscopic Telescope Array (NuSTAR) that can view the cosmos through the lens of hard X-rays. NuSTAR's mirrors will collect high energy X-ray photons emitted by cosmic sources, focusing the light into images 10 times sharper and 100 times more sensitive than any previous high energy X-ray telescope. It was quoted by the NuSTAR lead scientist that "the most important study will involve the cosmic X-ray background-understanding the individual phenomena that contribute to it."

Currently, intense X-rays are also produced in many high energy density (HED) laboratory facilities. They include X-rays from ultrafast beam irradiated clusters where high-pressure gas expands in vacuum creating clusters that are irradiated by a high-intensity laser. The hot dense plasma created in the interaction with the laser can provide a compact source of X-ray radiation, and energetic ions for fusion. Very recently, researchers working at the U.S. Department of Energy's (DOE) SLAC's Linac Coherent Light Source at National Accelerator Laboratory have used the world's most powerful X-ray laser to create and probe a 2-million-degree piece of matter in a controlled way for the first time. This feat, in *Nature* [26], takes scientists a significant step forward in understanding the most extreme matter found in the hearts of stars and giant planets, and could help experiments aimed at recreating the nuclear fusion process that powers the sun. The world's most powerful X-ray source, the Z facility at the US Sandia National Laboratories (SNL), offers a promising route to producing energy through controlled thermonuclear fusion. Our theoretical investigations will focus on interpretation and analysis of important parameters such as radiation output, charge-state distributions, and X-ray emission from high-temperature, high density laboratory Z-pinch plasmas that lead to significant advancements in the knowledge of the production and evolution of these plasmas. Understanding these parameters for plasmas in laboratory experiments using reliable theoretical modeling also provides insights into plasmas in astrophysical objects under similar conditions.

X-ray emissions in the K-shell from moderately high atomic number plasmas have generated substantial interest in recent years. However, the production of these hard photons from high Z materials comes with a price. There is substantial loss of radiative yield due to stripping through many electrons present in these materials to reach the H- or He-like ionization stages. Production of hard X-rays for materials with atomic numbers higher than Cu is very difficult, and theoretical predictions are more unreliable. Previous experimental efforts using Cu as a plasma pinch load on the Z machine at SNL are encouraging and may lead to further investigations of this element on the refurbished Z machine for achieving substantial radiative yields with photon energies higher than 5 keV. We have used a collisional radiative spectroscopic model to analyze the implosion dynamics of multiple high Z elements such as wire arrays of Stainless Steel (SS) and Cu on the Z accelerator at SNL. This approach combines the completeness of highly averaged Rydberg state models with the accuracy of detailed models for all important excited states. We have investigated the ionization dynamics and generated K- and L-shell spectra using the conditions in the Z accelerator, as calculated by a 1D (one dimensional) non-local thermodynamic equilibrium (LTE) radiation hydrodynamics model. These results are compared with K- and L-shell experimental spectra from Z. Theoretical K- and L-shell spectroscopy provides validation of atomic and plasma modeling when compared to available experimental data and also provides useful diagnostics for the plasma parameters. Our self-consistently generated non-LTE collisional radiative model employs an extensive atomic level structure and data for all dominant atomic processes that are necessary to model accurately the pinch dynamics and the spectroscopic details of the emitted radiation. X-ray emission spectra from the K- and especially complex L-shell of high Z (atomic number) ions provide valuable information with which to assess plasma conditions and X-ray performance in a variety of laboratory plasmas such as Z-pinch plasmas. Experiments at the Z facility can be used to test complex theories and models of X-ray production. The Z machine produces up to 250 TW and 1.8 MJ of radiation [15] by driving 20 MA with a 100 ns rise time through a nested cylindrical array of several hundred fine wires. Concentric gas-puff loads, such as Ar, Kr or DT can also be employed. The self-generated $\mathbf{j} \times \mathbf{B}$ force implodes the array, creating a high energy density, radiating plasma column on the axis of symmetry. Figure 5.1 shows a cross section of the Z machine and the pinch.

Typically there are two main classes of radiation sources that are studied on the Z machine. Low-to-mid atomic number loads with large diameter (40–80 mm) are employed to generate K- and L-shell emission in the 1–10 keV photon energy range. Compact high atomic number materials such as tungsten wire arrays provide powerful <1 keV soft X-ray sources for driving inertial confinement fusion (ICF) [22]. Our work is motivated by the need to analyze and interpret the experiments on the Z and the refurbished Z facilities at SNL using various load configurations. Space and time resolved spectroscopy is the most important diagnostic tool and comparison of these with synthetic spectra, produced from detailed atomic models, can produce a wealth of information about the plasma state. A test of the accuracy and reliability of our atomic model and simulation comes from the results of this comparison. The analysis used for Z-pinch plasmas can be used to study ICF and also astrophysical plasmas where laboratory measurements and simulations are the only means to interpret observed data.

Stainless Steel (SS) and Cu wire arrays are investigated as an intensive X-ray radiation source at the Z machine. The implosion dynamics of an array of Cu wires on the Z and/or ZR accelerator produces an abundance of radiation from the K- and L-shell ionization stages and provides reliable diagnostics capabilities [8]. These dynamic plasmas are inherently non-LTE, with opacity and other factors influencing the X-ray output. We will analyze the ionization dynamics and generate K- and

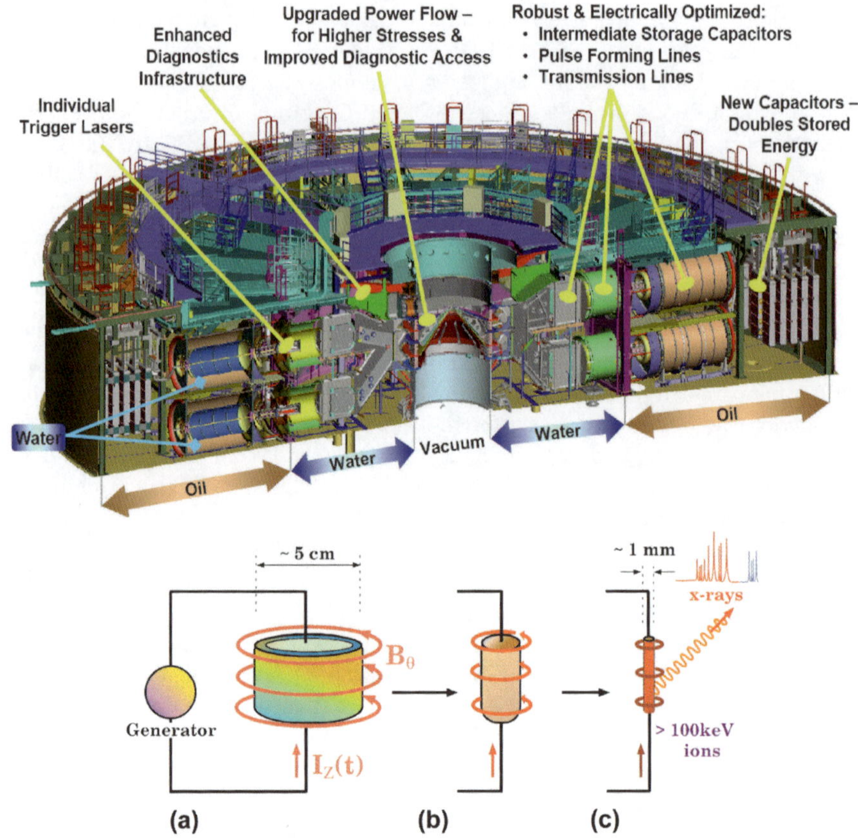

Fig. 5.1 (*Top*) Refurbished Z machine; (*bottom*) $\mathbf{J} \times \mathbf{B}$ force pinches wire array into a dense radiating plasma column

L-shell spectra using the conditions generated on Z and/or ZR, described by a 1D non-LTE radiation hydrodynamics model [25]. Although K-shell radiation is of extreme interest and K-shell photons for Cu have energies >5 keV, modeling L-shell spectra is more challenging due to the complex L-shell structure. The accuracy of predictions of K-shell energetics and diagnostics depend heavily on L-shell energetics. Analysis of L-shell spectra provides a wealth of information, as the X-ray spectra of these ions are rich in structure due to multiple overlapping satellite lines that accompany the resonance lines. Identification of these lines for diagnostic calculations poses challenges but diagnostic research involving L-shell ions, while not as extensive as that involving K-shell ions, provides potentially more rewards. Diagnostics of complex L-shell spectra open a new arena for validation of atomic and plasma models when compared with experiments. Our atomic model therefore includes the atomic data for all the relevant processes needed to accurately describe both L- as well as K-shell spectra [13].

Modeling of radiation from high atomic number species requires a detailed non-LTE kinetics model with accurate radiation transport. Plasmas created in Z-pinches are far from LTE, where the plasma properties such as internal energies and opacities depend only on the local temperature and density, and the level populations are described by statistical distributions of the excited states. This is especially true for the K-shell region. It is true that for low photon energies where free-free opacity is large, the plasma will approach LTE conditions. However, LTE atomic populations represent overionization of the plasma due mainly to neglect of radiative and dielectronic recombination, and also represent excessive excited state populations due chiefly to the neglect of spontaneous decay. The resulting LTE radiative emission will produce incorrect spectral characteristics. In pinch experiments, the dense plasma produced near stagnation is optically thick to line radiation and this radiation can significantly change the level populations by photoexcitation and photoionization. In the L- and K-shell photon energy range, a number of spectral lines will be optically thick, with optical depth greater than unity, especially near stagnation. In the K-shell region, the He line is quite thick, and the other lines typically exhibit moderate opacities. Thus a full collisional radiative model coupled with a non-local radiation field is necessary to describe the ionization dynamics. In addition, most of the energy deposited into a Z-pinch plasma is ultimately radiated away. Therefore, detailed radiation modeling, including accurate radiation transport, is necessary to obtain meaningful implosion dynamics.

The Cu nested wire arrays for Z1975 shots on the Z machine included 4 % Ni and additional radiation data from Fe and Cr were also observed in the spectra. The presence of Fe and Cr and additional Ni was due to the stainless steel (SS) cathode. Our atomic model thus included multi-material atomic data for a mixture of 93 % Cu, 4 % Ni, 2.4 % Fe, and 0.6 % Cr.

X-ray emissions from high atomic number materials often exhibit a substantial K-α radiation. K-α emission is associated with the $2p \rightarrow 1s$ inner-shell-electron radiative transitions in highly charged ions [3, 21]. These lines have the appearance of satellite lines in the X-ray spectra of laboratory and astrophysical plasmas. Besides photoexcitation and photoionization from hard K-shell radiation, K-α emission can originate from collisional processes involving hot electrons in the final phase of pinching plasmas such as Z-pinches, X-pinches, and gas-puffs [16]. The small amount of hot electrons that may be produced in these plasmas can alter the diagnostics [24]. There is evidence of K-α features of Cu, Ni, Fe, and Cr from the Z1975 data. As these K-α lines provide good diagnostics, we plan to investigate them in detail and the results of our findings will be reported in a future article.

5.2 Theoretical Model

Our theoretical atomic and plasma model has three essential components; (a) the kinetics atomic model, (b) radiation transport and (c) magneto-hydrodynamics. This is shown in Fig. 5.2. The Kinetics model embodies atomic structure and collisional

Fig. 5.2 Three critical components shown are solved simultaneously to generate radiation and spectroscopic data

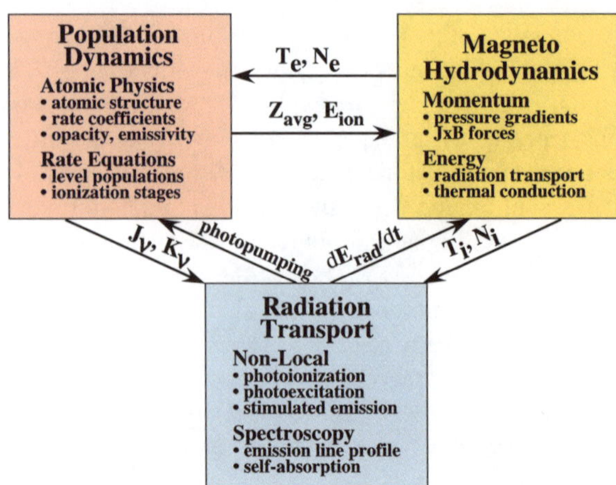

and radiative data. It generates the population densities for all levels of all ionization stages and provides opacities and emissivities for the Z-pinch plasmas. The hot and dense plasma produced in a Z-pinch is generally optically thick and thus the radiation emitted near the axis is attenuated in passing through the plasma much like the absorption of radiation from astrophysical objects when traveling through the atmosphere. It does then become necessary to employ a radiation transport scheme to estimate the continuum and line photon atenuation and deposition.

Magneto-hydrodynamics provides the evolution of the high energy density plasma and the plasma parameters such as internal energy, electron and ion temperatures and densities needed to obtain the level populations. In an optically thick plasma, non-local photopumping also affects populations. Thus all three components are fully dependent on each other and need to be coupled and solved simultaneously in order to predict accurate radiation and spectroscopic data from a Z-pinch implosion. The diagram in Fig. 5.2 illustrates this point. The detailed description of each of these components are given below.

5.2.1 Kinetic Models

Spectroscopic prediction of laboratory plasmas requires knowledge of atomic data and structure to evaluate the precise ionization dynamics and radiation transport. Typically, one of the three following models is valid depending on the plasma conditions.

5.2.1.1 Local Thermodynamic Equilibrium (LTE)

The LTE model describes a state of plasma in which population distribution is determined by the laws of statistical equilibrium except that radiation processes are not in a detailed balance. The LTE state is often found in laboratory plasmas of relatively high density and relatively low temperature when collisional processes with electrons from a Maxwellian distribution dominate in the rate equation and radiative processes do not affect population distributions. LTE conditions also hold deep in the interior of stellar atmospheres when collisions dominate and the photon mean-free-path is small.

The relative occupancies of the various states j of a given ion with energy E_j are simply proportional to the Boltzmann factor $\exp(-E_j/kT)$ where T is the electron temperature. The radiation field in LTE plasmas however, is approaching a Planck function and it depends not only on local plasma conditions but also on population distributions and atomic transition probabilities. Population distributions of LTE plasmas are entirely determined by local values of plasma conditions, that is by the Saha-Boltzmann equation as a function of *local* electron temperature T_e and density n_e. Using a simple formula for the collisional cross section of hydrogenic ions, a minimum electron density required to validate the LTE assumption between two levels in hydrogenic ions is roughly estimated by $n_e > 1.8 \times 10 T_e^{1/2} \Delta E_{ij}^3$, where ΔE is the transition energy between levels i and j, the energies are in electron volts and n_e is in cm^{-3}.

5.2.1.2 Coronal Equilibrium (CE)

When the plasma temperature is high and electron density is low the plasma state can be described by a CE model. These are the conditions exemplified by the plasma in the corona region in the Sun and this is reason for the name. Radiative decays are much larger than collisional excitations and deexcitations, and essentially all ions exist in the ground level. Distribution of atoms among various ionization stages is no longer determined by purely statistical considerations such as the Saha equation but is instead determined by balance between ionization and recombination processes. In a coronal model, electron density and radiation field are so low that collisional deexcitations and three body recombinations are insignificant. The collisional ionization or excitation is balanced by radiative recombination and dielectronic recombination or spontaneous decay, respectively. The application of the coronal model is limited to low electron densities, typically $n_e \leq 10^8$ cm^{-3}.

5.2.1.3 Collisional Radiative Equilibrium (CRE)

In the most general and applicable case for Z-pinch plasmas of intermediate to high densities, $10^{18} \leq n_i \leq 10^{22}$ cm^{-3}, the plasma ionization cannot be described by *coronal equilibrium*, in which radiative decay rates are much larger than collisional excitations and essentially all ions exist in the ground and $\Delta n = 0$ levels. Neither can the plasma be described as being in LTE, where all states of a given configuration are statistically populated. Here radiative processes compete with collisions and level populations for each ion must be determined by solving rate equations

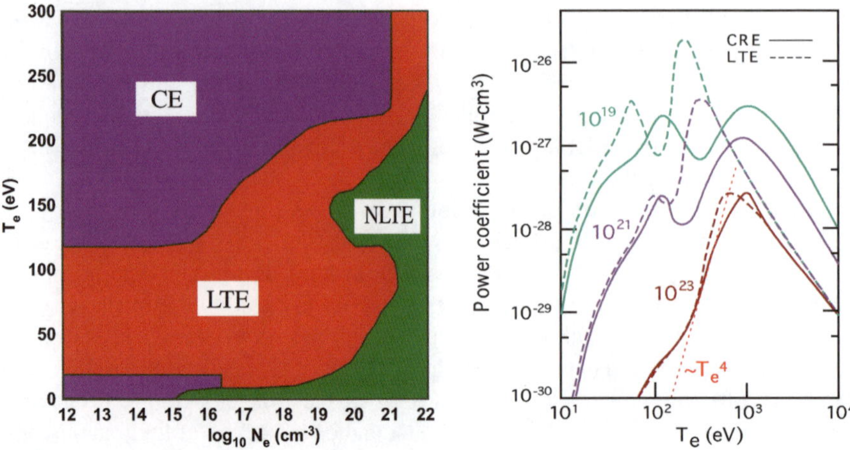

Fig. 5.3 (*Left*) Temperature and density regimes showing validity of CE, LTE and NLTE models for an aluminium plasma. (*Right*) Comparison of CRE and LTE power coefficients as function of temperature for several ion densities. At 10^{23} ions/cm^3 the CRE approaches the LTE predictions

including all important collisional and non-local radiative processes. This model is also the proper one for treating complex ions even at low densities. Figure 5.3 (left panel) shows the applicability of these models for plasma temperature and density regimes for an optically thin aluminum plasma [23]. In a collisional radiative equilibrium (CRE) model, atomic level populations are calculated by solving multi-level, atomic rate equations self-consistently with a radiation field, which is computed from a radiation transport model. This non-local radiation field changes population distributions and can cause spatial gradients in them. If collisional processes become dominant as the density increases, the population distribution and thus the radiation emission converges to the LTE solution. This is illustrated in Fig. 5.3 (right panel) where an identical level structure was used for an LTE calculation and for a CRE model to predict optically thin line emission from an aluminum plasma [6]. Solving for ionization balance for this CRE approximation involves not only the calculation of ion abundances, but also calculation of the populations of all important excited states. This calculation is complex, as it requires inversion of large matrices of atomic data involving a very large number of levels. For spectroscopic studies of K- and L-shell emission from a Z-pinch plasma we employ a full non-LTE (NLTE) CRE method for the application of our atomic model [12]; but a full time-dependent ion-equilibrium treatment of the atomic populations is also feasible. For K- and L-shell ions of moderate-Z elements, only at sufficiently high electron densities do the excited levels begin to come into statistical equilibrium. Thus a full non-LTE kinetics model is needed to investigate the electron density regimes found in laboratory plasmas. When the plasma ionization state responds to changes in hydrodynamic quantities, the plasma is in CRE and population densities are obtained by solving the corresponding equilibrium equations. For an ion in ionization stage Z and level j, the non-LTE kinetic rate equation appropriate for CRE is given by:

$$\frac{dn_{Z,j}}{dt} = 0 = +n_e \sum_{i<j} n_{Z,i} CX_{Z,i\uparrow Z,j} + n_e \sum_{i>j} n_{Z,i} CD_{Z,i\downarrow Z,j}$$
$$+ n_e^2 n_{Z+1,0} CR_{Z+1,0\downarrow Z,j} + \sum_{i>j} n_{Z,i} A_{Z,i\downarrow Z,j}$$
$$+ \sum_{i<j} n_{Z,i} PX_{Z,i\uparrow Z,j} + n_e n_{Z+1,0} RR_{Z+1,0\downarrow Z,j}$$
$$+ n_e n_{Z+1,0} DR_{Z+1,0\downarrow Z,j} - n_e n_{Z,j} \sum_{i>j} CX_{Z,j\uparrow Z,i}$$
$$- n_e n_{Z,j} \sum_{i<j} CD_{Z,j\downarrow Z,i} - n_e n_{Z,j} CI_{Z,j\uparrow Z+1,0}$$
$$- n_{Z,j} \sum_{i<j} A_{Z,j\downarrow Z,i} - n_{Z,j} \sum_{i>j} PX_{Z,j\uparrow Z,i} - n_{Z,j} PI_{Z,j\uparrow Z+1,0}$$

$$(5.1)$$

where the subscript arrow on the creation rates denote "from" and for the destruction rates it denotes "to". The nomenclature for the rate coefficients is: collisional excitation (CX); collisional ionization (CI); photoionization (PI); radiative decay (A); and photoexcitation (PX). The inverse processes of collisional deexcitation (CD), 3-body or collisional recombination (CR), and radiative recombination (RR) are determined by detailed balance from CX, CI, and PI, respectively. Dielectronic recombination is denoted by DR. Photoexcitation and photoionization can be non-local in the sense that the rate coefficient depends on the local radiation field that is determined by photon transport from all regions of the plasma [12].

Our atomic model has been constructed using a very detailed yet computationally efficient level structure employing a large number of excited states and the processes coupling these levels. DR is often the most important recombination process in laboratory and astrophysical plasmas [9–11]. Atomic structure data including energy levels and radiative transition probabilities as well as the collisional excitation and ionization rates and the collisional, radiative and dielectronic recombination rates were self-consistently generated using the Flexible Atomic Code (FAC) suite of codes [17, 18]. In FAC, structure calculations are based on relativistic configuration interactions with independent particle basis functions and relativistic effects are included using a Dirac Coulomb Hamiltonian. All ground levels and excited levels with $n \leq 4$ are kept at the fine-structure levels, while for $5 \leq n \leq 7$, the configuration-averaged level structures are used for all K- and L-shell ions. A more limited level structure is embedded in a configuration state model for the remainder of the ionization stages. These detailed atomic models obtained using self-consistently generated atomic data including low lying fine-structure Rydberg levels are crucial for proper spectroscopic diagnostics. Data for the atomic processes involved in level populations included in this calculation include:

(i) *Spontaneous radiative decay*—atomic structure data including energy levels and radiative transition rates are obtained using the FAC code where the bound states of the atomic system are calculated in the configuration mixing approximation with convenient specification of mixing schemes. The radial orbitals for the construction of basis states are derived from a modified self-consistent Dirac-Fock-Slater iteration on a fictitious mean configuration with fractional occupation numbers, representing the average electron cloud of the configurations included in the calculation. The radiative transition rates are calculated in the single multipole approximation with arbitrary ranks.

(ii) *Collisional excitation*—electron impact excitation cross sections from ground and all excited states and collisional coupling among all the excited states are calculated using FAC. Excitation cross sections for the most important $n = 2$–3 excitations are in good agreement with other published work. Extrapolations to high energies are obtained by using appropriate polynomial fitting formulae from FAC, but these are not adequate for non-allowed transitions.

(iii) *Collisional ionization*—the detailed level-to-level electron impact ionization cross sections for ground as well as all the excited levels in our model are obtained using FAC, incorporating the relativistic distorted-wave (DW) approximation, extending the factorization-interpolation method developed for the calculation of electron impact excitation cross sections. As noted in [18], the implementation allows for the most general configuration mixing to be included. Much more efficient methods based on the Coulomb-Born-Exchange (CBE) approximation or the Binary-Encounter-Dipole (BED) theory are also implemented.

(iv) *Photoionization*—the non-resonant photoionization cross sections are calculated using FAC where the continuum wavefunctions are obtained in the relativistic distorted-wave approximation. The non-relativistic multipole operators, rather than electric dipole (E1), are used for the calculation of bound-free differential oscillator strengths, although no interference between different multipoles is taken into account. In FAC, an efficient factorization-interpolation procedure is implemented by separating the coupling of the continuum electron from the bound states. FAC-generated near-threshold cross sections for both collisional and photoionization cross sections are not very accurate in many cases and therefore extrapolated values from higher energies are used to replace those cross sections.

(v) *Dielectronic recombination (DR)*—calculations of DR rates pose major challenges due to complexities involved in handling the complex atomic structures of large numbers of singly and doubly excited levels. DR involves competing processes of autoionization and radiative transitions. DR, often the dominant recombination process for ions in laboratory and astrophysical plasmas is also a significant contributor in kinetics calculations for level populations [9]. In FAC, DR data are calculated in the independent-process, isolated resonance approximation. In this calculation, $\Delta n = 0$ channels that contribute mainly and significantly to the low-temperature DR rates are included along with the $\Delta n = 1$ excitations.

(vi) *Photoexcitation*—when a photon from another region of the plasma is absorbed by a bound-bound process, the participating electron is placed into the corresponding excited level. The photoexcitation rate can be calculated from the radiative couplings $C_{kk'}$ which are defined as the probability that a photon emitted in zone k is absorbed in zone k' [1]. If $k_{\nu k}$ is the total opacity in zone k for frequency ν, and

$k_{vk}^{ii'm}$ is the partial opacity for transition $i - i'$ of material m, the photoexcitation rate for transition $i - i'$ can be written as:

$$w_{ii'mk} = \frac{4\pi \sum_v [k_{vk}^{ii'm}/k_{vk}] \sum_{k' \neq k} [C_{vk'k} J_{vk'} V_{k'} dv]}{[n_{km} f_{ikm} E_{ii'm} V_k]} \tag{5.2}$$

where $J_{vk'}$ is the total emissivity for frequency n in zone k', n_{km} is the ion density of species m in zone k, f_{ikm} is the atomic population of the lower level in zone k, $E_{ii'm}$ is the transition energy, and V_k is the volume of zone k.

The electrons in the plasma are assumed to have a Maxwellian velocity distribution and are the dominant species in populating the levels for the K- and L-shell spectra. Thus the rate coefficients for all the processes are obtained by integrating the cross sections over a Maxwellian electron distribution.

5.3 Radiation Transport and Hydrodynamics Simulation

Calculating X-ray and UV spectra of a hot plasma depends on the local atomic level populations. Atomic processes which determine the populations of the atomic levels are functions of the local electron temperature and density. For optically thick plasmas, the level populations depend on the radiation field arising from line transitions and radiative recombination processes, as pumping by photoionization and photoexcitation can significantly affect those population densities. This couples all the spatial zones in the plasma. Thus, for non-LTE dense plasmas, atomic kinetics must be coupled with radiation transport to include opacity effects. Because the populations determine the local radiation emissivities and opacities, and the radiation modifies the populations, an iterative process must be employed. In our model, a procedure is used where level populations are calculated using the radiation field from the previous iteration and these populations are used to calculate a new radiation field until convergence is reached.

The experiments of interest are the doubly nested wire array pinch, Z1975 for Cu and Z1860 for SS, performed on the refurbished Z (ZR) generator and Z581 for SS on the pre-refurbished Z at Sandia National Laboratories. For Z1975 the arrays, composed of a copper-nickel alloy wire, had a total mass of 2.57 mg and a 2:1 outer to inner mass and wire number ratio. The initial array radii were 65 and 32.5 mm and 20 mm in axial height. Emission lines of Fe and Cr were also observed in experimental spectra, and likely were due to scrape-off plasma from the feed electrode during the power flow.

The simulation presented below is performed with the one-dimensional DZAPP [14] code. The radial extent of the plasma is resolved into 30 Lagrangian zones. The magneto-hydrodynamics component solves for the radial momentum, ion and electron thermal energy, and the azimuthal magnetic field. A radiative loss term is included in the electron energy equation. The thermal conductivity for both ions and electrons and the resistivity for the diffusion of the magnetic field use the expressions of Braginskii [4] with a mean charge state. The plasma dynamics is coupled to a transmission line circuit model for the ZR generator at the plasma-vacuum interface. The circuit model includes current losses, treated as shunts, in the convolute and in the final feed with a right angle turn. Total energy conservation and Poynting's theorem are used to check the numerical solution for accuracy and found to be satisfied to better than 1 % (typically 0.1 % or less) throughout the simulation. The code is very nearly a "first-principles" code; the only phenomenological "adjustable" variable is an electron-ion thermal equilibration multiplier that artificially slows equilibration, lengthens the duration of stagnation, and prevents radiative collapse.

A large fraction of the plasma energy is radiated away in a typical Z-pinch. Thus, DZAPP treats the non-local thermodynamic equilibrium (non-LTE) atomic and radiation physics in a coupled, self-consistent manner. This method predicts the ionization populations more accurately. As noted above, lines of Cu, Ni, Fe and Cr were observed in the experimental spectra. The calculations are performed with the mixture: 93.0 % Cu, 4.0 % Ni, 2.4 % Fe, and 0.6 % Cr. The Ni abundance was chosen because the copper wires employed in the experiment had a nominal 4 % Ni component; the Fe and Cr abundances were selected to approximately match the respective He-α intensities of the experimental spectrum. The relative abundances are consistent with the Fe and Cr components coming from the stainless steel cathode. Detailed atomic models for Cu, Ni, Fe and Cr are used in the simulations. The model for Cu includes 779 atomic levels and 1609 transported emission lines. The models for Ni, Fe and Cr have 778, 776 and 774 levels, respectively, and, since they are minor constituents, 450 lines are transported for each. The radiation transport includes the effects of photoionization and photoexcitation. Non-local radiation couplings from each zone to every other zone are computed for each bound-bound, bound-free and free-free process using a probabilistic approach [1, 2, 5] where an escape probability is derived, integrated over the line profile, as a function of line-center optical depth, based on a frequency of maximum escape. Similarly, for bound-free continuum radiation arising from radiative recombination, an escape probability is derived, integrated over the bound-free profile, as a function of optical depth at the edge. Because the free-free emission and opacity have a gentle dependence on photon energy, a coarse multi-frequency treatment gives adequate accuracy for free-free escape probability. The probabilistic method of transport of both line and continuum radiation is an economic and efficient way for simulation of most laboratory and astrophysical plasmas. Inner-shell opacities, which are important in the cool, dense plasma regions, are also included in our model. The positions of the ionization-dependent absorption edges are taken from the Hartree-Fock calculations of Clementi and Roetti. Since the local emissivities and opacities are functions of the atomic populations, and the populations depend on photon transport, such as the effects of photopumping by the radiation field, an iterative process is employed on each call to the non-LTE solver to obtain convergence in the level populations and radiation field. Such calls are made approximately every 100 MHD steps, or more often if the total internal energy changes by more than a fixed amount (typically 5 %) in any zone. The total internal energy consists of the ion and electron thermal energy plus the ionization/excitation energy of the population levels. The radiative loss term from each zone for the electron energy equation is also computed during this call. In between these calls, the radiative loss and excitation/ionization energy per unit mass are kept constant. At the end of each MHD step the spectral energy distribution of the radiated energy escaping the whole plasma is constructed. Each bound-bound transition is taken to have a Voigt profile evaluated at the local ion temperature of the emitting region, and 21 photon energies are used to resolve the line. Hundreds of energies are used to describe the free-bound recombination and free-free emissions. Upon completion of the entire simulation the synthetic time-integrated spectrum is calculated from the accumulated totals of these spectral distributions.

For simulations of Z1860 and Z581 SS spectra, the atomic level structures of the component materials Fe, Ni, Cr, and Mn were employed, analogous to the Z1975 modeling. We also used the same 1D radiation hydrodynamics method described above for Z1975. 1D simulations of Z-pinches cannot reproduce the multidimensional "structure" in the density and temperatures from plasma instabilities. For some cases such as for high Z ions, the K-shell radiation comes mostly from "hot spots" in the plasma. Nevertheless, in some recent experiments, fairly uniform temperatures and densities are observed along the axis at peak implosion and appear to be nearly 1D in nature. For these experiments, including some of the stainless steel cases, the simulated spectra can come close to the measured spectra. Multidimensional structure is seen in these cases, but (a) they are energy-rich, so atomic populations are less sensitive to plasma details, and (b) the structuring is more pronounced in lower density regions of the plasma that do not contribute as much to the radiation.

For very dense and optically thick plasmas, the profile-averaged method may not be adequate to resolve spectral lines and a multi-frequency method may be necessary. In the multi-frequency case, each line is resolved using multiple frequencies which resolve the line profile. The photons generated in each zone, for each frequency v, attenuate as: $dI_v/d\tau_v = -I_v + S_v$, where I_v, S_v, and τ_v are the intensity, source function and the optical depth respectively. The radiative coupling $C_{kk'}$ discussed above can be used to calculate the radiative heating and cooling in each spatial zone, the radiation escaping from the plasma, and the photopumping of the observed levels [5].

5.4 Spectral Line Intensities

5.4.1 Energetics and Spectroscopic Simulated Data for Cu

Figure 5.4 presents some overall properties of the simulation results for Z1975. The load current is that through the wire array and its peak value is about 5 MA less than the current before the convolute due to the shunts in the circuit model. The breakdown and ablation phases of the wires is not addressed in the present model. Instead we take the initial density distribution to be composed of two Gaussians, one centered at the outer and the other at the inner array radius. This profile represents an early radial expansion of each array as the rising current initiates heating

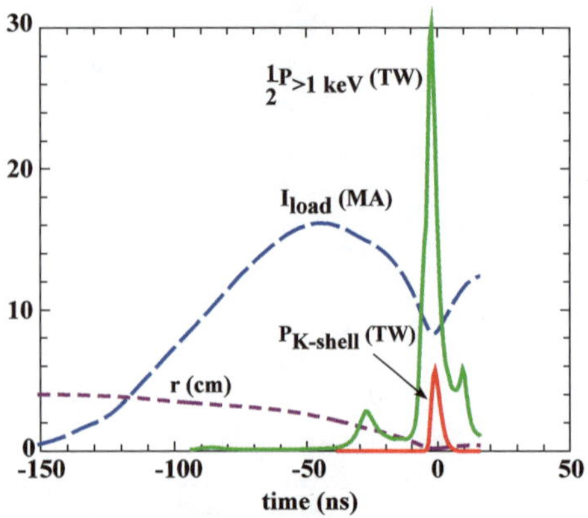

Fig. 5.4 DZAPP simulation of current profile, radius, total and K-shell radiative powers as a function of time for Z1975 data

of the wires. The denoted radius in Fig. 5.4 is the location of the plasma-vacuum interface, initially at 4 cm. The fiducial time ($t = 0$) is set at the peak of the K-shell radiation pulse, which comprises photons above 5 keV. The experiment provided a K-shell yield of 24 kJ while the integral of the simulated pulse has 27 kJ. The larger pulse contains the total radiation, which is predominantly from the L-shell like ions of Cu and Ni. The total simulated yield is 477 kJ, notably above the measured value of 258 kJ. This difference arises from two limitations of the model. First the atomic model for the components does not contain a sufficient number of levels in the low ionization stages, because it emphasizes the level structure of the L- and K-shell ions. Hence the simulated pinch, which radiates its coupled energy, emits via L-shell ions rather than through lower ionization stages. Second, the 1D character of the simulation means that all the load material implodes together. In the experiment there is axial structure due to trailing mass, which radiates at a low temperature. Despite these limitations, the agreement of the simulated spectra with the data discussed below indicates that the temporal simulation accurately accounts for some aspects of the implosion physics. These findings are also in good agreement with the investigations of Hansen et al. for the same Z1975 shot [20]. Figure 5.5 presents snapshots of the plasma dynamics before ($t = -4$ ns), at ($t = 0$ ns), and following ($t = 8$ ns) peak power as a function of the radius r. The ion density n_i of Fig. 5.5(a) shows the maintenance of a dense shell during the stagnation. Inside of this shell the plasma has a high electron temperature T_e and ion temperature T_i shown in Figs. 5.5(c) and 5.5(d). At each time the dense shell is at the same location as the rapid falloff in T_e, T_i and the mean ionization state Z bar in Fig. 5.5(b). The primary character of the stagnation is a hot core from which the K-shell emission arises, and an outer, high density shell comprised of L-shell ions. The maximum in radiative emission occurs near the time of peak compression (minimum radius), as seen in this figure. At peak compression, the density reaches about 10^{20} cm^{-3} and the ion temperature in this model reaches almost 500 keV. The average charge is about 27 near the peak compression.

Time-integrated synthetic spectra were generated for K- and L-shell Cu, Ni, Fe and Cr ions from the simulation of shot Z1975 and compared with data. We em-

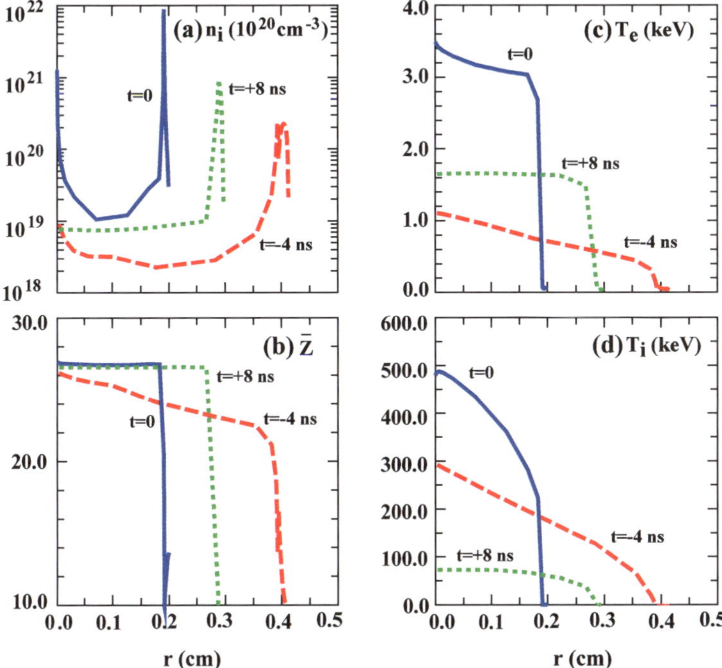

Fig. 5.5 Plasma properties as a function of radius at three different times centering on time of peak K-shell emission power. (**a**) ion densities, (**b**) average charge state, (**c**) electron temperatures and (**d**) ion temperature

phasize that the synthetic K- and L-shell spectra are time-integrals of the non-LTE emission (including photopumping and radiation transport) self-consistently calculated throughout the pinch. Thus, these spectra are not the result of a post-process, but represent the integral of the radiated power that cools the electrons throughout the simulation. In Fig. 5.6, we show a comparison of K-shell spectra with strong He-like and H-like lines of Cu and Ni, the components in the nested wire arrays and some He-like lines for Fe and Cr that are coming from the cathode. Our 1D non-LTE collisional radiative simulation of the K-shell spectra is in good agreement with the experimental data, except for several strong satellite like features present in the data but which are missing in our simulated spectra. Most of these strong overlapping lines may be due to K-α emissions from Cu, Ni, Fe, and Cr. The spectral feature of K-α emission lines from Cu ions in the energy range of 8.05 to 8.15 keV overlaps with the Ni Ly-α lines in this figure. There are also features around 6.3 and 7.6 keV that could correspond to inner-shell K-α emission from Fe and Ni ions, respectively. However, since we do not include any contributions due to K-α emission in our simulation, our spectrum does not show the these features.

The K-shell spectrum contains the Rydberg series of lines from H-like and He-like Cu, resulting from the decay of np ($n < 7$) states to the 1s orbitals. The strongest line in the spectrum is the Cu He-α complex, consisting of the closely-spaced

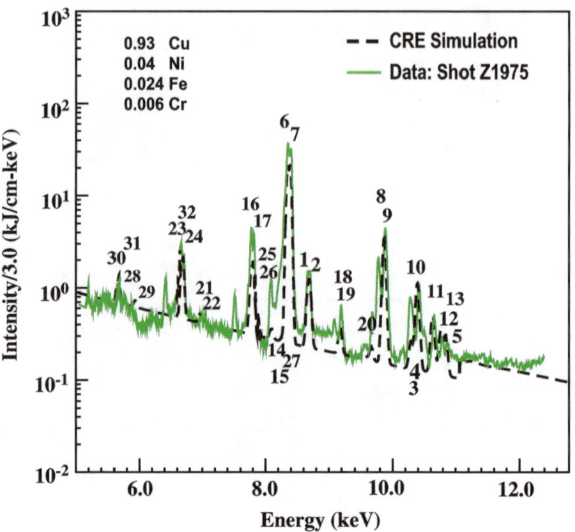

Fig. 5.6 Time-integrated K-shell spectra of Cu, Ni, Fe, and Cr compared to Z1975 data. The lines are identified in Table 5.1

1s2p ^{1}P$_1$ → 1s^2 ^{1}S$_0$ line and the density-sensitive 1s2p ^{3}P$_1$ → 1s^2 ^{1}S$_0$ intercombination line. In time-dependent spectra, the He-α line can exhibit self-reversal; the optical depth at line center becomes large near peak implosion, and peak power is radiated in the wings of the line. This does not normally occur with the intercombination line. It sometimes happens that the overlap of the self-reversed He-α and intercombination lines makes it appear that the latter line is stronger. There is good agreement in the Cu He-like Rydberg lines out through $n = 6$. The Ly-α Cu line consisting of contributions from 2p ^{2}P$_{1/2}$ and 2p ^{2}P$_{3/2}$ → 1s ^{2}S$_{1/2}$ also shows good agreement, whereas the higher-order lines do not. The He-α lines of Fe and Cr match the data reasonably well, as does the Ly-α line of Fe, but the Ni He-α line intensities are lower compared to the data.

In Fig. 5.7, we compare part of our simulated L-shell spectra with the data of Cu Z1975. At present, the model underpredicts the continuum emission seen in the data in this region, which could indicate that there is a greater quantity of cooler trailing material in the experiment than in the model. In order to compare modeled line emission with the data, we subtracted the continuum background from the data. This was accomplished by first calculating a local minimum for each frequency point in the spectrum (defined as the minimum value over the adjacent 30–50 spectral frequencies). Then this quantity was subtracted from the spectral intensity at that point, and the resulting spectrum was normalized. The strong lines in this figure are from $n = 3$–2 transitions of Li-like and Be-like Cu lines in the energy range of 1.3 to 1.5 keV. Our simulated spectra are in good agreement with the data except for the central strong 1s^{2}3d ^{2}D$_{5/2}$ → 1s^{2}3p ^{2}P$_{3/2}$ Li-like line. This disagreement is somewhat disturbing given the fact that most of the other lines in the data of similar intensities are in excellent agreement. However, the strong Li-like line is most likely to be affected by opacity in the experimental data. Further, the time-integrated L-shell spectra of single and nested Cu wire arrays shot Z1268 [7], show that this

Table 5.1 K-shell line identification for Cu, Ni, Fe and Cr

Ion stage	Index	Ion	Upper level	Lower level	E (keV)	A^r (s^{-1})
H-like Cu	1	CuXXIX	$2p\,^2P_{1/2}$	$1s\,^2S_{1/2}$	8.666	4.42E+14
	2	CuXXIX	$2p\,^2P_{3/2}$	$1s\,^2S_{1/2}$	8.699	4.47E+14
	3	CuXXIX	$3p\,^2P_{1/2}$	$1s\,^2S_{1/2}$	10.281	1.16E+14
	4	CuXXIX	$3p\,^2P_{3/2}$	$1s\,^2S_{1/2}$	10.291	1.20E+14
	5	CuXXIX	$4p$	$1s\,^2S_{1/2}$	10.848	4.83E+13
He-like Cu	6	CuXXVIII	$1s2p\,^3P_1$	$1s^2\,^1S_0$	8.345	9.51E+13
	7	CuXXVIII	$1s2p\,^1P_1$	$1s^2\,^1S_0$	8.390	7.05E+14
	8	CuXXVIII	$1s3p\,^3P_1$	$1s^2\,^1S_0$	9.859	2.75E+13
	9	CuXXVIII	$1s3p\,^1P_1$	$1s^2\,^1S_0$	9.872	1.91E+14
	10	CuXXVIII	$1s4p\,^3P_1$	$1s^2\,^1S_0$	10.389	2.25E+13
	11	CuXXVIII	$1s\,n=5$	$1s^2\,^1S_0$	10.633	1.37E+12
	12	CuXXVIII	$1s\,n=6$	$1s^2\,^1S_0$	10.764	5.52E+11
	13	CuXXVIII	$1s\,n=7$	$1s^2\,^1S_0$	10.842	2.58E+11
H-like Ni	14	NiXXVIII	$2p\,^2P_{1/2}$	$1s\,^2S_{1/2}$	8.072	3.84E+14
	15	NiXXVIII	$2p\,^2P_{3/2}$	$1s\,^2S_{1/2}$	8.103	3.88E+14
He-like Ni	16	NiXXVII	$1s2p\,^3P_1$	$1s^2\,^1S_0$	7.764	7.38E+13
	17	NiXXVII	$1s2p\,^1P_1$	$1s^2\,^1S_0$	7.805	6.18E+14
	18	NiXXVII	$1s3p\,^3P_1$	$1s^2\,^1S_0$	9.171	2.15E+13
	19	NiXXVII	$1s3p\,^1P_1$	$1s^2\,^1S_0$	9.182	1.68E+14
	20	NiXXVII	$1s4p$	$1s^2\,^1S_0$	9.663	9.04E+12
H-like Fe	21	FeXXVI	$2p\,^2P_{1/2}$	$1s\,^2S_{1/2}$	6.950	2.86E+14
	22	FeXXVI	$2p\,^2P_{3/2}$	$1s\,^2S_{1/2}$	6.976	2.88E+14
He-like Fe	23	FeXXV	$1s2p\,^3P_1$	$1s^2\,^1S_0$	6.661	4.20E+13
	24	FeXXV	$1s2p\,^1P_1$	$1s^2\,^1S_0$	6.699	4.68E+14
	25	FeXXV	$1s3p\,^3P_1$	$1s^2\,^1S_0$	7.869	1.25E+13
	26	FeXXV	$1s3p\,^1P_1$	$1s^2\,^1S_0$	7.879	1.27E+14
	27	FeXXV	$1s4p$	$1s^2\,^1S_0$	8.292	1.44E+13
H-like Cr	28	CrXXIV	$2p\,^2P_{1/2}$	$1s\,^2S_{1/2}$	5.914	2.08E+14
	29	CrXXIV	$2p\,^2P_{1/2}$	$1s\,^2S_{1/2}$	5.931	2.09E+14
He-like Cr	30	CrXXIII	$1s2p\,^3P_1$	$1s^2\,^1S_0$	5.645	2.19E+13
	31	CrXXIII	$1s2p\,^1P_1$	$1s^2\,^1S_0$	5.681	3.44E+14
	32	CrXXIII	$1s3p\,^1P_1$	$1s^2\,^1S_0$	6.677	6.65E+12

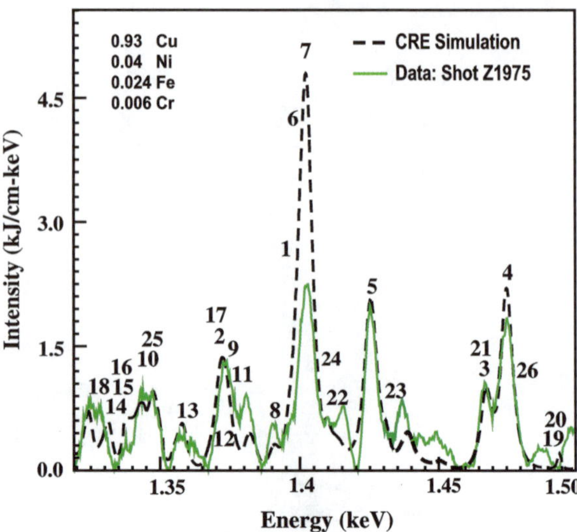

Fig. 5.7 Time-integrated Cu L-shell spectra compared to Z1975 data. The lines are identified in Table 5.2

line is much stronger than the adjacent Li-like and Be-like lines as predicted by our modeling and has overall good agreement with our findings of the relative intensities of the Li-like $1s^2 3d\ ^2D_{5/2} \rightarrow 1s^2 3p\ ^2P_{3/2}$ and $1s^2 3d\ ^2D_{3/2} \rightarrow 1s^2 3d\ ^2D_{1/2}$ lines in Fig. 5.7. The electron temperature and density quoted in their analysis of highly ionized L-shell species were 1.1 keV and 7×10^{20} cm^3 respectively [7]. This suggests that the spectral data were obtained from line-outs near the axis and thus the electron temperature was higher in comparison to that of the Z1975 data in Fig. 5.7. In the cooler region away from the axis, as the temperature becomes lower, the strong Li-like $1s^2 3d\ ^2D_{5/2} \rightarrow 1s^2 2p\ ^2P_{3/2}$ line becomes less intense in comparison to other strong lines. This is a possible explanation for the discrepancy between the spectral features of data and modeling of this line in Fig. 5.7.

5.4.2 X-Ray Stainless Steel Spectra

The atomic, kinetic and radiation transport models described above were used in this investigation to analyze Z-pinch stainless steel (SS) data obtained from experiments on the Z machine at SNL. Recent shots on the refurbished Z machine at SNL are exceeding previous record high K-shell X-ray powers and yields. Stainless steel spectra are sensitive to the accuracy of the atomic modeling. The time-integrated spectra are also sensitive to the implosion history because they depend on the time-dependent details of the imploding plasma over the full history of the pinch. In this case, we solve for implosion history dependent, i.e., non-equilibrium atomic populations, although these often approach collisional radiative equilibrium values. Even if ground state populations diverge appreciably from CRE values, the corresponding excited state populations remain close to CRE. The SS shot Z1860 [27] on refurbished ZR machine with a nested wire array, 65.0 on 32.5 mm, 200 on 100 wires and total mass of 2.5 mg, has K-shell outputs that exceeds the best

Table 5.2 Cu L-shell line identification

Ion stage	Index	Ion	Upper level	Lower level	E (keV)	A^r (s^{-1})
Li-like	1	CuXXVII	$1s^23s\,^2S_{1/2}$	$1s^22p\,^2P_{1/2}$	1.396	1.49E+12
	2	CuXXVII	$1s^23s\,^2S_{1/2}$	$1s^22p\,^2P_{3/2}$	1.371	3.16E+12
	3	CuXXVII	$1s^23p\,^2P_{1/2}$	$1s^22s\,^2S_{1/2}$	1.467	1.22E+13
	4	CuXXVII	$1s^23p\,^2P_{3/2}$	$1s^22s\,^2S_{1/2}$	1.475	1.18E+13
	5	CuXXVII	$1s^23d\,^2D_{3/2}$	$1s^22p\,^2P_{1/2}$	1.425	2.97E+13
	6	CuXXVII	$1s^23d\,^2D_{3/2}$	$1s^22p\,^2P_{3/2}$	1.400	5.80E+12
	7	CuXXVII	$1s^23d\,^2D_{5/2}$	$1s^22p\,^2P_{3/2}$	1.402	3.50E+13
Be-like	8	CuXXVI	$1s^22s3d\,^3D_2$	$1s^22s2p\,^3P_1$	1.390	2.77E+13
	9	CuXXVI	$1s^22s3d\,^3D_3$	$1s^22s2p\,^3P_2$	1.373	3.57E+13
	10	CuXXVI	$1s^22s3d\,^1D_2$	$1s^22s2p\,^1P_1$	1.343	2.65E+13
	11	CuXXVI	$1s^22p3d\,^3P_0$	$1s^22p^2\,^3P_1$	1.382	2.66E+13
	12	CuXXVI	$1s^22p3d\,^3D_3$	$1s^22p^2\,^3P_2$	1.371	3.33E+13
	13	CuXXVI	$1s^22p3d\,^1F_3$	$1s^22p^2\,^1D_2$	1.358	5.33E+13
B-like	14	CuXXV	$1s^22s^23d\,^2D_{5/2}$	$1s^22s^22p\,^2P_{3/2}$	1.331	3.05E+13
	15	CuXXV	$1s^22s^23d\,^4D_{7/2}$	$1s^22s^22p\,^4P_{5/2}$	1.334	1.41E+13
	16	CuXXV	$1s^22s2p3d\,^2F$	$1s^22s2p^2\,^2D_{5/2}$	1.332	2.10E+13
C-like	17	CuXXIV	$1s^22s2p^23p$	$1s^22s^22p^2\,^3P_2$	1.369	1.30E+12
N-like	18	CuXXIII	$1s^22s2p^33p$	$1s^22s^22p^3\,^4S_{3/2}$	1.325	5.77E+11
O-like	19	CuXXII	$1s^22s^22p^34s$	$1s^22s^22p^4\,^3P_0$	1.491	3.51E+11
	20	CuXXII	$1s^22s^22p^34d$	$1s^22s^22p^4\,^1D_2$	1.494	1.28E+12
	21	CuXXII	$1s^22s^22p^34d$	$1s^22s^22p^4\,^1S_0$	1.467	2.60E+11
F-like	22	CuXXI	$1s^22s^22p^44d$	$1s^22s^22p^5\,^2P_{1/2}$	1.410	7.75E+11
	23	CuXXI	$1s^22s^22p^44d$	$1s^22s^22p^5\,^2P_{3/2}$	1.431	1.56E+12
	24	CuXXI	$1s^22s2p^54d$	$1s^22s2p^6\,^2S_{1/2}$	1.409	9.39E+11
Ne-like	25	CuXX	$1s^22s^22p^54d$	$1s^22s^22p^6\,^1S_0$	1.344	1.05E+12
	26	CuXX	$1s^22s2p^64p$	$1s^22s^22p^6\,^1S_0$	1.478	9.34E+11

previously obtained yield at $\sim$7 keV from a nested wire array. More typical 55 mm nested SS loads produced $\sim$50 kJ K-shell yield and $\sim$10 TW K-shell power. In the left panel of Fig. 5.8 a comparison of our multi-frequency generated K-shell SS spectrum with the time-integrated Z1860 data is shown. The experimental spectrum was normalized to the measured power. The spectrum was generated by post-processing simulation-derived hydrodynamic profiles with a full multi-frequency radiation transport model, and then summing the weighted results at a series of times near peak calculated emission. It is apparent that the intensities of the He-α and He-intercombination lines of Fe, Cr, Ni, and Mn, the constituents of SS, are not in the ratios of the compositions of the materials. While the intensities of Fe and Ni are approximately in the ratios as they exist in SS, the Cr lines are somewhat stronger than this. One explanation could be that Cr is somewhat easier to ionize

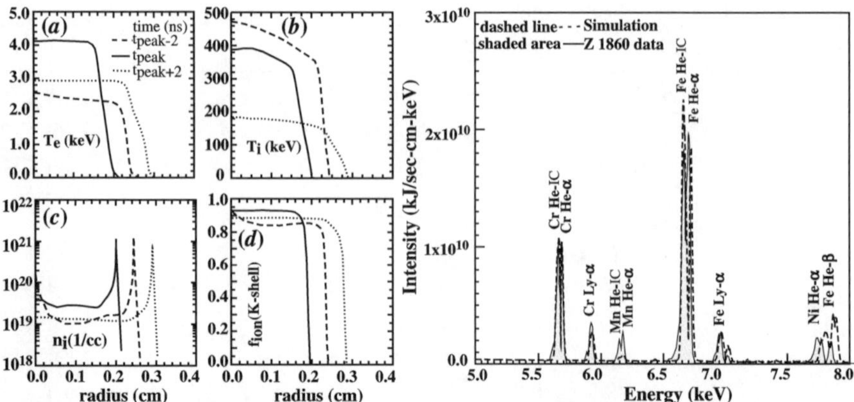

Fig. 5.8 (*Left*) Plasma properties as a function of radius at three different times centering around the time of peak K-shell emission power. (**a**), (**b**), (**c**), and (**d**) show the electron temperatures, ion temperatures, ion densities and the K-shell fractional ion populations. (*Right*) Comparison of K-shell SS spectrum with shot Z1860 data. To obtain the multi-frequency K-shell spectrum, hydrodynamic profiles near the time of peak radiative (K-shell) emission (note the units on the y-axis) were obtained from the probabilistic simulation, and post-processed with the multi-frequency radiation transport algorithm

than Fe and Ni. In the right panel of Fig. 5.8, profiles of electron and ion temperatures, ion densities and the populations of the K-shell (H- and He-like) for shot Z581 are plotted as functions of radius at the time of peak radiative emission, and for times 2.0 ns before and after the peak. The maximum in radiative emission occurs near the time of peak compression (minimum radius), as seen in this figure. At peak compression, the density reaches about 3×10^{19} cm^{-3} and the ion temperature in this model is almost 400 keV. The K-shell fraction is a maximum (about 93 %) near this time.

In Fig. 5.9 we show a time-integrated experimental Z-pinch spectrum for SNL shot Z581 [28] in the energy range between 1.05 and 1.20 keV. We have identified some of the strong Li-like transitions in this figure. These data are compared with a time-integrated synthetic spectrum computed from a 1D time-dependent radiation magnetohydrodynamic simulation of Z581. It is important to mention that most spectra available to us until fairly recently have been time-integrated. Now that time-resolved spectra are becoming available, interesting features, including time-dependent Doppler splitting, are observed. There are a number of disadvantages in using time-integrated data. For example, density and temperature information can be obtained from certain K-shell line ratios, but using integrated spectra, these determinations can be inaccurate or misleading.

A detailed-configuration FAC model, as described above, was used for Fe. In order to show the significance of proper atomic and radiation physics in the spectral simulations, we also compare the data in Fig. 5.9 with LTE and optically thin

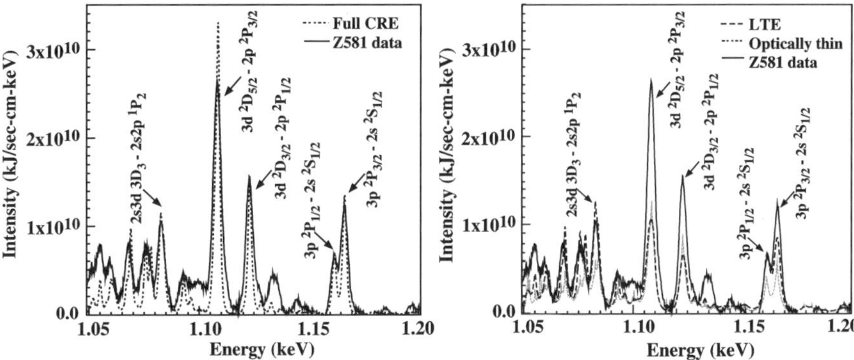

Fig. 5.9 (*Left*) Li-like Fe spectra for a full SS Li-like Fe simulation compared with shot Z581. (*Right*) LTE and Optically thin Li-like Fe simulated spectra compared with shot Z581

models. This comparison clearly shows the inadequacies of these modelings compared to a full CRE calculation with radiation transport shown in the left panel of Fig. 5.9.

This same analysis can be used to predict and analyze Fe spectra from astrophysical objects. In X-ray astronomy we generally are faced with more unknowns than in the laboratory, so beyond diagnostics of e.g. temperature, density, and Doppler shifts, we are also interested in element abundances. Absolute abundances are usually measured from line-to-continuum ratios, at least in relatively quiescent objects such as supernova remnants (SNR) or clusters of galaxies that only evolve on timescales of many years. Relative abundances involve line-to-line comparisons. The left panel in Fig. 5.10 shows the Cassiopeia A SNR with explosion. This was the target of a 1 million second observation with *Chandra*, performed over 3 weeks in 2004 with $\sim 3 \times 10^8$ photons collected. The CCD spectra shown in the right panel of Fig. 5.10 appear to be almost pure Fe with no Si, S, Ar, and Ca emission lines that usually accompany Fe lines. This implies that this plasma must have originated in the very center of the explosion and has been thrown to the outer layers of the explosion by instabilities in the core. We can quantify this by measuring the departure from collisional ionization equilibrium (CIE) by measuring ratios of Fe XXV to lower charge states, Fe XXIV, Fe XXII, etc. at the temperature given by the continuum. This represents heating by the shock followed by partial relaxation to CIE. By estimating the time since shock passage using the observed spectral signature and modeled evolution for the SNR, we can build up a precise picture of the element abundance distribution produced by the explosion. Currently analysis of around 6000 spectra from different regions of the SNR is under way. Although the same analysis used for Z-pinches can be used to study ICF and astrophysical plasmas, spectroscopic simulations provide different kinds of knowledge for different plasma regimes and diagnostic properties that are relevant for laboratory or astrophysical plasmas.

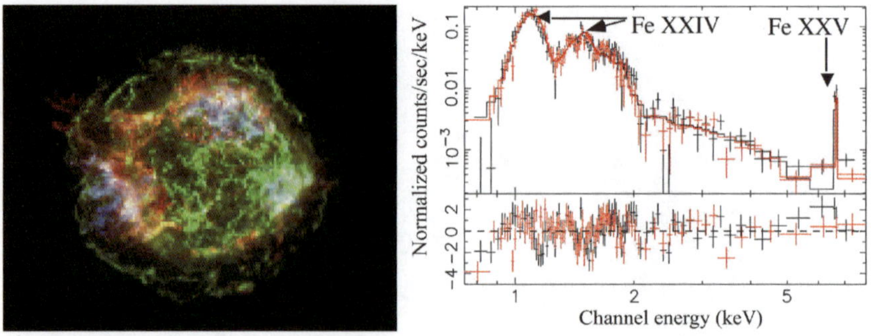

Fig. 5.10 (*Left*) Cassiopeia A SNR 1 million sec observation with Chandra. Fe in *blue*, Si in *red* and continuum in *green*. (*Right*) The 2000 (*red*) and 2002 (*black*) fitted epoch ACIS spectra of the diffuse Fe cloud with a single-temperature model

5.5 Summary and Conclusions

The reliability and scalability of atomic data used in the ionization dynamics calculations of K- and L-shell moderate-Z elements are of great concern. It is crucial to include an adequate number of excited levels per ion to generate accurate synthetic spectra with which to diagnose plasma conditions. We have constructed a collisional radiative model including a large and adequate number of excited levels and atomic rates of all dominant processes and opacity effects by including radiation transport for non-LTE plasmas. Self-consistent coupling of detailed radiation transport provides accurate photoionization and photoexcitation modifications to the local atomic populations. We have presented the synthetic time-integrated K- and L-shell radiation from a Cu and Fe Z-pinch plasma using a 1D non-LTE collisional radiative model. Comparisons of our predictions with time-integrated Z1975 and Z581 data from the Z generator show excellent agreement. Since the nested wire arrays contained some Ni and radiation from the SS cathode material was also observed in the data, our Z1975 simulation included a detailed atomic model for Cu as well as Ni, Fe, and Cr. Calculating the X-ray spectra of hot plasmas requires knowledge of atomic data and structure to evaluate the precise ionization dynamics and it is imperative that the data needed for the CRE model be generated using state-of-the art computer codes. Due to the enormous quantity of the data involved for many atomic levels, especially for complex ions, simple approximations and crude atomic models are often employed for analysis. A major issue of this investigation is how to construct an atomic model containing a detailed yet manageable number of levels that will adequately describe the plasma [19]. Self-consistent coupling of detailed radiation transport provides accurate photoionization and photoexcitation modifications to the local atomic populations. Although the Ni Ly-α line overlaps with some of the Ne-like and F-like K-α lines, we believe that due to the relatively very small concentration (4 %) of Ni in the wires, the strong line in the range 8.05 to 8.15 keV is due to Cu K-α lines. More work needs to be done to validate this further. We note that time-integrated spectra from a Z-pinch embody the history of the plasma dynamics throughout the stagnation period. The good agreement of our simulated spectra, using the 1D MHD DZAPP code and self-consistently generated data from state-of-the-art codes such as FAC for both K- and L-shell line emissions, with experimental data is very encouraging. We will continue to analyze both Cu and Fe spectral data from new experiments that are planned in the near future on Z.

Acknowledgements   This work was supported by the US Department of Energy/NNSA. Sandia National Laboratories is a multi-program laboratory managed and operated by Sandia Corporation, a wholly owned subsidiary of Lockheed Martin Corporation, for the U.S. Department of Energy's National Nuclear Security Administration under contract DE-AC04-94AL85000.

References

1. J.P. Apruzese, J. Quant. Spectrosc. Radiat. Transf. **34**, 447 (1985)
2. J.P. Apruzese, J. Davis, K.G. Whitney, J.W. Thornhill, P.C. Kepple, R.W. Clark, C. Deeney, C.A. Coverdale, T.W.L. Sanford, Phys. Plasmas **9**(5), 2411 (2002)
3. P. Beiersdorfer, T. Phillips, V.L. Jacobs, K.W. Hill, M. Bitter, S. von Goeler, S.M. Kahn, Astrophys. J. **409**, 846 (1993)
4. S.I. Braginskii, Rev. Plasma Phys. **1**, 205 (1965)
5. R.W. Clark, J. Davis, J.P. Apruzese, J.L. Giuliani, J. Quant. Spectrosc. Radiat. Transf. **53**(3), 307 (1995)
6. R. Clark, J. Davis, M. Blaha, J.L. Giuliani, Laser Part. Beams **19**, 557–577 (2001)
7. C.A. Coverdale, B. Jones, P.D. LePell, C. Deeney, A.S. Safronova, V.L. Kantsyrev, D. Fedin, N. Ouart, V. Ivanov, J. Chittenden, V. Nalajala, S. Pokola, I. Shrestha, in *The 6th International Conference on Dense Z-Pinches*. AIP Conference Proceedings, vol. 808 (2006), p. 45
8. C.A. Coverdale, B. Jones, D.J. Ampleford, J. Chittenden, C. Jennings, J.W. Thornhill, J.P. Apruzese, R.W. Clark, K.G. Whitney, A. Dasgupta, J. Davis, J. Guiliani, P.D. LePell, C. Deeney, D.B. Sinars, M.E. Cuneo, High Energy Density Phys. **6**, 143 (2010)
9. A. Dasgupta, K.G. Whitney, M. Blaha, M. Buie, Phys. Rev. A **46**, 5973 (1992)
10. A. Dasgupta, K.G. Whitney, H.L. Zhang, D.H. Sampson, Phys. Rev. E **55**, 3460 (1997)
11. A. Dasgupta, J. Davis, R.W. Clark, J.W. Thornhill, J.L. Giuliani, K.G. Whitney, in *The 7th International Conference on Dense Z-Pinches*. AIP Conference Proceedings, vol. 1088 (2009), p. 37
12. A. Dasgupta, J.G. Giuliani, J. Davis, R.W. Clark, C.A. Coverdale, B. Jones, D. Ampleford, IEEE Trans. Plasma Sci. **38**(4), 598 (2010), Special Issue Z-Pinch Plasmas
13. A. Dasgupta, R.W. Clark, N.D. Ouart, J.L. Giuliani, W. Thornhill, J. Davis, B. Jones, D.J. Ampleford, S.B. Hansen, C.A. Coverdale, High Energy Density Phys. **8**, 284 (2012)
14. J. Davis, R.W. Clark, J.L. Giuliani, J.W. Thornhill, C. Deeny, IEEE Trans. Plasma Sci. **26**, 1192 (1998)
15. C. Deeney, M.R. Douglas, R.B. Spielman, T.J. Nash, D.L. Peterson, P. L'Eplattenier, G.A. Chandler, J.F. Seamen, K.W. Struve, Phys. Rev. Lett. **81**, 4883 (1998)
16. H. Gabriel, K.J.H. Philips, Mon. Not. R. Astron. Soc. **189**, 319 (1979)
17. M.F. Gu, Astrophys. J. **590**, 1131 (2003)
18. M.F. Gu, Can. J. Phys. **86**, 675 (2008)
19. S.B. Hansen, J. Bauche, C. Bauche-Arnoult, M.F. Gu, High Energy Density Phys. **3**, 109 (2007)
20. S.B. Hansen, B. Jones, J.L. Giuliani, J.P. Apruzese, J.W. Thornhill, H.A. Scott, D.J. Ampleford, C.A. Jennings, C.A. Coverdale, M.E. Cuneo, G.A. Rochau, J.E. Bailey, A. Dasgupta, R.W. Clark, J. Davis, High Energy Density Phys. **7**, 303 (2011)
21. V.L. Jacobs, G.A. Doschek, J.F. Seely, R.D. Cowan, Phys. Rev. A **39**, 2411 (1989)

22. B. Jones, C.A. Coverdale, C. Deeney, D.B. Sinars, E.M. Waisman, M.E. Cuneo, D.J. Ampleford, D. LePell, K.R. Cochrane, J.W. Thornhill, J.P. Apruzese, A. Dasgupta, K.G. Whitney, R.W. Clark, J.P. Chittenden, Phys. Plasmas **15**, 122703 (2008)
23. R. Rodríguez, J.M. Gil, R. Florido, J.G. Rubiano, P. Martel, E. Mínguez, in *34th EPS Conference on Plasma Physics*, Warsaw, 2–6 July 2007. ECA, vol. 31F (2007), p. 2.092
24. F.B. Rosmej, J. Phys. B, At. Mol. Opt. Phys. **30**, L819–L828 (1997)
25. J.W. Thornhill, A.L. Velikovich, R.W. Clark, J.P. Apruzese, J. Davis, K.G. Whitney, P.L. Coleman, C.A. Coverdale, C. Deeney, B.M. Jones, P.D. Lepell, IEEE Trans. Plasma Sci. **34**, 2377 (2006)
26. S.M. Vinko, O. Ciricosta, B.I. Cho, K. Engelhorn, H.K. Chung, C.R. Brown, T. Burian, J. Chalupský, R.W. Falcone, C. Graves, V. Hájková, A. Higginbotham, L. Juha, J. Krzywinski, H.J. Lee, M. Messerschmidt, C.D. Murphy, Y. Ping, A. Scherz, W. Schlotter, S. Toleikis, J.J. Turner, L. Vysin, T. Wang, M.B. Wu, U. Zastrau, D. Zhu, R.W. Lee et al., Nature **482**, 59 (2012)
27. Z1860 data obtained from SNL (courtesy of Jones B., Ampleford D.)
28. Z581 data obtained from SNL (courtesy of Coverdale C.)

Chapter 6
Dielectronic Satellites and Auger Electron Heating: Irradiation of Solids by Intense XUV-Free Electron Laser Radiation

F. Petitdemange and F.B. Rosmej

Abstract Auger electron heating of dense matter that is driven by irradiation of solids with intense XUV/X-ray Free Electron Laser radiation permits studying the dielectronic capture channels that are usually closed in standard atomic physics investigations (like, e.g., in accelerators, EBITS, tokamaks, astrophysics). After disintegration of crystalline order dense strongly coupled plasma is formed where dielectronic capture coupled to excited states is identified as the primary population source of autoionizing hole states. We demonstrate that excited states coupling of the inverse Auger effect changes entirely the known picture of dielectronic satellite emission.

6.1 Introduction

The analysis of the emission from the solar corona has lead to the conclusion that dielectronic recombination is of great importance for the ionization balance of atoms and ions and Burgess [1] elaborated the first quantitative calculations. It was quickly realized that the associated radiative transitions, so called dielectronic satellite transitions, can be directly observed in the spectral distribution and Gabriel has introduced these transitions as a sensitive method to determine the electron temperature [4]. In low-density plasmas this method approaches the ideal picture of a temperature diagnostic, i.e. selected line intensity ratios depend only on temperature but not on other plasma parameters. It turns out that also in high-density plasmas, this method is applicable (although more complex compared to the low-density plasma

F. Petitdemange · F.B. Rosmej (✉)
Sorbonne Universités, Pierre et Marie Curie, UMR 7605, LULI, case 128, 4 place Jussieu, 75252 Paris Cedex 05, France
e-mail: frank.rosmej@courriel.upmc.fr

F. Petitdemange
e-mail: frederick.petitdemange@etu.upmc.fr

F. Petitdemange · F.B. Rosmej
LULI, CEA, CNRS, Physique Atomique dans les Plasmas Denses—PAPD, Ecole Polytechnique, route de Saclay, 91128 Palaiseau Cedex, France

M. Mohan (ed.), *New Trends in Atomic and Molecular Physics*,
Springer Series on Atomic, Optical, and Plasma Physics 76,
DOI 10.1007/978-3-642-38167-6_6, © Springer-Verlag Berlin Heidelberg 2013

case) and represents now one of the most powerful methods for electron temperature determination in hot plasmas. Since then, important steps forward have been made to employ dielectronic satellite emission as a unique method to characterize also complex plasma phenomena like, e.g., suprathermal electrons, charge exchange, non-equilibrium effects, inhomogeneities, see review by Rosmej [10].

With the emergence of the 4th generation light sources (XUV and X-ray Free Electron Lasers) matter can be created under extreme conditions. As the principle laser absorption mechanism is photoionization of inner-shells, matter evolution passes via the creation of multiple excited autoionizing states. The subsequent radiative emission of these states provides therefore a great challenge to investigate matter under extreme conditions never obtained in laboratories so far: Auger electron heating of conduction band electrons and subsequent disintegration of crystalline order has been identified via the radiative emission of L-shell hole states [5]; K-shell hollow ion emission is proposed to provide high time resolution down to the autoionizing time scale of some 10 fs [13, 17]. This manifests the unique and important role of dielectronic satellite emission for the world wide emerging 4th generation light sources too; recent advances have been reviewed by Rosmej [11].

6.2 Basic Principles of Dielectronic Satellite Emission

Dielectronic recombination is a term used in atomic physics to describe the ensemble of various different processes, namely dielectronic capture, autoionization, radiative decay and corresponding dielectronic satellite emission. The ionic fractions are strongly linked to dielectronic recombination and the dielectronic satellite emission has outstanding importance for plasma diagnostics.

Let us consider the basic principles of dielectronic satellite emission via an example: the dielectronic $1s2l3l'$ satellites of Li-like ions. As the Li-like states $1s2l3l'$ are located above the ionization limit of the He-like ground state, a non-radiative decay to the He-like ground state (autoionization) is possible:

$$\text{autoionization:} \quad 1s2l3l' \rightarrow 1s^2 + e$$

By first quantum mechanical principles, the reverse process, so called "dielectronic capture" or "inverse Auger effect" must exist:

$$\text{dielectronic capture:} \quad 1s^2 + e \rightarrow 1s2l3l'$$

Another important excitation channel for satellite transitions is electron collisional excitation from inner-shells. Concerning the above-discussed example the excitation channels are:

$$\text{inner-shell excitation:} \quad 1s^2 2l + e \rightarrow 1s2l3l' + e$$

$$\text{inner-shell excitation:} \quad 1s^2 3l' + e \rightarrow 1s2l3l' + e$$

The inner-shell excitation channel is important for satellite transitions with low autoionizing rates but high radiative decay rates. It drives satellite intensities that allow an advanced characterization of the plasma, e.g., the characterization of suprathermal electrons [10] or non-equilibrium phenomena (ionizing plasma, recombining plasma [4]). We note that for electron temperature measurements, the inner-shell excitation channel should be weak.

The radiative decay channels are given by:

$$\text{radiative decay:} \quad 1\text{s}2\text{l}3\text{l}' + e \rightarrow \begin{cases} 1\text{s}^2 3\text{l}' + \hbar\omega_{\text{He}_\alpha\text{-satellite}} \\ 1\text{s}^2 2\text{l} + \hbar\omega_{\text{He}_\beta\text{-satellite}} \end{cases}$$

The emitted photon is called a "satellite". The satellite transition is of similar nature as the resonance transition $\text{He}_\alpha = 1\text{s}2\text{p} \, ^1\text{P}_1 \rightarrow 1\text{s}^2 \, ^1\text{S}_0 + \hbar\omega_{\text{He}_\alpha}$ and $\text{He}_\beta = 1\text{s}3\text{p} \, ^1\text{P}_1 \rightarrow 1\text{s}^2 \, ^1\text{S}_0 + \hbar\omega_{\text{He}_\beta}$ except the circumstance that an additional electron (the so called "spectator" electron) is present in the quantum shell $n = 3$ and $n = 2$ respectively. As the spectator electron screens the nuclear charge, the satellite transitions are mostly located on the long wavelength side of the corresponding resonance line.

However, due to intermediate coupling effects and configuration interaction, also satellites on the short wavelengths side exist [12]. As the number of possible angular momentum couplings increases rapidly with the number of electrons, usually numerous satellite transitions are located near the resonance line that often cannot be resolved spectrally even with high-resolution spectroscopy.

The genius idea of Gabriel [4] is to obtain the electron temperature with the help of satellite transitions. In low-density plasma, the intensity of the resonance line is given by

$$I^{\text{res}}_{k',j'i'} = n_e n_{k'} \frac{A_{j'i'}}{\sum_{l'} A_{j'l'}} \langle C_{k'j'} \rangle \tag{6.1}$$

where n_e is the free electron density, $n_{k'}$ the ground state density from which electron collisional excitation proceeds (k' is the 1s^2-level in our example), $A_{j'i'}$ is the transition probability of the resonance transition $j' \rightarrow i'$ (the sum over A in the denominator accounts for possible branching ratio effects) and $\langle C_{k'j'} \rangle$ is the electron collisional excitation rate coefficient from level k' to level j'. The intensity of a satellite transition with a large autoionizing rate (and negligible collisional channel) is given by

$$I^{\text{sat}}_{k,ji} = n_e n_k \frac{A_{ji}}{\sum_l A_{jl} + \sum_m \Gamma_{jm}} \langle D_{kj} \rangle \tag{6.2}$$

A_{ji} is the transition probability of the particular satellite transition and $\langle D_{kj} \rangle$ is the dielectronic capture rate coefficient from level k to the level j. The sums over the radiative decay rates and autoionizing rates account for possible branching ratio effects (in our example only $m = k$ exist, however, a particular upper level $1\text{s}2\text{l}3\text{l}'$ may have more than one radiative decay possibilities $j \rightarrow l$). As both intensities (Eqs. (6.1), (6.2)) are proportional to the electron density n_e and to the same ground state density ($k' = k$), the intensity ratio is a function of the electron temperature only, because the rate coefficients $\langle C \rangle$ and $\langle D \rangle$ depend only on the electron temperature but not on density:

$$\frac{I^{\text{sat}}_{k,ji}}{I^{\text{res}}_{k',j'i'}} = G(T_e) \tag{6.3}$$

The dielectronic capture rate is an analytical function and given by (assuming a Maxwellian electron energy distribution function)

$$\langle D_{kj} \rangle = \alpha \, \Gamma_{jk} \frac{g_j}{g_k} \frac{\exp(-\frac{E_{kj}}{kT_e})}{(kT_e)^{3/2}} \tag{6.4}$$

$\alpha = 1.656 \times 10^{-22}$ cm$^3 \cdot$ s^{-1}, g_j and g_k are the statistical weights of the states j and k, Γ_{jk} is the autoionizing rate in [s^{-1}], E_{kj} is the dielectronic capture energy in [eV] and kT_e is the electron temperature in [eV]. The intensity of a satellite transition can therefore be written as

$$I^{\text{sat}}_{k,ji} = \alpha n_e n_k \frac{Q_{k,ji}}{g_k} \frac{\exp(-\frac{E_{kj}}{kT_e})}{(kT_e)^{3/2}} \tag{6.5}$$

$Q_{k,ji}$ is the so called "dielectronic satellite intensity factor" and given by

$$Q_{k,ji} = \frac{g_j A_{ji} \Gamma_{jk}}{\sum_l A_{jl} + \sum_m \Gamma_{jm}} \tag{6.6}$$

The calculation of the dielectronic satellite intensity factors $Q_{k,ji}$ requests multi-configuration relativistic atomic structure calculations that have to include (in particular for the configurations to be discussed below) intermediate coupling and configuration interaction.

Figure 6.1 shows the simulation of the spectral distribution $I(\omega)$ of the He$_\beta$ resonance line (1s3p ^{1}P$_1$–1s^2 ^{1}S$_0$) and corresponding satellites (1s2l3l'–1s^{2}2l) calculated according [10]

$$I(\omega) = \sum_{i=1}^{N} \sum_{j=1}^{N} \frac{\hbar \omega_{ji}}{4\pi} n_j A_{ji} \varphi_{ji}(\omega_{ji}, \omega) \tag{6.7}$$

N is the maximum number of states included in the atomic kinetic population equations, n_j is the atomic population density of level j, ω_{ji} is the angular frequency for the transition $j \to i$, A_{ji} is the corresponding spontaneous radiative decay and φ_{ji} the associated line profile. The atomic population density is obtained from a set of differential equations (e.g. [11]):

$$n_j \sum_{k=1}^{N} W_{jk} = \sum_{i=1}^{N} n_i W_{ij} \tag{6.8}$$

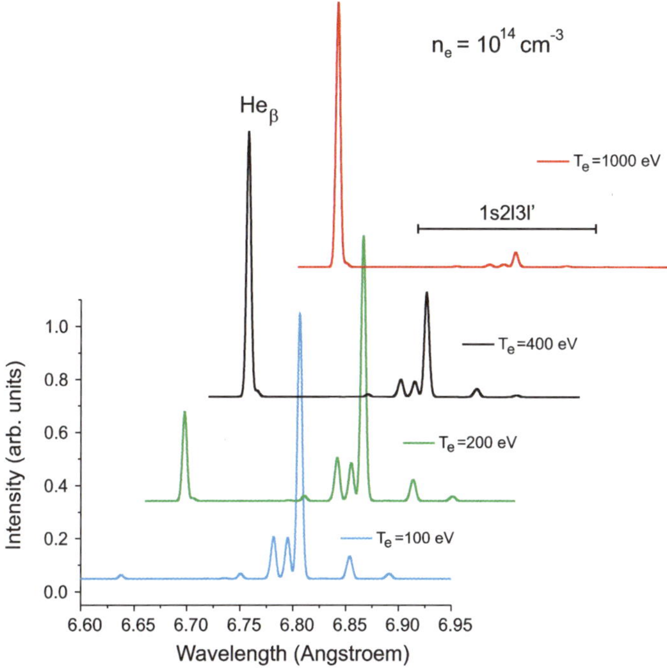

Fig. 6.1 Temperature sensitivity of He_β satellites 1s2l3l' of aluminum for an electron density of $n_e = 10^{14}$ cm^{-3}

The W-matrix for an optically thin plasma is given by

$$W_{ij} = A_{ij} + C_{ij} + I_{ij} + T_{ij} + R_{ij} + DC_{ij} + \Gamma_{ij} \qquad (6.9)$$

where A represents the spontaneous radiative emission rate, C denotes the collisional excitation/deexcitation rate, I is the ionization rate, T is the 3-body recombination rate, R is the radiative recombination rate, DC is the dielectronic (radiationless) capture rate, and Γ is the autoionization rate. The matrix elements (rate coefficients) for the inverse processes are obtained by the application of the principle of detailed balance. In the case on non-Maxwellian plasmas, the principle of micro-reversibility has to be employed. The interesting reader is referred to [14] for more details concerning non-Maxwellian population kinetics. If a particular transition does not occur, the corresponding rate is set equal to zero. The set of population equation (6.8) is sometimes also called "Rate Equations". The level populations n_j are also called "non-LTE level populations", the term "dynamic level populations" is often used in the framework of line profile calculations.

Equation (6.8) indicates equilibrium between depopulating (left hand side of Eq. (6.8)) and populating (right hand side of Eq. (6.8)) processes. This means that the level population densities are in equilibrium in the sense that they do not change in time. Weather the level population densities are in statistical equilibrium remains open. Therefore, the approach according Eqs. (6.8) is also called "collisional-radiative" model.

Figure 6.1 shows that the temperature variation ranges from negligible resonance line emission ($kT_e < 100$ eV) to negligible satellite emission ($kT_e > 1000$ eV). The satellite emission is driven by dielectronic capture and by inner-shell collisional excitation and forms different emission groups $\beta_1, \ldots, \beta_7$ [16]. At densities of $n_e = 10^{14}$ cm^{-3}, the low density limit is achieved (Corona limit) and the spectral distribution does not depend on the density variations. As can be seen from Fig. 6.1, the spectral distribution shows a strong sensitivity to electron temperature that is essentially due to the exponential factor in Eq. (6.5).

6.3 Dielectronic Capture in Dense Plasmas

Although the use of dielectronic satellite emission for electron temperature plasma diagnostics has been developed for the low density Corona limit, simulations show that strong temperature sensitivities are even maintained in dense plasmas. Moreover, modern simulations of detailed atomic population kinetics of autoionizing states allow extending the satellite diagnostics also to plasma density.

In dense plasmas, collisions between the autoionizing levels become of increasing importance (compared to the radiative decay and autoionizing rates) and population is effectively transferred between the autoionizing levels of a particular configuration. These mixing effects result in characteristic changes of the satellite spectral distribution (total contour).

Figure 6.2 shows the principle mechanism for the 1s2l3l'-configurations (indicated numerical values are for Si). In low-density plasmas, only those autoionizing levels are strongly populated that have a high autoionizing rate because in this case the dielectronic capture rate is large, see Eq. (6.4). This results in a high intensity of satellite transitions that do have high autoionizing rates and high radiative decay rates. Contrary, satellite transitions with high radiative decay rates but low autoionizing rates have small intensities as the dielectronic capture is small, Fig. 6.2(a). In high-density plasmas, population is transferred from highly populated levels to low populated ones, resulting in a density dependent change of satellite line intensity, Fig. 6.2(b). These characteristic changes of the spectral distribution can in turn be used for density diagnostics: this has first been explored by Vinogradov et al. [19] via the 2l2l'-satellites near Ly$_\alpha$ and has later been extended to the 1s2l2l'-satellites [7] near He$_\alpha$ and the 1s2l3l' [12] near He$_\beta$.

Figure 6.3 shows a strong density sensitivity (over many orders of magnitude) of the spectral distribution of the He$_\beta$ satellites of aluminum for an electron temperature of $kT_e = 100$ eV. A strong variation of intensity of different groups (in particular between ($\beta_1, \beta_2, \beta_3, \beta_5$) and β_4) is observed. At densities above 10^{22} cm^{-3}, the spectral distribution does not show anymore strong variations (except line shapes) because the Boltzmann limit is achieved. For densities below 10^{18} cm^{-3}, the Corona limit is achieved and the spectral distribution does not show anymore significant variation with respect to electron density.

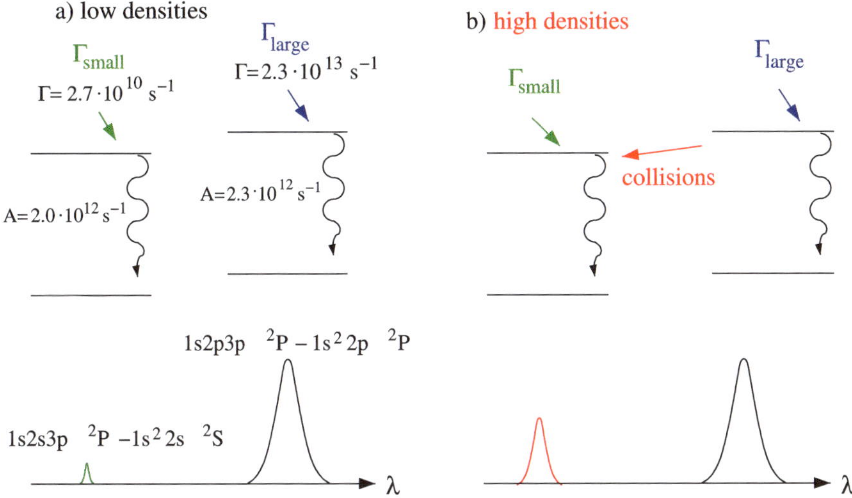

Fig. 6.2 Collisional transfer of population between autoionizing states. For illustration numerical values for A and Γ are indicated for the 1s2l3l'-configurations of Li-like silicon

Figure 6.4 demonstrates the temperature sensitivity of the spectral distribution in a dense aluminum plasma: $n_e = 10^{22}$ cm^{-3}. Although the sensitivity is not so much pronounced than for the low density limit (see Fig. 6.1) the sensitivity is still strong and permits to obtain valuable information about electron temperature also in dense plasmas.

Density diagnostic that employs dielectronic satellite transitions has the great advantage that emission is usually confined to the high temperature high-density volume elements only thereby providing an inherent space resolution. It also provides an effective time resolution: dielectronic satellite emission is not sensitive to radiative recombination regimes (that usually takes place at low densities, e.g., in dense laser produced plasmas) and a characteristic time scale that is dictated by the large autoionization rate rather than the radiative decay rate [10]. Therefore, in time-integrated spectra, emission from high density is not much masked by low-density emission. This is a very important circumstance when dealing with emission induced from the 4th generation light sources: the pulse duration of some 10 fs is much too short for time resolved diagnostics (current limits of streak camera are about 0.5 ps) and time and space resolution must be realized with "internal atomic physics properties". For a detailed discussion on these subjects the reader is referred to the recent review articles and references therein [10, 11].

Standard Stark broadening analysis of the dielectronic satellite transitions (that neglect interference effects and ions dynamics) has been proposed for advanced density diagnostics [20]. It has been understood only recently, that quantum mechanical interference effects are very important in order to predict accurately the shape of the satellite emission groups for density diagnostics [11].

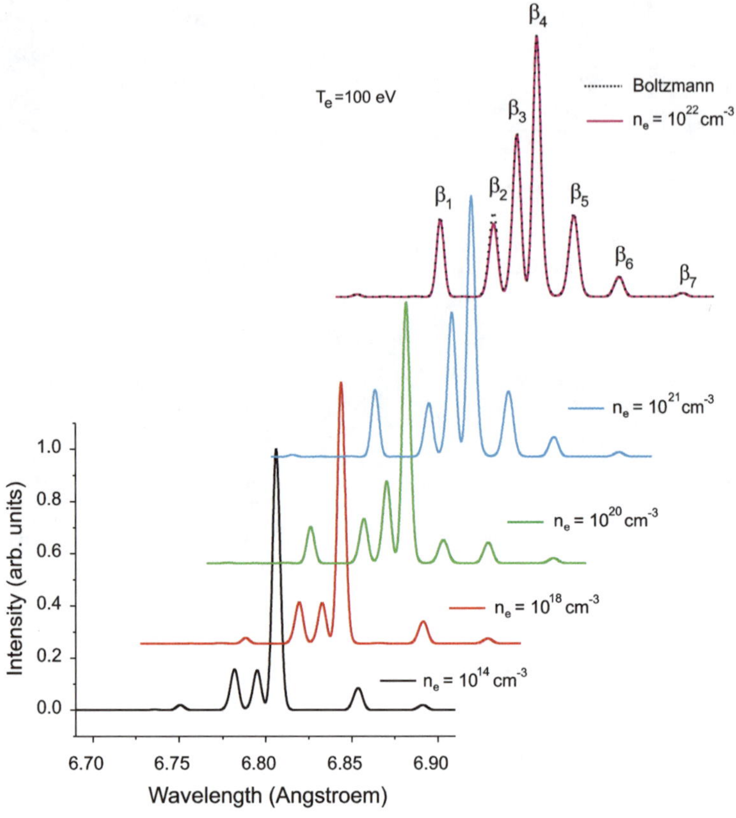

Fig. 6.3 Density sensitivity of the He$_\beta$ satellites 1s2l3l$'$ of aluminum for an electron temperature $kT_e = 100$ eV. Stark effect is not included in order to visualize better atomic population effects

6.4 Dielectronic Satellites Coupled to Excited States

In traditional plasma spectroscopy, ground state population plays the dominating role for the radiative properties. In high density plasmas, however, excited state populations become of increasing importance. For several important cases, relevant elementary processes are much stronger from excited states than from ground states and highly excited state population may therefore drive an emission that dominates over the corresponding traditional ground state driven emission.

High density effects on dielectronic satellites may not only provoke transfer of populations between the autoionizing states as discussed above, but also enlarge the number of states "k" (Eq. (6.2)) into which dielectronic capture can occur. This has been explored by means of the 1s3lnl'-satellites near He$_\beta$ [15]. Let us consider the 1s3l3l$'$-configuration as an example. From a simple hydrogen-like estimation of ionization energies it can be seen that the 1s3l3l$'$ configuration lies not only above the He-like ground state but also above the 1s2l-states (the ionization energy for the two 3l-electron is about $2Z^2$ Ry$/3^2$ being smaller than the ionization potential

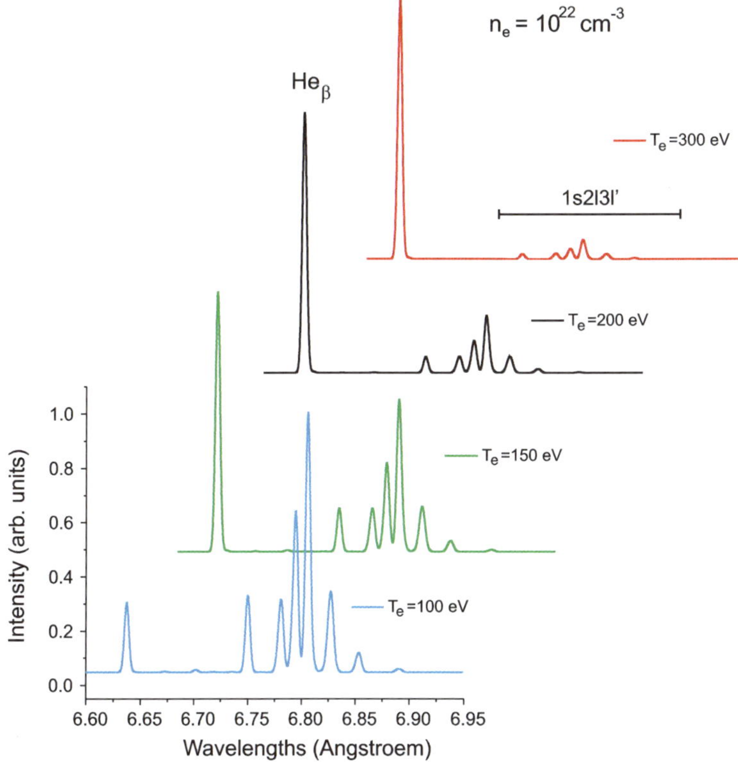

Fig. 6.4 Temperature sensitivity of the He$_\beta$ satellites 1s2l3l' of aluminum for an electron density of $n_e = 10^{22}$ cm^{-3}. Although the atomic population kinetics is fully collisional-radiative due to high density, a strong temperature sensitivity for the satellite emission is maintained

$\simeq Z^2$ Ry/2^2 of one 2l-electron):

$$\text{autoionization:} \quad 1s2l3l' \rightarrow \begin{Bmatrix} 1s^2 + e \\ 1s2l + e \end{Bmatrix}$$

By first quantum mechanical principles, for each channel the reverse process must exist:

$$\text{dielectronic capture:} \quad \begin{Bmatrix} 1s^2 + e \\ 1s2l + e \end{Bmatrix} \rightarrow 1s2l3l'$$

The first channel is the usual ground state coupling ($k = 1s^2$ in Eq. (6.5)), the second channel is the so called "excited states coupling", $k = 1s2l$ in Eq. (6.5) (note, that for the above example the excited states coupling involves many different states: $k = 1s2s\ ^3S_1$, $1s2s\ ^1S_1$, $1s2p\ ^3P_0$, $1s2p\ ^3P_1$, $1s2p\ ^3P_2$, $1s2p\ ^1P_1$). In low-density plasmas, the excited state population is negligible (compared to the ground state

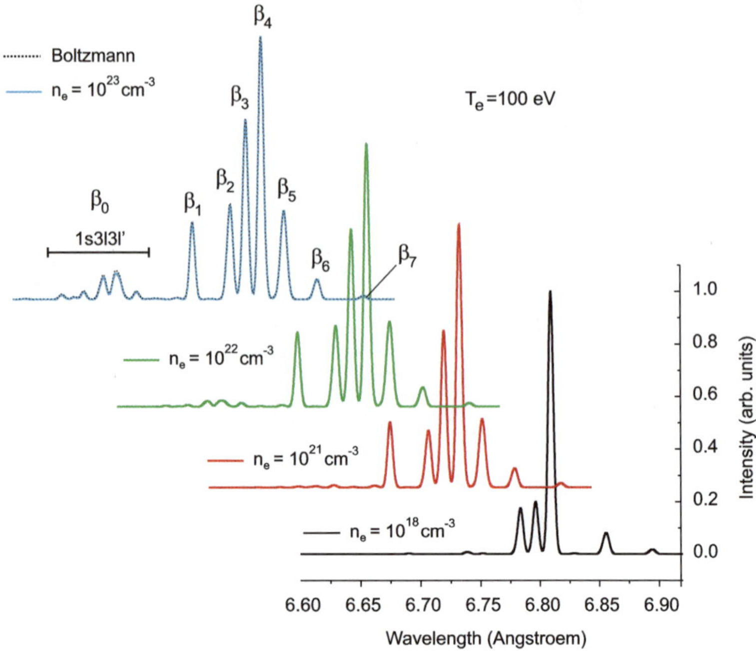

Fig. 6.5 Density sensitivity of the He$_\beta$ satellites 1s3l3l' + 1s2l3l' of aluminum for an electron temperature of $kT_e = 100$ eV. Stark effect is not included for better demonstration of atomic population effects

population) and corresponding excited state coupling is equally negligible as has been shown in connection to satellite emission in magnetic fusion plasmas [8]. In high-density plasmas, however, excited state population increases considerably (up to the Boltzmann limit) thereby driving satellite intensities to an observable level [15].

Figure 6.5 shows the simulations of the He$_\beta$ satellite emission that include the configurations 1s3l3l' and 1s2l3l'. It should be emphasized, that all LSJ-split levels for the 1s3l3l' and 1s2l3l' configuration are included in a collisional radiative approach (the included elementary processes are dielectronic capture, autoionization, radiative decay, electron collisional excitation/de-excitation, ionization, three-body recombination, radiative recombination). The figure shows that the Boltzmann limit for the configuration 1s2l3l' and 1s3l3l' are quite different: 1s2l3l'-satellites approach their Boltzmann limit at about 10^{22} cm^{-3}, whereas the 1s3l3l'-satellites are still far from their respective Boltzmann limit. In fact, due to the excited states coupling, only at densities above 10^{21} cm^{-3} a significant intensity of the 1s3l3l'-satellites is observed. At densities above 10^{23} cm^{-3}, also the 1s3l3l'-satellites approach the Boltzmann limit (compare the dotted curve and the blue curve in Fig. 6.5). The reason for this different behavior is connected to the large autoionizing rates that are coupled to excited states. Table 6.1 depicts the autoionization rates for aluminum and titanium. Excited states autoionizing rates are about two or-

Table 6.1 Averaged autoionization rates from levels K^1M^2 to K^2 and K^1L^1

Autoionization from K^1M^2 to level	Aluminum $\Gamma\,(\mathrm{s}^{-1})$	Titanium $\Gamma\,(\mathrm{s}^{-1})$
$1s^2$	$3.9 \cdot 10^{11}$	$5.2 \cdot 10^{11}$
$1s^1 2s^1\ ^3S_1$	$2.4 \cdot 10^{13}$	$2.9 \cdot 10^{13}$
$1s^1 2s^1\ ^3P_0$	$1.3 \cdot 10^{13}$	$1.6 \cdot 10^{13}$
$1s^1 2p^1\ ^1P_1$	$4.0 \cdot 10^{13}$	$4.8 \cdot 10^{13}$
$1s^1 2p^1\ ^1S_0$	$8.9 \cdot 10^{12}$	$1.0 \cdot 10^{13}$
$1s^1 2p^1\ ^3P_2$	$6.6 \cdot 10^{13}$	$8.0 \cdot 10^{13}$
$1s^1 2p^1\ ^3P_1$	$4.3 \cdot 10^{13}$	$5.0 \cdot 10^{13}$

ders of magnitude larger than corresponding ground state autoionization rates. This has two effects: first, the large excited state autoionization rate drives large dielectronic capture resulting in an important excited state contribution to the emission as discussed above. Second, the Boltzmann limit is shifted to higher values as the collisional rate needs also to compete with the autoionizing rate of excited states.

Figure 6.6 shows the spectral simulations of $1s3l3l'$ and $1s2l3l'$-satellite for titanium. It can also be seen for this chemical element, that $\beta_{1,2,3,5}$-satellites rise in

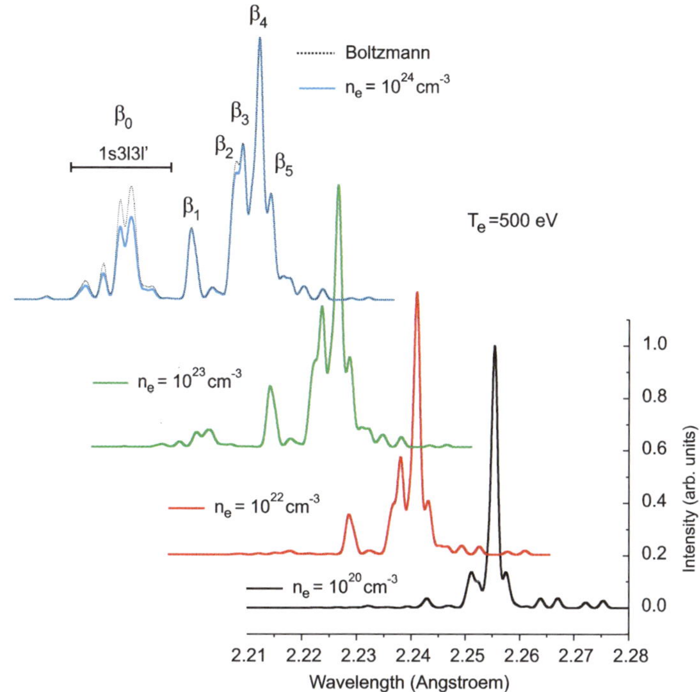

Fig. 6.6 Density sensitivity of the He_β satellites $1s3l3l' + 1s2l3l'$ of titanium for an electron temperature of $kT_e = 500\ \mathrm{eV}$

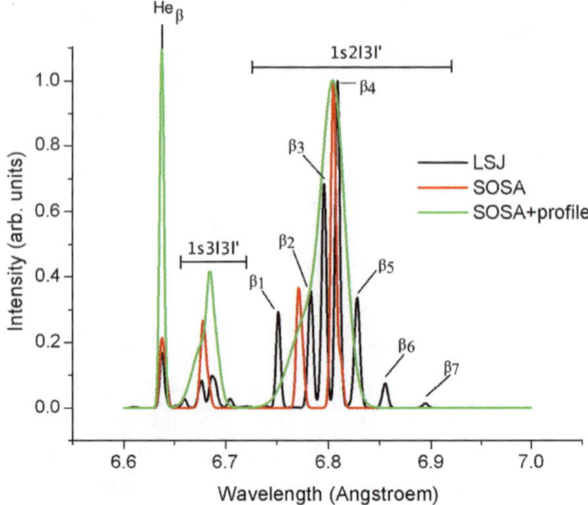

Fig. 6.7 Comparison of the spectral distribution of He$_\beta$ satellites 1s3l3l$'$ + 1s2l3l$'$ of aluminum for $kT_e = 100$ eV, $n_e = 10^{23}$ cm^{-3} calculated in the SOSA approximation (*red and green curves*), and full LSJ-split atomic level system (*black curve*)

intensity relative to the β_4-satellites with increasing density. For titanium, even at electron densities of $n_e = 10^{24}$ cm^{-3}, the 1s3l3l$'$-satellites are not yet driven to their Boltzmann limit (compare the blue curve and the dotted curve near the β_0-group in Fig. 6.6).

6.5 Spectroscopic Precision and Spin-Orbit Split Arrays

Dielectronic satellite emission provides outstanding diagnostic possibilities, however, the number of autoionizing levels is usually very large. Therefore, simplifications in the atomic population kinetics level system are introduced. A widely spread method is the "Unresolved Transition" approach, where nLSJ-split level structure is replaced by an nl-split level structure. It introduces serious errors in the spectral distribution, as intermediate coupling and corresponding level splittings are not taken into account.

Figure 6.7 compares the spectral distribution of the He$_\beta$ resonance line emission and its dielectronic satellites 1s3l3l$'$ and 1s2l3l$'$ for different approximations. The black curve (designated as "LSJ") includes all nLSJ-split levels and its corresponding atomic data and rate coefficient in a collisional-radiative approach that has been described above (Eqs. (6.7)–(6.9)). The red curve is calculated in the so called SOSA-approximation (spin-orbit split array) that involves only a relativistic nl-level structure, e.g., the 1s2l3l$'$-levels are replaced by

$$1s_{1/2}2s_{1/2}3s_{1/2}, \quad 1s_{1/2}2s_{1/2}3p_{1/2}, \quad 1s_{1/2}2s_{1/2}3p_{3/2}, \quad 1s_{1/2}2s_{1/2}3d_{3/2},$$

$$1s_{1/2}2s_{1/2}3d_{5/2}, \quad 1s_{1/2}2p_{1/2}3s_{1/2}, \quad 1s_{1/2}2p_{1/2}3p_{1/2}, \quad 1s_{1/2}2p_{1/2}3p_{3/2},$$

$$1s_{1/2}2p_{1/2}3d_{3/2}, \quad 1s_{1/2}2p_{1/2}3d_{5/2}, \quad 1s_{1/2}2p_{3/2}3s_{1/2}, \quad 1s_{1/2}2p_{3/2}3p_{1/2},$$

$$1s_{1/2}2p_{3/2}3p_{3/2}, \quad 1s_{1/2}2p_{3/2}3d_{3/2}, \quad 1s_{1/2}2p_{1/2}3d_{5/2}.$$

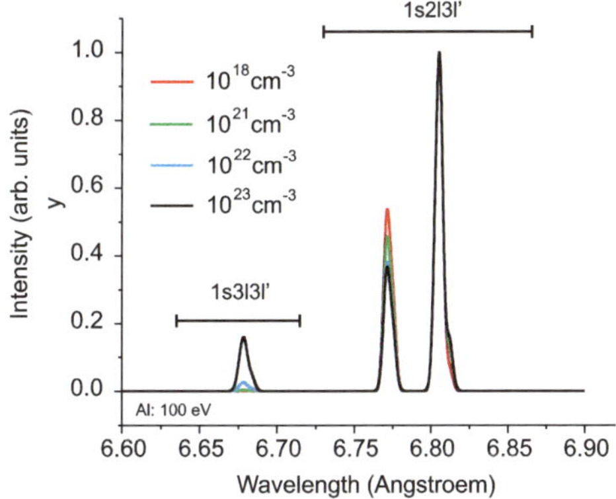

Fig. 6.8 Spectral distribution of He$_\beta$ satellites 1s3l3l' + 1s2l3l' of aluminum for $kT_e = 100$ eV and various electrons densities calculated in the SOSA

This means the 66 LSJ-split levels are replaced by 15 levels. Moreover, intermediate coupling effects are not included in this approach. The green curve (designated as "SOSA + profile") is calculated in the SOSA-approximation taking into account the reduction of the levels by averaging each atomic data (rate and population) for each configuration and by using corrected gaussian line profile:

$$\varphi^G(\omega) = \frac{1}{\sqrt{\pi \times \sigma^2}} \times \exp\left[-\frac{(\hbar \times \omega - \overline{E_{ij}})^2}{\sigma^2} \right] \tag{6.10}$$

where $\overline{E_{ij}}$ is the averaged energy of the transition between the level j to the level i calculated from the corresponding LSJ-split levels and σ^2 is the variance taking into account the Doppler broadening and the energy distribution of the LSJ transitions. The variance is given by

$$\sigma^2 = \hbar^2 \omega_{ji}^2 \frac{2kT_i}{Mc^2} + \left(\overline{E_{ij}^2} - \overline{E_{ij}}^2 \right) \tag{6.11}$$

where T_i is the ion temperature, M is the mass of the ion. This induces larger lines than for LSJ-calculus, for example, we see one large line for $1s^2nl \rightarrow 1s2l3l'$ transition instead of seven. It can be seen from Fig. 6.7, the SOSA-approach reduces considerably the number of emission features for the 1s3l3l' and 1s2l3l'-satellites. The distortion of the spectral distribution is so strong that the spectroscopic precision (that is necessary to employ satellite transitions for diagnostic purposes, see, e.g. Figs. 6.1, 6.3, 6.4, 6.5, 6.6) is evidently not anymore maintained.

Figure 6.8 shows the density sensitivity of the spectral distribution in the SOSA-approximation. It can be seen that for higher densities, the emission group near 6.77 Å decreases relative to the emission group near 6.81 Å. Quite the opposite is observed for the exact nLSJ-split levels system (see Figs. 6.3, 6.5, 6.6). Therefore, the change of the spectral distribution in the SOSA-approximation can even not be used to roughly estimate the density effects.

The complete distortion of the emission groups (Fig. 6.7) and the wrong density dependencies (Fig. 6.8) implies equally, that Stark broadening is equally not useful

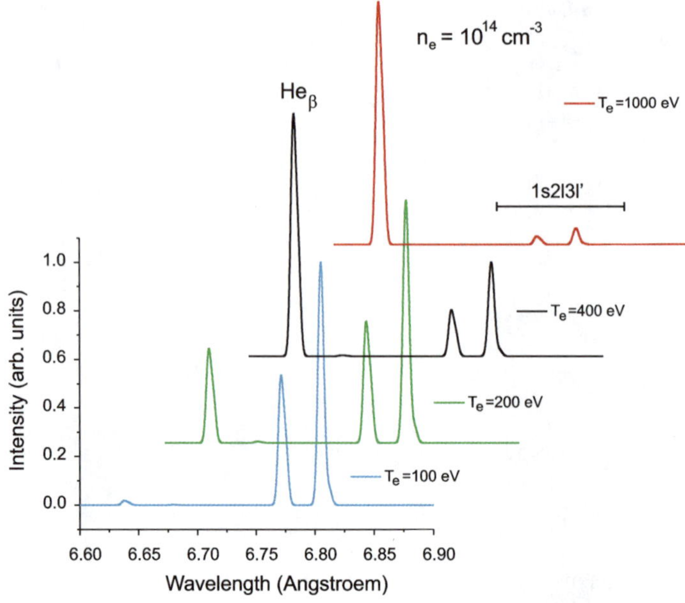

Fig. 6.9 Temperature sensitivity of the He$_\beta$ satellites 1s2l3l$'$ of aluminum for an electron density of $n_e = 10^{14}$ cm^{-3} in the SOSA approximation

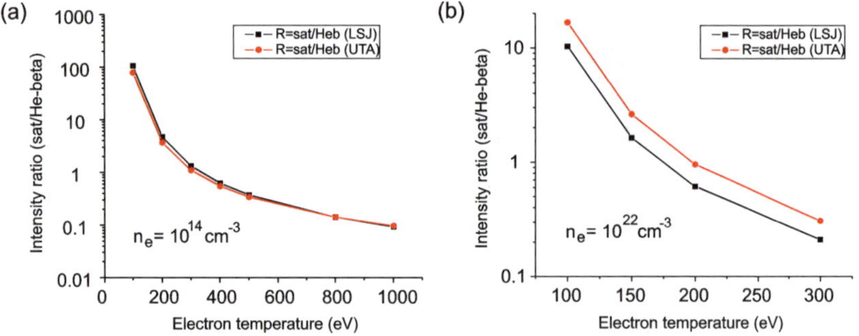

Fig. 6.10 Temperature dependent ratio (1s2l3l$'$-satellites/He$_\beta$) for the exact nLSJ-split level system and in SOSA-approximation, (**a**) for an electron density of $n_e = 10^{14}$ cm^{-3}, (**b**) for an electron density of $n_e = 10^{22}$ cm^{-3}

for density determination in the SOSA-approximation. The overall broadening of the satellite emission is twofold: the Stark broadening of a single transition and the distribution of the oscillator strengths over wavelengths [9, 11].

Figure 6.9 shows the temperature dependence of the spectral distribution in the SOSA-approximation. Compared to Fig. 6.1, the temperature sensitivity might differ not too much from the exact calculations.

In order to quantify this observation, we present in Fig. 6.10 the ratio of the total 1s2l3l$'$-satellite emission in SOSA-approximation with the exact nLSJ-level system.

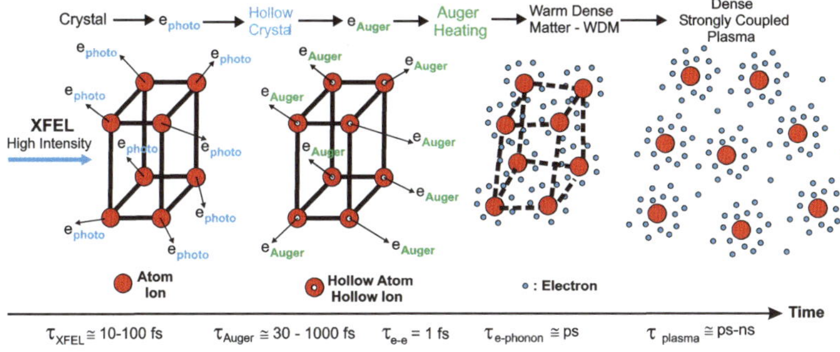

Fig. 6.11 Schematic temporal evolution of a solid crystal after irradiation with intense XUV/X-ray Free Electron Laser, τ_{XFEL} is the pulse duration of the laser pulse, τ_{Auger} is the time scale for disintegration of an autoionizing state, τ_{e-e} is the electron equilibration time in the conduction band, $\tau_{e-phonon}$ is the characteristic time scale for the electron-phonon coupling and τ_{plasma} is the live time of the dense plasma

Figure 6.10(a) presents the intensity ratios for an electron density of $n_e = 10^{14}$ cm^{-3} (corona limit). It can be seen that for almost all temperatures a reasonable agreement for the total intensity ratio to the resonance line is obtained in the corona limit (note that the agreement concerns only the *total* satellite intensity and *the corona limit* as different groups are not correctly described, see discussions in connection with Figs. 6.7 and 6.8). Figure 6.10(b) presents the ratio for an electron density of $n_e = 10^{22}$ cm^{-3}. The deviations are strong for all temperatures (overestimating the temperature for a given experimental ratio).

6.6 Excited States Coupling Following Auger Electron Heating

The interaction of XUV/X-ray Free Electron Laser radiation with solids allows matter heating at near solid density due to the short (10 fs) pulse duration. Excited state coupling effects are therefore strongly pronounced changing the entire picture of the spectral distribution.

6.6.1 Auger Electron Heating Driven by Intense XUV/X-Ray Free Electron Laser Radiation

Interaction of intense XUV/X-ray Free Electron Laser radiation with matter is quite different from those of optical laser radiation. The primary absorption mechanism is photoionization of inner-shells thereby creating multiple excited autoionizing states. If laser intensities exceed certain threshold values [11], creation of hole states via the laser radiation is more effective than depopulating elementary processes like, e.g. radiative decay, autoionization and usual collisions. Consequently, an almost instantaneous transformation of the solid crystal into a "hollow crystal" can be achieved. The hollow crystal decays rapidly because the Auger effect has a characteristic time scale of some 10 fs, see Fig. 6.11.

At high intensities almost every atom in the crystal is photoionized resulting in a burst of Auger electrons. Auger electrons carry an important amount of kinetic

Table 6.2 Values of the coefficients to compute the energy of the Auger electron energies (in keV) for some transitions. The fitting values for LMM and MNN select one branch only (center)

	KLL	LMM	MNN
α	$1.559 \cdot 10^{-2}$	$6.97 \cdot 10^{-3}$	$4.95 \cdot 10^{-3}$
σ	1.248	10.262	32.181
β	1.824	1.617	1.495

Table 6.3 Comparison of the Auger electron energy measured and calculated from the analytical formula for aluminum

	E_{Auger} measured (keV)	E_{Auger} fit (keV)	Accuracy
KLL	1.38	1.40	1.5 %
LMM	0.040	0.035	12.5 %

energy E_{Auger} as can be seen from the following analytic approximation:

$$E_{\text{Auger}} = \alpha \times (Z_n - \sigma)^\beta \tag{6.12}$$

where the coefficients, given in Table 6.2, are obtained from a fit to experimental data [3].

The analytic formula provides results with a precision better than 15 %, Table 6.3 depicts some values for aluminum.

The crystal is therefore effectively heated by Auger electron as has been demonstrated for the first time via the analysis of complex dielectronic satellite spectra [5]. The release of the potential energy of the hollow crystal drives therefore a material heating (driven by Auger electrons) that disintegrates crystalline order and transforms the solid into Warm Dense Matter and, later on, into a strongly coupled plasma.

Figure 6.11 summarizes schematically the important steps in the temporal evolution after irradiation of solids with intense XUV/X-ray radiation: irradiation of the solid with intense XUV/X-ray FEL radiation, photoionization from inner-shells of almost all atoms, Auger decay of the hollow crystal structure, Auger electron heating and creation of Warm Dense Matter, strongly coupled plasma.

If almost all atoms are ejecting Auger electrons, the average electron thermal energy is of the same order (see also Eq. (6.13)) as the dielectronic capture energy itself. Therefore, all excitation channels from ground and excited states provide therefore effective dielectronic capture in the near solid density plasma.

6.6.2 Dielectronic Satellite Intensity Coupled to Ground and Excited States

In a single approximation, the intensity of a satellite transition due to dielectronic capture is given by

$$I_{ji}^{\text{sat}} = \sum_k I_{k,ji}^{\text{sat}} = \alpha n_e \sum_k n_k \frac{Q_{k,ji}}{g_k} \frac{\exp(-\frac{E_{kj}}{kT_e})}{(kT_e)^{3/2}} \tag{6.13}$$

Numerical calculations show [15, 18] that autoionizing rates related to excited states can be orders of magnitude larger than those coupled to ground states. It is therefore helpful to separate the ground state contribution from the excited state contribution:

$$I_{ji}^{\text{sat}} = \alpha n_e n_{\text{gr}} \frac{Q_{\text{gr},ji}}{g_{\text{gr}}} \frac{\exp(-\frac{E_{\text{gr},j}}{kT_e})}{(kT_e)^{3/2}} + \alpha n_e \sum_{\substack{k \in \text{ excited} \\ \text{states}}} n_k \frac{Q_{k,ji}}{g_k} \frac{\exp(-\frac{E_{kj}}{kT_e})}{(kT_e)^{3/2}} \qquad (6.14)$$

with

$$Q_{\text{gr},ji} = \frac{g_j A_{ji} \Gamma_{j,\text{gr}}}{\sum_l A_{jl} + \Gamma_{j,\text{gr}} + \sum_{\substack{m \in \text{ excited} \\ \text{states}}} \Gamma_{jm}} \qquad (6.15)$$

and

$$Q_{k,ji} = \frac{g_j A_{ji} \Gamma_{jk}}{\sum_l A_{jl} + \Gamma_{j,\text{gr}} + \sum_{\substack{m \in \text{ excited} \\ \text{states}}} \Gamma_{jm}} \qquad (6.16)$$

Equations (6.15) and (6.16) show that order of magnitude differences in autoionizing rates translates directly in order of magnitude different intensity factors Q of the respective satellite transitions:

$$\text{if } \Gamma_{j,\text{gr}} \ll \Gamma_{jk}: \quad Q_{\text{gr},ji} \ll Q_{k,ji} \qquad (6.17)$$

Relation (6.17) has direct implications for the satellite line intensity. The relative importance of ground and excited state coupled intensities is given by the ratio

$$R_{\text{gr},k} = \frac{g_k n_{\text{gr}} Q_{\text{gr},ji}}{g_{\text{gr}} n_k Q_{k,ji}} \exp\left(-\frac{E_{\text{gr},j} - E_{kj}}{kT_e}\right) \qquad (6.18)$$

If the excited state population per statistical weight (i.e., n_k/g_k) is of the same order like the ground state population per statistical weight (i.e., $n_{\text{gr}}/g_{\text{gr}}$) and relation (6.17) holds, excited states coupling might be of great importance. In this parameter regime, the simple exponential temperature dependence of the satellite intensity is seriously altered: the excited state populations n_k are strongly density dependent thereby changing the entire picture of satellite emission originally developed by Gabriel [4].

6.7 Dielectronic Satellites Coupled to L-Shell Excited States

The following paragraph considers the specific properties of L-shell excited states coupling. Particular emphasize is put to the "Coster-Kronig Transition" that involves a 2s–2p transition when the Auger electron is ejected.

Let us consider L-shell dielectronic satellite emission $K^2L^7M^2 \rightarrow K^2L^8M^1 + \hbar\omega_{\text{sat}}$ of aluminum that is relevant to experiments when intense XUV-FEL radiation interacts with solid aluminum [5]: photoionization creates L-hole states and Auger

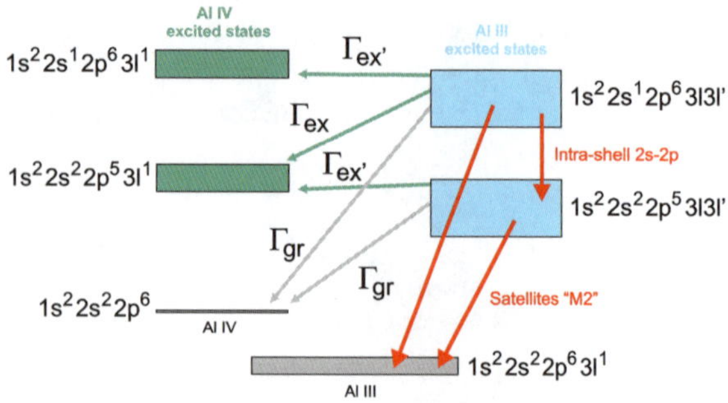

Fig. 6.12 Schematic energy level diagram of Al III and Al IV including autoionizing hole states. Possible autoionization channels coupled to ground (Γ_{gr}) and excited states (Γ_{ex} and $\Gamma_{ex'}$) are indicated. Radiative transitions (satellites) in Al III are indicated with red flashes

Table 6.4 Averaged autoionizing rates Γ in [s^{-1}] of Al III related to ground and excited states

	$1s^2 2s^2 2p^5 3l3l'$	$1s^2 2s^1 2p^6 3l3l'$
$1s^2 2s^2 2p^6$	$2.2 \cdot 10^{12}$	$1.1 \cdot 10^{12}$
$1s^2 2s^2 2p^5 3l$	$2.1 \cdot 10^{13}$	$5.4 \cdot 10^{14}$
$1s^2 2s^1 2p^6 3l$	–	$2.1 \cdot 10^{13}$

electron heating of the conduction band electrons drives dielectronic recombination in the strongly coupled plasma (see Fig. 6.11). Figure 6.12 shows the relevant energy level diagram.

Figure 6.12 shows that autoionization is not only related to the ground state but to excited states too (note that in solid physics, the Auger effect that involves intra-shell transition is called "Coster-Kronig" [2] that is a special case of the general excited states coupling effect described above). In dense plasmas the excited states are strongly populated via electron collisional excitation: $K^2 L^8 + e \rightarrow K^2 L^7 M^1 + e$. This opens up the possibility to proceed towards dielectronic capture from the excited states if the energy level structure energetically does permit this channel. The level diagram depicted in Fig. 6.12 shows interesting features: the levels $1s^2 2s^1 2p^6 3l3l'$ are not only coupled to the ground state $1s^2 2s^2 2p^6$ (Γ_{gr}) and excited states $1s^2 2s^2 2p^5 3l$ (Γ_{ex}) but even partially to the states $1s^2 2s^1 2p^6 3l$ ($\Gamma_{ex'}$) too. Similar relations hold true for the $1s^2 2s^2 2p^5 3l3l'$-levels: ground state coupling and partially excited states $1s^2 2s^1 2p^6 3l$ ($\Gamma_{ex'}$) coupling. Table 6.4 illustrates that autoionizing rates from excited states are even more important than those from ground states: about 1–2 orders of magnitude.

Table 6.5 shows the relevant spontaneous radiative decay rates that are 2–5 orders of magnitude smaller than autoionizing rates. The statistically averaged data in Tables 6.4, 6.5 have been calculated with the FAC code [6] employing a multi-configuration relativistic atomic structure, fine structure (LSJ-split levels), intermediate coupling and configuration interaction.

The data depicted in Tables 6.4 and 6.5 imply that characteristic line emission from hole states that are produced by dielectronic capture from the ground state are barely visible due to a non-favorable satellite intensity factor (as discussed in connection with Eqs. (6.15) (6.16)). In a single level approximation, the line intensity

Table 6.5 Averaged spontaneous decay rates A in $[\mathrm{s}^{-1}]$ of Al III

	$1\mathrm{s}^2 2\mathrm{s}^2 2\mathrm{p}^5 3l3l'$	$1\mathrm{s}^2 2\mathrm{s}^1 2\mathrm{p}^6 3l3l'$
$1\mathrm{s}^2 2\mathrm{s}^2 2\mathrm{p}^6 3l$	$3.8 \cdot 10^9$	$2.4 \cdot 10^9$
$1\mathrm{s}^2 2\mathrm{s}^2 2\mathrm{p}^5 3l3l'$	–	$4.1 \cdot 10^{10}$

is given by

$$
I_{ji}^{\mathrm{sat}}\left(\mathrm{K}^2\mathrm{L}^7\mathrm{M}^2\right) = \alpha n_e n_{\mathrm{K}^2\mathrm{L}^8} \frac{Q_{\mathrm{K}^2\mathrm{L}^8,ji}}{g_{\mathrm{K}^2\mathrm{L}^8}} \frac{\exp\left(-\dfrac{E_{\mathrm{K}^2\mathrm{L}^8,j}}{kT_e}\right)}{(kT_e)^{3/2}}
$$

$$
+ \alpha n_e \sum_{\mathrm{K}^2\mathrm{L}^7\mathrm{M}^1} n_{\mathrm{K}^2\mathrm{L}^7\mathrm{M}^1} \frac{Q_{\mathrm{K}^2\mathrm{L}^7\mathrm{M}^1,ji}}{g_{\mathrm{K}^2\mathrm{L}^7\mathrm{M}^1}} \frac{\exp\left(-\dfrac{E_{\mathrm{K}^2\mathrm{L}^7\mathrm{M}^1,j}}{kT_e}\right)}{(kT_e)^{3/2}} \qquad (6.19)
$$

As the excited states autoionizing rates are much larger than the radiative decay rates and the ground state autoionizing states, the first term in Eq. (6.19) is small. The second term is almost independent of the autoionizing rates as the ground state autoionization rate is small compared to the excited states one:

$$
Q_{k,ji} = \frac{g_j A_{ji} \Gamma_{jk}}{\sum_l A_{jl} + \Gamma_{j,\mathrm{gr}} + \sum_{m \in \text{ excited} \atop \text{states}} \Gamma_{jm}} \simeq g_j A_{ji} \qquad (6.20)
$$

The satellite intensity is therefore given by

$$
I_{ji}^{\mathrm{sat}}\left(\mathrm{K}^2\mathrm{L}^7\mathrm{M}^2\right) \simeq \alpha n_e \sum_{\mathrm{K}^2\mathrm{L}^7\mathrm{M}^1} n_{\mathrm{K}^2\mathrm{L}^7\mathrm{M}^1} \frac{g_j A_{ji}}{g_{\mathrm{K}^2\mathrm{L}^7\mathrm{M}^1}} \frac{\exp\left(-\dfrac{E_{\mathrm{K}^2\mathrm{L}^7\mathrm{M}^1,j}}{kT_e}\right)}{(kT_e)^{3/2}} \qquad (6.21)
$$

The dependence on the excited state density $n(\mathrm{K}^2\mathrm{L}^7\mathrm{M}^1)$ implies that the intensity (Eq. (6.21)) depends strongly on the electron density and not only on temperature as originally proposed by Gabriel [4] (see also discussion related to Figs. 6.5, 6.6). This is demonstrated with simulations of the characteristic line emission from $\mathrm{K}^2\mathrm{L}^7\mathrm{M}^2$ hole states together with the corresponding resonance line emission $\mathrm{K}^2\mathrm{L}^7\mathrm{M}^1$–$\mathrm{K}^2\mathrm{L}^8$, Fig. 6.13. Simulations have been carried out employing all LSJ-split levels of the $\mathrm{K}^2\mathrm{L}^8$-configuration (1 level), $\mathrm{K}^2\mathrm{L}^7\mathrm{M}^1$-configuration (36 levels), $\mathrm{K}^2\mathrm{L}^8\mathrm{M}^1$-configuration (5 levels), $\mathrm{K}^2\mathrm{L}^7\mathrm{M}^2$-configuration (237 levels) and including intermediate coupling and configuration interaction. Corresponding atomic population kinetics include electron collisional excitation/deexcitation, ionization/three body recombination, spontaneous radiative decay, autoionization and dielectronic capture.

Figure 6.13 shows the spectral range near the resonance transitions 2p–3d and 2p–3s (indicated by dashed lines) for different electron densities, other emission features are due to characteristic line emission from the hollow ion configuration $\mathrm{K}^2\mathrm{L}^7\mathrm{M}^2$. At low electron densities, $n_e = 10^{19}$ cm^{-3}, satellite emission is barely visible. With increasing electron density, the satellite emission rises considerably due to excited state coupling effects (see discussion of Eqs. (6.19) (6.20) (6.21)).

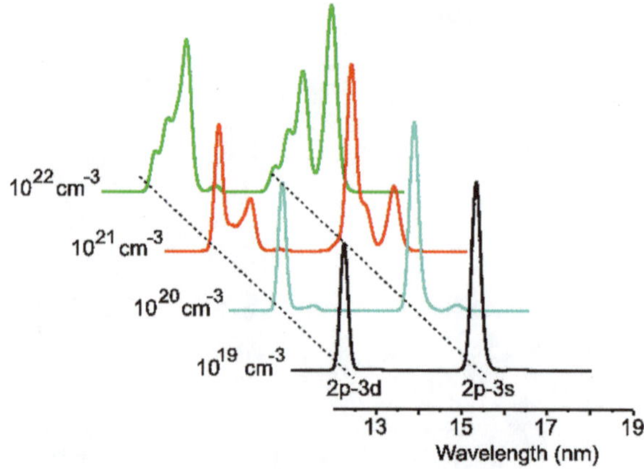

Fig. 6.13 Simulations of the spectral emission of Al IV and Al III (normalized to peak) for different electron densities, $kT_e = 10$ eV

We note that also collisional redistribution between the autoionizing levels leads to changes of the spectral distribution of satellite transitions (included in the present simulations and to be discussed in the next paragraph, see also Fig. 6.2), however, this concerns essentially deformations of the spectral distribution and not much an overall drastic intensity increase as observed in Fig. 6.13.

With respect to the overall temporal evolution of matter irradiated by XFEL radiation (Fig. 6.11), Fig. 6.13 demonstrates (see curve for $n_e = 10^{22}$ cm^{-3}) that the characteristic line emission from hole states might be even more important than usual resonance line emission. This behavior is quite different from the example of 1s3l3l'-satellites discussed in Sect. 6.6.2 where 1s3l3l'-satellite emission is strong but never really dominating. The so called L-shell satellite emission discussed via a particular example (Fig. 6.13) plays therefore an exceptional role to explore radiative properties of high-density matter under extreme conditions.

6.8 Collisional Redistribution Between Autoionizing States That Are Coupled to M-Shell Excited States

Excited states coupling does seriously influence on the way how the atomic system is approaching the Boltzmann limit. For M-shell configurations, the Coster-Kronig transition type transitions lead to significant changes over well-established criteria to achieve local thermodynamic equilibrium.

Let us now proceed to the study of collisional redistribution effects between the autoionizing states. The basic principle of this effect is depicted in Fig. 6.2. This picture is altered when excited states coupling is important: satellite emission might be provoked from excited states that have negligible coupling to the ground state.

Figure 6.14 shows the spectral distribution of the $K^2L^7M^2$-satellites for different electron densities, $kT_e = 20$ eV. In order to identify the critical density for which the spectral distribution is almost close to the Boltzmann case, the black dotted curve

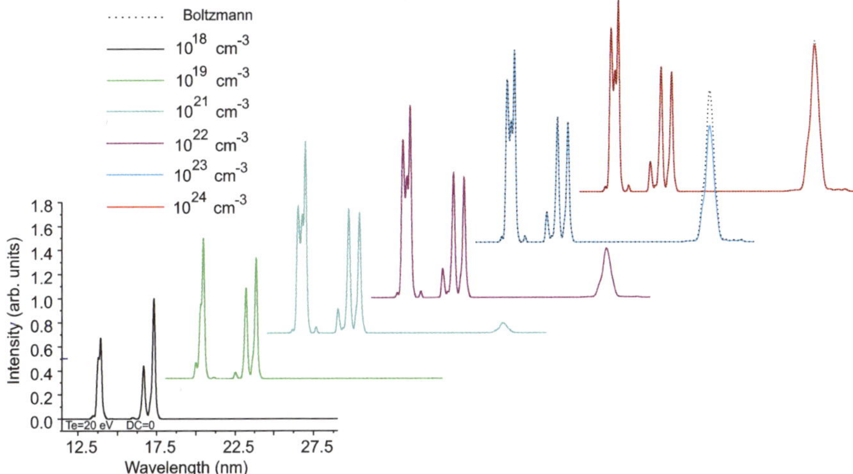

Fig. 6.14 Simulation of the spectral distribution of $K^2L^7M^2$-satellites excited by inner-shell electron collisions only: $K^2L^8M^1 + e \rightarrow K^2L^7M^2 + e$. Different electron densities are depicted, $kT_e = 20$ eV. Also indicated is the Boltzmann limit of the spectral distribution (*dotted curves*). Stark effect is not included for better demonstration of atomic population effects

is calculated for the Boltzmann limit. The simulations in Fig. 6.14 have been carried out assuming inner-shell excitation only: $K^2L^8M^1 + e \rightarrow K^2L^7M^2 + e$ (means the dielectronic capture channel is artificially switched off). It can be seen that the Boltzmann limit is only achieved for rather high electron densities (larger than about 10^{24} cm^{-3}). The reason for this large critical density is due to the large autoionizing rates to excited states (see Table 6.4): the non-radiative decay has to be compensated by electron collisions in the fine structure of the autoionizing states.

Figure 6.15 shows the same calculations like in Fig. 6.14, however, both excitation channels, dielectronic capture ($K^2L^8 + e \rightarrow K^2L^7M^2$ and $K^2L^7M^1 + e \rightarrow K^2L^7M^2$) and inner-shell excitation ($K^2L^8M^1 + e \rightarrow K^2L^7M^2 + e$) are included in the simulation. It can clearly be seen that the Boltzmann limit is already obtained for a much lower critical electron density (about 10^{21} cm^{-3}).

These results lead us to the conclusion that inverse Auger channel (dielectronic capture) is a mechanism to achieve a Boltzmann distribution under less stringent conditions. The mechanism to drive the spectral distribution towards a Boltzmann one due to inverse Auger effect can be analytically explained as follows. The detailed balance in steady state leads to (G, S indicates ground and single excited states)

$$n_e \times \sum_{j \in G,S} n_j \times C_{ji} + n_e \times \sum_{j' \in \text{ satellite}} n_{j'} \times C_{j'i} + n_e \times \sum_{l > i} n_l \times A_{li}$$

$$+ n_e \times \sum_{k} n_k \times DC_{ki} = n_i \times \sum_{l' < i} A_{il'} + n_i \times \sum_{k'} \Gamma_{ik'} + n_i \times n_e \times \sum_{j''} C_{ij''}$$

$$(6.22)$$

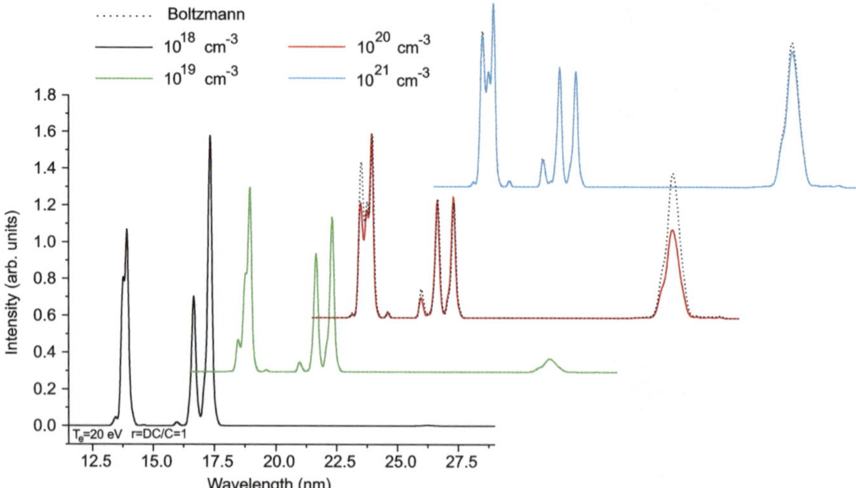

Fig. 6.15 Simulation of the spectral distribution of $K^2L^7M^2$-satellites excited by inner-shell electron collisions $K^2L^8M^1 + e \rightarrow K^2L^7M^2 + e$ and dielectronic capture $K^2L^7M^1 + e \rightarrow K^2L^7M^2$ assuming equal ground state populations. Different electron densities are depicted, $kT_e = 20$ eV. Also indicated is the Boltzmann limit of the spectral distribution (*dotted curves*). Stark effect is not included for better demonstration of atomic population effects

Using average rates, we find:

$$n_e \times n_k \times DC_{ki} \simeq n_i \times \Gamma_{ik} \tag{6.23}$$

which connects the multiple excited states "i" to the resonant level "k" via the Saha-Boltzmann relation. As a consequence, if the levels k (ground state and singly excited states) have a Boltzmann population, then the multiple excited states have a Boltzmann distribution too. To complete this scheme we remember that single excited states are not autoionizing, therefore the Boltzmann limit is achieved for orders of magnitude lower densities. This is exactly what is reflected in Figs. 6.14 and 6.15.

Excited states coupling drives therefore spectral distributions of satellite transitions that have critical densities (means densities for which the distribution is almost a Boltzmann one) much different from those where only ground state coupling is important. This has important impact for the study of matter under extreme conditions never obtained in laboratories so far.

6.9 Conclusion

High intensity short pulse XUV/X-FEL Free Electron Laser radiation provides to the scientific community outstanding tools to investigate matter under extreme conditions never obtained in laboratories so far. After a general introduction to dielectronic satellite emission we have presented novel effects in the solid-to-plasma transition considering multiple excited states and the corresponding dielectronic satellite emission. Employing a detailed LSJ-split atomic level structure it has been demonstrated that excited states coupling effects alter the entire picture of dielectronic satellite emission. We identified critical densities orders of magnitude different from those when excited states coupling is neglected. Detailed simulations carried our for aluminum L-shell satellites explored the great challenges to study not only extreme states of matter but exotic ones (like, e.g. hollow crystals, Auger electron heating) too.

Acknowledgements Support from the project "Émergence-2010: *Métaux transparents créés sous irradiations intenses émises par un laser XUV/X à électrons libres*" of the University Pierre and Marie Curie and the "Extreme Matter Institute EMMI" is greatly appreciated.

References

1. A. Burgess, Dielectronic recombination and the temperature of the solar corona. Astrophys. J. **139**, 776 (1964)
2. D. Coster, R. De L. Kronig, New type of Auger effect and its influence on the X-ray spectrum. Physica **2**, 13 (1935)
3. Evans Analytical Group (2012), http://www.eaglabs.com/training/tutorials/aes_theory_tutorial/energies.php
4. A.H. Gabriel, Dielectronic satellite spectra for highly charged He-like ion lines. Mon. Not. R. Astron. Soc. **160**, 99 (1972)
5. E. Galtier, F.B. Rosmej, D. Riley, T. Dzelzainis, F.Y. Khattak, P. Heimann, R.W. Lee, S.M. Vinko, T. Whitcher, B. Nagler, A. Nelson, J.S. Wark, T. Tschentscher, S. Toleikis, R. Faustlin, R. Sobierajski, M. Jurek, L. Juha, J. Chalupsky, V. Hajkova, M. Kozlova, J. Krzywinski, Decay of the crystalline order and equilibration during solid-to-plasma transition induced by 20-fs microfocused 92 eV Free Electron Laser Pulses. Phys. Rev. Lett. **106**, 164801 (2011)
6. M.F. Gu, The flexible atomic code FAC. Can. J. Phys. **86**, 675 (2008)
7. V.L. Jacobs, M. Blaha, Effects of angular-momentum-changing collisions on dielectronic satellite spectra. Phys. Rev. A **21**, 525 (1980)
8. F.B. Rosmej, K_β-line emission in fusion plasmas. Phys. Rev. E **58**, R32 (1998). Rapid Communication
9. F.B. Rosmej, An alternative method to determine atomic radiation. Europhys. Lett. **76**, 1081 (2006)
10. F.B. Rosmej, X-ray emission spectroscopy and diagnostics of non-equilibrium fusion and laser produced plasmas, in *Highly Charged Ions*, ed. by Y. Zou, R. Hutton (Taylor & Francis, London, 2012). ISBN:9781420079043
11. F.B. Rosmej, Exotic states of high density matter driven by intense XUV/X-ray Free Electron Laser, in *Free Electron Laser*, ed. by S. Varró (InTech, Rijeka, 2012). ISBN:9789535102793
12. F.B. Rosmej, J. Abdallah Jr., Blue satellite structure near He_α and He_β and redistribution of level populations. Phys. Lett. A **245**, 548 (1998)
13. F.B. Rosmej, R.W. Lee, Hollow ion emission driven by pulsed X-ray radiation fields. Europhys. Lett. **77**, 24001 (2007)
14. F.B. Rosmej, V.S. Lisitsa, Non-equilibrium radiative properties in fluctuating plasmas. Plasma Phys. Rep. **37**, 521 (2011)
15. F.B. Rosmej, A.Ya. Faenov, T.A. Pikuz, F. Flora, P. Di Lazzaro, S. Bollanti, N. Lizi, T. Letardi, A. Reale, L. Palladino, O. Batani, S. Bossi, A. Bornardinello, A. Scafati, L. Reale, Line formation of high intensity He_β-Rydberg dielectronic satellites $1s3lnl'$ in laser produced plasmas. J. Phys. B, At. Mol. Opt. Phys. **31**, L921 (1998)

16. F.B. Rosmej, A. Calisti, B. Talin, R. Stamm, D.H. Hoffmann, W. Süß, M. Geißel, A.Ya. Faenov, T.A. Pikuz, Observation of two-electron transitions in dense non Maxwellian laser produced plasmas and their use as diagnostic reference lines. J. Quant. Spectrosc. Radiat. Transf. **71**, 639 (2001)

17. F.B. Rosmej, R.W. Lee, D.H.G. Schneider, Fast X-ray emission switches driven by intense X-ray Free Electron Laser radiation. High Energy Density Phys. **3**, 218 (2007)

18. F.B. Rosmej, F. Petitdemange, E. Galtier, Identification of Auger electron heating and inverse Auger effect in experiments irradiating solids with XUV Free Electron Laser Radiation at intensities higher than 10^{16} W/cm^2, in *SPIE International Symposium on Optical Engineering and Applications "X-Ray Lasers and Coherent X-Ray Sources: Development and Applications (OP321)"*. Proceedings SPIE, vol. 8240-2 (2011). Invited talk

19. V.A. Vinogradov, I.Yu. Skobelev, E.A. Yukov, Effect of collisions on the intensities of the dielectronic satellites of resonance lines of hydrogenlike ions. Sov. Phys. JETP **45**, 925 (1977)

20. L.A. Woltz, V.L. Jacobs, C.F. Hooper, R.C. Mancini, Effects of electric microfields on argon dielectronic satellite spectra in laser-produced plasmas. Phys. Rev. A **44**, 1281 (1991)

Chapter 7
The Iron Project: Photoionization and Photoexcitation of Fe XVII in Solar Opacity

Sultana N. Nahar

Abstract Opacity is a fundamental quantity for plasmas and gives a measure of radiation transport. It is caused by the absorption and emission of photons by the constituent elements of the plasma and hence depends mainly on the atomic processes of photoexcitation and photoionization. It is also affected by photon scatterings. Monochromatic opacity at a particular frequency, $\kappa(\nu)$, is obtained from oscillator strengths (f) for bound-bound transitions and photoionization cross sections (σ_{PI}). However, the total monochromatic opacity depends on the summed contributions of all possible transitions from all ionization stages of all elements in the plasma. Calculation of accurate atomic parameters for such a large number of transitions has been the main problem for obtaining accurate opacities. The overall mean opacity, such as Rosseland mean opacity (κ_R), depends also on the physical conditions, such as temperature and density, elemental abundances and equation-of-state such as local thermodynamic equilibrium (LTE) of the plasmas.

In this report, I will illustrate the necessity for high-precision atomic calculations for the radiative processes of photoexcitation and photoionization in order to resolve some perplexing astrophysical problems, particularly solar abundances and opacities. Fe XVII is most abundant iron ion in the solar convection zone. I will present new results on oscillator strengths and new features in high energy photoionization cross sections of Fe XVII which give clear indication of the reason for discrepancy between measured and theoretically predicted abundances and on how the discrepancy is resolved or reduced.

7.1 Introduction

The fundamental quantity opacity (κ) is needed for studying various quantities such as elemental abundances, diffusion of radiation, optical depth, speed of seismic waves, stellar pulsations etc. of astrophysical and laboratory plasmas. As the radiation propagates, it looses energy as well as slows down by absorption and emission

S.N. Nahar (✉)
Department of Astronomy, The Ohio State University, Columbus, OH 43210, USA
e-mail: nahar@astronomy.ohio-state.edu

M. Mohan (ed.), *New Trends in Atomic and Molecular Physics*,
Springer Series on Atomic, Optical, and Plasma Physics 76,
DOI 10.1007/978-3-642-38167-6_7, © Springer-Verlag Berlin Heidelberg 2013

by the constituent elements. The resultant effect is the opacity such that more radiation means less opacity and vice versa. Because of opacity the high energy gamma radiation produced by the nuclear fusion in the core of the sun takes over a million years to travel to the surface and escape as optical or low energy photons. Opacity depends on the atomic process of photoexcitations, photoionization and photon scattering.

For the photon-ion interactions, calculations of opacity depends on the oscillator strengths and photoionization cross sections. Consideration of these processes require large amount of atomic data for all possible radiative transitions. Currently available atomic data for all ions are not accurate and complete enough to compute accurate opacities for various astrophysical problems.

7.1.1 The Opacity Project and the Iron Project

Prior to the Opacity Project (OP) [26, 28, 29], there were large discrepancies between astrophysical observations and theoretical predictions obtained using existing opacities for plasmas. The earlier opacities using the atomic data from simple approximations were incorrect by factors of 2 to 5 resulting in inaccurate stellar models. The international collaborative effort, the Opacity Project, of about 25 scientists from 6 countries was initiated in 1982 in order to study in detail and obtain accurate atomic data of radiative processes with large number of states and apply them for computation of accurate opacities. The first systematic and detailed studies were carried out for the radiative processes of photoexcitation and photoionization for all astrophysically abundant atoms and ions from hydrogen to iron. Computations were carried out in ab initio close coupling approximation and using R-matrix method. Large amount of atomic data for energy levels, oscillator strengths and photoionization cross sections were obtained and are made available at data base TOPbase [31] at CDS. The monochromatic opacities and Rosseland mean opacities under the OP are available at the OPServer [22] at the Ohio Supercomputer Center (OSC). New features in photoionization cross sections were revealed. Many long standing astrophysical problems were solved. These atomic data have continued to solve many astrophysical problems. However, a large part of the data are not precise and complete enough for various diagnostics and astrophysical problems.

Following the successful collaboration of the OP, the international collaboration of the Iron Project (IP) [6] was initiated to focus on collisional processes as well as radiative processes, but mainly for the astrophysically abundant iron and iron-peak elements. Work under IP emphasizes the relativistic effects and achievement of higher accuracy. The large amount of atomic data from the IP are available at TIPbase [30] at CDS, at *Atomic Data and Nuclear Data Tables* database and at the NORAD-Atomic-Data website [21]. A new project, RMAX, was also initiated under the IP to focus on the X-ray astrophysics.

The OP team made extensive extension of the existing R-matrix codes for the radiative processes [2]. The computations under the OP were carried out in non-relativistic LS coupling approach. The OP R-matrix IP were extended to include

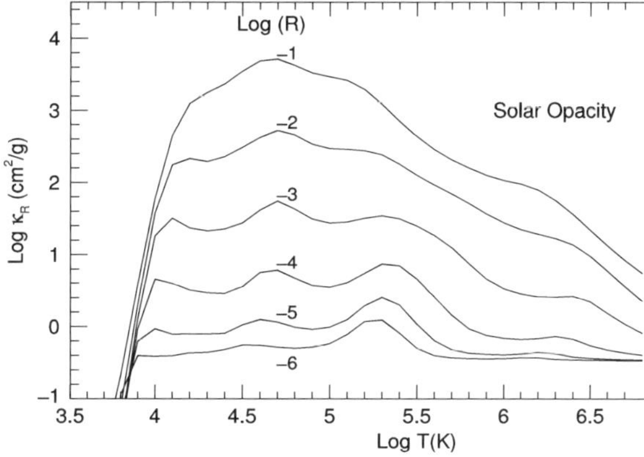

Fig. 7.1 Opacities in temperature-density regimes throughout the solar interior. κ is the Rosseland mean opacity and $R = \rho(g/cc)/T_6^3$ is the temperature-density parameter with $T_6 = 10^6$ K, i.e. $T_6 = T * 10^{-6}$. For the sun, $-6 \leq R \leq -1$. The four bumps in the curves of various R represent H-, He-, Z-, and inner-shell bumps, that is, higher opacities due to excitation/ionization of the atomic species at those temperatures [27]

relativistic fine structure effects in Breit-Pauli approximation under the IP and the new method of computation is known as the Breit-Pauli R-matrix or BPRM method [3]. Several extension and modifications of the codes were made by the Ohio state team of the IP (e.g. [4, 19]).

Considerable advances in computational capabilities for higher accuracy is continuing. Higher order relativistic corrections have been added to the BPRM method [4]. Theoretical spectroscopy has been developed for identification of large number of fine structure levels for all practical purposes [10, 15]. This enables calculation of accurate oscillator strengths for larger number of transitions than considered before. We are also finding existence of extensive and dominant resonant features in the high energy photoionization cross sections (e.g. [13]) as well as important fine structure effects in low energy region [19]. We will illustrate results from recent calculations showing more complete atomic data and new features and considerable improvement in accuracy in monochromatic opacities, particularly Fe XVII opacity in the sun.

7.2 High Temperature Plasmas in Solar Corona and Abundances

The solar elemental abundances are known within 10 % uncertainty as: H—90 % (by number) and 70% (by mass fraction), He—10 % (by number) and 28 % (by mass), and metals (all elements heavier than helium)—2 % (by mass) (e.g. [23]). Using the OP data and standard mixture of solar elemental abundances, solar opacity were calculated under the OP as shown in Fig. 7.1 [27]. The curves of various R in the figure show about four bumps or kinks, representing higher opacities. These are

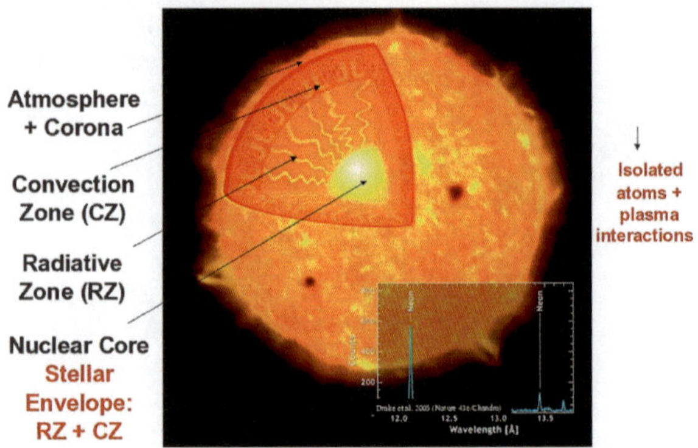

Fig. 7.2 Solar interior is defined from the core to the end of the convection zone. At the convection zone plasma bubbles and the radiation escapes

due to excitation/ionization, and hence higher absorption of photons, of different atomic species at those temperatures. The first bump is the H-bump, the second one is the He-bump, the third one the Z-bump (sum of all elements heavier than H and He) and the fourth one is due to inner-shell excitation/ionization bump. The general features of these solar opacity curves are in agreement with those by the other opacity by OPAL [7, 8].

However, these do not solve require precision for elemental abundances for various applications. For example, the recent determination of solar abundances of light elements, from measurements and 3D hydro NLTE models, show 30–40 % lower abundances of C, N, O, Ne, Ar than the standard abundances; these contradict the accurate helioseismology data. One major problem has been the discrepancy between the observed and predicted boundary between the solar radiative zone and the convection zone, R_{CZ}. Radiation from the solar nuclear core travels through the radiative zone to the convection zone where the phase changes as the plasma bubbles and beyond it the photons escape. From helioseismology, the distance of boundary R_{CZ}, relative to the solar radius, is accurately measured to be 0.713 (Fig. 7.2). R_{CZ} can be calculated from opacity through optical depth and elemental abundances in the solar plasma. However, the calculated boundary R_{CZ} is 0.726, is a much larger value than the measured value. The plasma in the convection zone is in a HED (high energy density) condition where the temperature is $T_e \sim 193$ eV (about 2.2 MK) and the electron density is $n_e \sim 10^{23}/\mathrm{cm}^3$. At this condition, the abundant elements are O, Ne, especially Fe ions in the ionic states of Fe XVII, Fe XVIII, and Fe XIX, Fe XVII the most.

The plasma in the solar convection zone has been under study to resolve the discrepancy. With technological advances, the laboratory set-up zeta pinch or Z-pinch machines, which can confine plasma through intense magnetic field, at the Sandia National Lab (SNL) is now able to create this plasma condition and study the radiation transmission, the inverse of opacity. The measurements will enable calibration of the theoretical calculations of basic parameters that govern the opacity. Bailey et al. [1] reported achieving HED plasmas at temperature ($T > 10^6$ K) and density ($N > 10^{20}$ cm^{-3}) similar to those in the convection zone. At the Z-pinch set-up the plasma is created with laser heating and compress or implode the plasma by a gigantic magnetic field created by about 27 MAmp current passing through the coil around it. The measurement at SNL Z-pinch machine showed that the measured iron opacity is much higher than the prediction obtained using the radiative atomic data, oscillator strengths for photoexcitation and cross sections for photoionization, from the Opacity Project.

7.3 Photoexcitation, Photoionization and Opacity

Opacity $\kappa(\nu)$ depends on oscillator strengths of photoexcitation,

$$X^{+Z} + h\nu \rightleftharpoons X^{+Z*}$$

where X^{+Z} is the ion with charge Z and the inverse arrow represents de-excitation. The emitted or absorbed photon $(h\nu)$ from the transition is observed as a spectral line. The relevant atomic parameters for the direct and inverse processes are oscillator strength (f) and radiative decay rate (A-value). f_{ij} is related to $\kappa(\nu)$ as

$$\kappa_\nu(i \to j) = \frac{\pi e^2}{mc} N_i f_{ij} \phi_\nu \tag{7.1}$$

where N_i is the ion density in state i, ϕ_ν is a profile factor which can be Gaussian, Lorentzian, or combination of both over a small wavelength range.

Photoionization can be direct when an electron is ejected with absorption of a photon,

$$X^{+Z} + h\nu \rightleftharpoons X^{+Z+1} + \varepsilon$$

where the inverse process is radiative recombination (RR). Photoionization also occurs via an intermediate doubly excited autoionizing state:

$$e + X^{+Z} \rightleftharpoons \left(X^{+Z-1}\right)^{**} \rightleftharpoons \begin{cases} e + X^{+Z} & \text{AI} \\ X^{+Z-1} + h\nu & \text{DR} \end{cases}$$

A colliding electron excites the target and attaches to form the short-lived doubly excited autoionizing state which leads either to autoionization (AI), where the electron goes free and target drops to ground state, or to dielectronic recombination (DR), where the electron gets bound by emission of a photon. The inverse of DR is photoionization. The autoionizing state manifests as an enhancement or resonance in the process and can be seen in absorption spectra. The atomic parameters for these processes are photoionization cross sections (σ_{PI}) and recombination rates. $\kappa(\nu)$ for photoionization is obtained from σ_{PI} as

$$\kappa_\nu = N_i \sigma_{\text{PI}}(\nu) \tag{7.2}$$

κ_ν also has contributions from two other processes, inverse bremsstrahlung or free-free (ff) scattering and photon-electron scattering. Bremsstrahlung refers to the radiation emitted by a charged particle accelerated in an electromagnetic field. The inverse process is where a free electron and an ion can absorb a photon in free-free interaction, that is,

$$h\nu + \left[X_1^+ + e(\varepsilon)\right] \to X_2^+ + e(\varepsilon') \tag{7.3}$$

Explicit calculations for the free-free scattering cross sections may be done using the elastic scattering matrix elements for electron impact excitation of ions. An approximate expression for the free-free opacity is given by

$$\kappa_\nu^{\text{ff}}(1,2) = 3.7 \times 10^8 N_e N_i g_{\text{ff}} \frac{Z^2}{T^{1/2}\nu^3} \tag{7.4}$$

where g_{ff} is a Gaunt factor.

The photon-electron scattering, giving some contributions to opacity, can be of two type, Thomson scattering, when the electron is free and Rayleigh scattering when the electron is bound to an atomic or molecular species. κ is related to Thomson scattering cross section σ^{Th} as,

$$\kappa(sc) = N_e \sigma^{\text{Th}} = N_e \frac{8\pi e^4}{3m^2 c^4} = 6.65 \times 10^{-25} \text{ cm}^2/\text{g} \tag{7.5}$$

To Rayleigh scattering cross section σ^R, opacity is related as

$$\kappa_\nu^R = n_i \sigma_\nu^R \approx n_i f_t \sigma^{\text{Th}} \left(\frac{\nu}{\nu_I}\right)^4 \tag{7.6}$$

where n_i is the density of the atomic or molecular species, $h\nu_I$ is the binding energy and f_t is the total oscillator strength associated with the bound electron, i.e. the sum of all possible transitions, such as the Lyman series of transitions $1s \rightarrow np$ in hydrogen.

To find the average opacity, such as Rosseland opacity κ_R, we need elemental abundances pertinent to the plasma conditions of temperature and density. These are obtained from proper equation-of-state (EOS) which gives the ionization fractions and level populations of each ion of an element in levels with non-negligible occupation probability. For example, for plasmas in local thermodynamic equilibrium (LTE) Saha equation is the EOS. However, Saha equation is not applicable in non-LTE condition. Some details of the opacity calculations can be found in [23].

Rosseland mean opacity $\kappa_R(T,\rho)$ is the harmonic mean of monochromatic opacity averaged over the Planck function, $g(u)$,

$$\frac{1}{\kappa_R} = \frac{\int_0^\infty \frac{1}{\kappa_\nu} g(u)du}{\int_0^\infty g(u)du}, \tag{7.7}$$

where $g(u)$ is given by

$$g(u) = \frac{15}{4\pi^4} \frac{u^4 e^{-u}}{(1-e^{-u})^2}, \quad u = \frac{h\nu}{kT} \tag{7.8}$$

$g(u)$, for an astrophysical state is calculated with different chemical compositions H (X), He (Y) and metals (Z), such that

$$X + Y + Z = 1 \tag{7.9}$$

7.4 Theoretical Approach: Breit-Pauli R-Matrix Method

As mentioned above, the relativistic Breit-Pauli R-matrix method with close coupling (CC) approximation is used to calculate the oscillator strengths and photoionization cross sections. (Details are given in [23]). The BPRM Hamiltonian is given by

$$H^{\mathrm{BP}} = H^{\mathrm{NR}} + H^{\mathrm{mass}} + H^{\mathrm{Dar}} + H^{\mathrm{so}} \tag{7.10}$$

$$+ \frac{1}{2} \sum_{i \neq j}^{N} \left[g_{ij}\left(so + so'\right) + g_{ij}\left(ss'\right) + g_{ij}\left(css'\right) + g_{ij}(d) + g_{ij}\left(oo'\right) \right] \tag{7.11}$$

where H^{NR} is the nonrelativistic Hamiltonian,

$$H^{\mathrm{NR}} = \sum_{i=1}^{N} \left\{ -\nabla_i^2 - \frac{2Z}{r_i} + \sum_{j>i}^{N} \frac{2}{r_{ij}} \right\} \tag{7.12}$$

and the one-body relativistic correction terms are mass correction, Darwin and spin-orbit interaction terms respectively,

$$H^{\mathrm{mass}} = -\frac{\alpha^2}{4} \sum_i p_i^4, \qquad H^{\mathrm{Dar}} = \frac{\alpha^2}{4} \sum_i \nabla^2 \left(\frac{Z}{r_i}\right), \qquad H^{\mathrm{so}} = \left[\frac{Ze^2 \hbar^2}{2m^2 c^2 r^3} \right] \tag{7.13}$$

The spin-orbit interaction H^{so} splits LS energy in to fine structure levels. Rest of the terms are two-body interaction terms where the notation are s for spin and a prime indicates 'other', o for orbit, c for contraction, and d for Darwin. These terms are much weaker but can become important for weak transitions and when relativistic effects are important. The first two two-body terms comprise the Breit interaction

$$H^B = \sum_{i>j} \left[g_{ij}\left(so + so'\right) + g_{ij}\left(ss'\right) \right] \tag{7.14}$$

which contributes more relatively to the other terms. In the latest development of the BPRM codes, Breit interaction has been included [4].

In CC approximation, the wave function expansion is expressed as a core or target ion of N-electrons interacting with the $(N + 1)$ interacting electron. The total wave function expansion is expressed as:

$$\Psi_E(e + \mathrm{ion}) = A \sum_i^N \chi_i(\mathrm{ion})\theta_i + \sum_j c_j \Phi_j(e + \mathrm{ion}) \tag{7.15}$$

where χ_i is the target ion or core wave function which includes excitations. θ_i is interacting electron wave function (continuum or bound), and Φ_j is a correlation function of $(e \mid \mathrm{ion})$. The complex resonant structures in the atomic processes are included through coupling of bound and continuum channels with core excitations.

The target wave functions χ_i are obtained from atomic structure calculations, e.g. SUPERSTRUCTURE (SS) [5, 18].

Substitution of $\Psi_E(e + \text{ion})$ in $H\Psi_E = E\Psi_E$ results in a set of coupled equations which are solved by the R-matrix method. In the R-matrix method (Fig. 7.5), the space is divided in two regions, the inner and the outer regions, of a sphere of radius r_a with the ion at the center. r_a, the R-matrix boundary, is chosen large enough for electron-electron interaction potential to be zero outside the boundary. The wave function at $r > r_a$ is Coulombic due to perturbation from the long-range multipole potentials. In the inner region, the partial wave function of the interacting electron is expanded in terms of a basis set, called the R-matrix basis, and are made continuous at the boundary by matching with the Coulomb functions outside the boundary. For negative energy, the solution is a bound $(e + \text{ion})$ states, Ψ_B and for positive energy, the solution is a continuum state, Ψ_F.

The oscillator strengths and photoionization cross sections are obtained from the line strength $\mathbf{S}$. It depends on the transitions matrix elements with dipole operator $\mathbf{D} = \sum_i \mathbf{r}_i$ for photoexcitation and photoionization:

$$\langle \Psi_B \| \mathbf{D} \| \Psi_{B'} \rangle, \qquad \langle \Psi_B \| \mathbf{D} \| \Psi_F \rangle \tag{7.16}$$

respectively. The reduced tensor $\|\mathbf{D}\|$ gives 3-j symbols for angular momenta on simplification and the generalized line strength as

$$S = \left| \left\langle \Psi_f \sum_{j=1}^{N+1} r_j \Psi_i \right\rangle \right|^2 \tag{7.17}$$

The oscillator strength (f_{ij}) and radiative decay rate (A_{ji}) for the bound-bound transition are obtained as

$$f_{ij} = \left[\frac{E_{ji}}{3g_i} \right] S, \qquad A_{ji}(\text{sec}^{-1}) = \left[0.8032 \times 10^{10} \frac{E_{ji}^3}{3g_j} \right] S \tag{7.18}$$

The photoionization cross section, σ_{PI}, is obtained as

$$\sigma_{\text{PI}} = \left[\frac{4\pi}{3c} \frac{1}{g_i} \right] \omega S, \tag{7.19}$$

where ω is the incident photon energy in Rydberg unit. With consideration of relativistic fine structure effects, BPRM method enables extensive sets of E1 transitions ($\Delta j = 0, \pm 1$, $\Delta L = 0, \pm 1, \pm 2$, parity π changes) with same spin-multiplicity ($\Delta S = 0$) and intercombination ($\Delta S \neq 0$). On the contrary LS coupling allows only same spin-multiplicity transitions.

Under the IP, large sets of various forbidden transitions are also considered. They are obtained in Breit-Pauli approximation but using atomic structure calculations. Due to small f-values, typically the radiative decay rates, A-values, are calculated for the forbidden transitions. These transitions are mainly

(i) electric quadrupole (E2) transitions ($\Delta J = 0, \pm 1, \pm 2$, parity does not change)

$$A_{ji}^{\text{E2}} = 2.6733 \times 10^3 \frac{E_{ij}^5}{g_j} S^{\text{E2}}(i, j) \text{ s}^{-1}, \tag{7.20}$$

(ii) magnetic dipole (M1) transitions ($\Delta J = 0, \pm 1$, parity does not change),

$$A_{ji}^{\mathrm{M1}} = 3.5644 \times 10^4 \frac{E_{ij}^3}{g_j} S^{\mathrm{M1}}(i, j)\ \mathrm{s}^{-1}, \tag{7.21}$$

(iii) electric octupole (E3) transitions ($\Delta J = \pm 2, \pm 3$, parity changes) and

$$A_{ji}^{\mathrm{E3}} = 1.2050 \times 10^{-3} \frac{E_{ij}^7}{g_j} S^{\mathrm{E3}}(i, j)\ \mathrm{s}^{-1}, \tag{7.22}$$

(iv) magnetic quadrupole (M2) transitions ($\Delta J = \pm 2$, parity changes) and

$$A_{ji}^{\mathrm{M2}} = 2.3727 \times 10^{-2}\ \mathrm{s}^{-1} \frac{E_{ij}^5}{g_j} S^{\mathrm{M2}}(i, j). \tag{7.23}$$

Some details of the forbidden transitions are given in [11, 18].

7.5 Results and Discussions

The accuracy and completeness of oscillator strengths for bound-bound transitions and new features in photoionization affecting the plasma opacities are discussed with examples in the two subsections below. Results on opacity is discussed in the third subsection.

7.5.1 Energy Levels and Oscillator Strengths

In order to obtain accurate solar opacity at the convection zone, we have been focusing on the radiative data for the most abundant iron ions, Fe XVII–Fe XIX. The oscillator strengths of these ions have already been obtained [12, 14, 18]. We have completed the photoionization data for Fe XVII recently [20]. Being complex with strong electron-electron correlation, computation of radiative data for each iron ion is a challenge. Another major task is the spectroscopic identification of the large number of energy levels and transitions that BPRM method can compute. BPRM identification is quite different from atomic structure calculations where identification of an energy level is based on the percentage contributions of the configurations. The BPRM Identification is carried out using a method based on the channel contributions in the outer region of the R-matrix method, quantum defect theory, and algebraic algorithms [10, 15].

BPRM calculations for Fe XVII transitions under the IP resulted in a total of 490 fine structure energy levels of total angular momenta $0 \leq J \leq 7$ of even and odd parities with $2 \leq n \leq 10$, $0 \leq l \leq 8$, $0 \leq L \leq 8$, and singlet and triplet multiplicities. These energies agree within 1 % with all available 52 measured energies. The calculated levels yielded to over 2.6×10^4 allowed (E1) transitions that are of dipole and intercombination type. In addition a set of 2312 forbidden transitions of

Table 7.1 Example set of dipole allowed and intercombination transitions in Fe XVII. The $g:I$ indices refer to the statistical weight:energy level index. The notation $a(b)$ means $a \times 10^{b}$

C_i	C_j	T_i	T_j	$g_i:I_i$	$g_j:I_j$	$\lambda_{ij}/\text{Å}$	f	$A\cdot s$
2p6	2s22p53s	$^1S^e$	$^3P^o$	1:1	3:1	17.1	1.223(−1)	9.35(11)
2p6	2s22p63s	$^1S^e$	$^1P^o$	1:1	3:2	16.8	1.008(−1)	7.96(11)
2p6	2s22p53d	$^1S^e$	$^3P^o$	1:1	3:3	15.4	8.136(−3)	7.58(10)
2p6	2s22p53d	$^1S^e$	$^3D^o$	1:1	3:4	15.3	6.208(−1)	5.93(12)
2p6	2s22p53d	$^1S^e$	$^1P^o$	1:1	3:5	15.0	2.314	2.28(13)
2p6	2s2p63p	$^1S^e$	$^3P^o$	1:1	3:6	13.9	3.501(−2)	4.03(11)
2p6	2s2p63p	$^1S^e$	$^1P^o$	1:1	3:7	13.8	2.835(−1)	3.30(12)
2p6	2s22p54s	$^1S^e$	$^3P^o$	1:1	3:8	12.7	2.286(−2)	3.16(11)
2p6	2s22p54s	$^1S^e$	$^1P^o$	1:1	3:9	12.5	1.758(−2)	2.49(11)
2p6	2s22p54d	$^1S^e$	$^3P^o$	1:1	3:10	12.3	3.281(−3)	4.81(10)
2p6	2s22p54d	$^1S^e$	$^3D^o$	1:1	3:11	12.3	3.594(−1)	5.31(12)
2p53s	2s22p53p	$^3P^o$	$^3P^e$	3:1	1:2	296.0	3.354(−2)	7.66(09)
2p53s	2s22p53p	$^3P^o$	$^3P^e$	3:1	3:4	262.7	5.893(−5)	5.70(06)
2p53s	2s22p53p	$^3P^o$	$^3P^e$	5:1	3:4	252.5	4.985(−3)	8.69(08)
2p53s	2s22p53p	$^3P^o$	$^3P^e$	3:1	5:2	340.4	9.075(−2)	3.13(09)
2p53s	2s22p53p	$^3P^o$	$^3P^e$	5:1	5:2	323.5	6.913(−2)	4.41(09)
LS		$^3P^o$	$^3P^e$	9	9		8.959(−2)	6.71(09)
2p63s	2s22p53p	$^1P^o$	$^3P^e$	3:2	1:2	413.8	9.557(−3)	1.12(09)
2p63s	2s22p53p	$^1P^o$	$^3P^e$	3:2	3:4	351.6	4.162(−2)	2.25(09)
2p63s	2s22p53p	$^1P^o$	$^3P^e$	3:2	5:2	506.3	1.464(−3)	2.29(07)
2p53p	2s22p53d	$^3P^e$	$^3P^o$	1:2	3:3	369.5	9.560(−3)	1.56(08)
2p53p	2s22p53d	$^3P^e$	$^3P^o$	3:4	1:2	457.5	1.443(−3)	1.38(08)
2p53p	2s22p53d	$^3P^e$	$^3P^o$	3:4	3:3	439.0	1.262(−3)	4.37(07)
2p53p	2s22p53d	$^3P^e$	$^3P^o$	3:4	5:2	415.7	5.280(−4)	1.22(07)
2p53p	2s22p53d	$^3P^e$	$^3P^o$	5:2	3:3	317.7	9.904(−3)	1.09(09)
2p53p	2s22p53d	$^3P^e$	$^3P^o$	5:2	5:2	305.4	5.093(−2)	3.64(09)
LS		$^3P^e$	$^3P^o$	9	9		3.145(−2)	1.64(09)

types E2, M1, E3, and M2 were also obtained. These transitions correspond to an order of magnitude more atomic data available data for Fe XVII. Table 7.1 gives an example set of Fe XXVII transitions that have been identified and processed in the standard NIST format.

The effect of accuracy and more complete set of transitions on opacities was tested for Fe IV in a plasma with temperature $\log T(K) = 4.5$ and electron density $\log N_e(\text{cm}^{-3}) = 17.0$, where Fe IV dominates the iron opacity. It was found by Nahar and Pradhan [16] that iron opacity κ_ν for that plasma depends primarily on oscillator strengths, and varies over orders of magnitude between 500–4000 Å. Significant differences and shifts of opacity was found with the opacity obtained using from earlier oscillator strengths calculated under the OP.

7.5.2 High Energy Photoionization

Most of the opacity caused by photoionization is through the photon absorption at resonant energies. The resonances in photoionization are introduced as the core goes through excitations. The OP work computed the resonances in photoionization of the ground and many excited states of many atomic systems for the first time. However, the work considered only the low energy resonances with the assumption that they are more dominant due to strong couplings of channels, and that couplings become weaker with higher energy and photoionization becomes featureless and decays smoothly. However, later investigations under the IP revealed that resonances due to excitations of the high lying core states of ions, such as for Fe XXI, are much stronger and hence play a very important role for opacities [13].

We found extensive dominance of high-peak resonances in the high energy photoionization of Fe XVII in our recent investigations [20, 33]. There is a large energy gap of about 47 Rd between the core states of $n = 2$ and $n = 3$ complexes. Such large gap is often taken as vanished impact of the channels with $n = 3$ core levels with the lower ones. Hence earlier calculations considered only 3 levels of the $n = 2$ complex [24]. It is also known that the $n = 3$ excitations do not form any bound state of Fe XVII. However, large radiative decay rates of $n = 3$ levels indicated possibility of high autoionizing resonances and were found to be the fact in the 60 level calculations as shown in Fig. 7.3.

Figure 7.3 presents σ_{PI} of the (a) ground level $2s^2 2p^6(^1S_0)$ and (b) an excited level $2s^2 2p^5 3d(^3P^o_0)$ of Fe XVII. The resonances due to $n = 3$ levels of the core do not have much impact, except appearing as some weak resonances, on the ground level. While resonances introduced by the core excitations to $n = 2$ levels are important for the ground level, excitations to $n = 3$ levels are more important for the excited level photoionization. The high energy region of σ_{PI} of the excited level is filled with high-peak resonances.

Most of these resonances in Fig. 7.3 are Rydberg series autoionizing resonances. They form at energies E_p,

$$(E_t - E_p) = z^2/v^2$$

where E_t is an excited core state or threshold and v is the effective quantum number of the state. These resonances are usually narrow and more common. However, photoionization of excited levels is characterized by wide PEC (photoexcitation-of-core) or Seaton resonances. These resonances occur when the core goes through a dipole allowed transition while the outer electron remains as a spectator. The state is followed by ionization via the outer electron while the core drops down to the ground state. Since the ground level of core Fe XVIII is $2s^2 2p^5(^2P^o_{3/2})$, a PEC resonance can form at the excited energy threshold of the core when it goes through

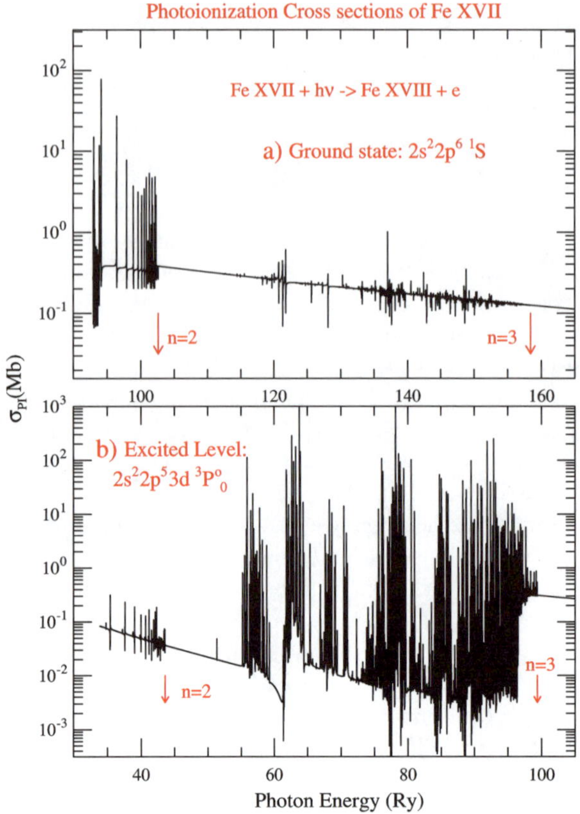

Fig. 7.3 Photoionization cross section σ_{PI} of Fe XVII with arrows pointing energy limits of $n = 2$ and 3 core states. *Top*: ground level, $n = 2$ resonances are important and for an excited level. *Bottom*: $n = 3$ resonances are important

the dipole transition, for example, $2s^22p^5(^2P^o_{3/2})$–$2s^22p^43s$ $(^2P_{3/2})$. The resonant phenomena was first explained by Seaton in [32].

Figure 7.4 presents σ_{PI} of the excited level, $2s^22p^57f(^3D_1)$, of Fe XVII. The 60 core levels includes 29 possible dipole allowed transitions for the core ground level and each corresponds to a Seaton resonance. The cross sections are delineated at 24,653 energies; the energy intervals are chosen so as to resolve resonance profiles in so far as practical. The overlapping Seaton resonances have been pointed by couple of arrows in the figure. As the figure shows, Seaton resonances are strong which increase the background cross sections by orders of magnitude. The shape and strength of PEC resonances depend on the interference of core excitations and overlapping Rydberg series of resonances. These dominating features, especially those due to $\Delta n = 3$–2 core transitions, show large enhancement of photon absorptions related to the opacities. These are non-existent in the available data and thereby grossly underestimating the opacity.

The impact of the myriad $\Delta n(3 - 2)$ PECs in Fig. 7.4b and 7.4c, can be seen to dominate the distribution of oscillator strength over a large region $\sim$56–66 Ry, or $>$100 eV, due to the L-shell excitation array $2p \rightarrow 3d$. The cumulative resonance

Fig. 7.4 Photoionization cross section (σ_{PI}) of excited level $2p^37f(^3D_1)$ over various ranges of energy (**a, b, c**) while the bottom panel (**d**) presents the total range. PEC resonances appear, positions pointed by *arrows*, occur at energies of dipole transitions in the core. They are high-peak, wider and enhance the background σ_{PI} by orders of magnitude, and will affect photoionization and recombination rates, especially of high temperature plasmas

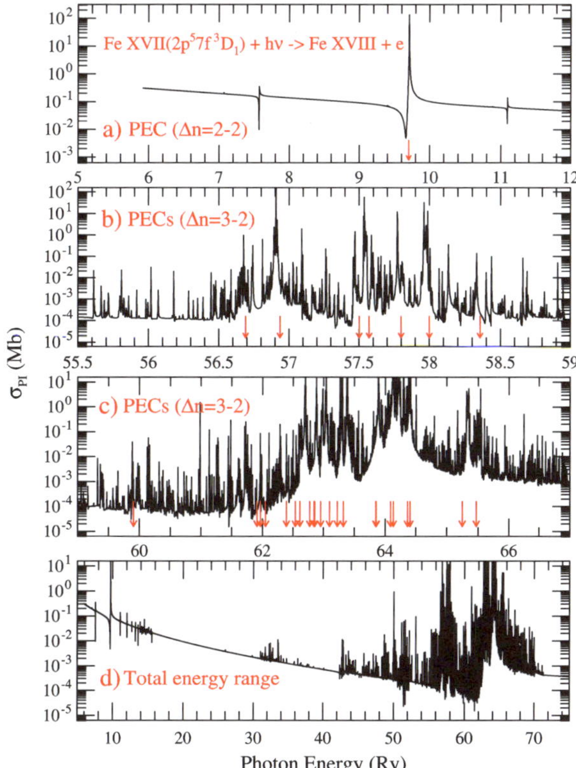

(or bound-free continuum) oscillator strength corresponding to Fig. 7.4 is related to the photoionization cross section (viz. [17]) as

$$f_r\left[^3D_1 \longrightarrow \varepsilon SLJ : {}^3(P, D, F)\,(J = 0, 1, 2)^o\right]$$
$$= \left[\frac{1}{4\pi\alpha a_o^2}\right] \int_0^{\varepsilon_0} \sigma_{PI}\left(2s^2 2p^5 7f\,^3D_1\right)d\varepsilon, \tag{7.24}$$

where ε is the energy relative to ionization threshold and up to $\varepsilon_o \sim 80$ Ry. An integration over the range shown in Fig. 7.4 yields the *partial* resonance oscillator strength including all of the PECs and the non-PEC resonances due to coupling of all 60 Fe XVIII levels. The sum over the oscillator strengths corresponding to the 30 PECs gives a total $f_{PEC} = 2.31$. The integrated resonance oscillator strength f_r is found to be 4.38. Hence, the 30 PEC resonances involving transitions up to the $n = 3$ levels of the core ion Fe XVIII contribute over half of all the continuum bound-free oscillator strength in photoionization of *any* excited state of Fe XVII. Although we chose an excited level to demonstrate the quantitative effect of PEC resonances; they manifest themselves in photoionization of most levels.

σ_{PI} of two levels of Fe XVII obtained from 60CC BPRM calculations and from the OP are compared in Fig. 7.5. Panels (a, b) shows σ_{PI} of level $2p^53p(^1P)$ and panels (c, d) of level $2p^53d(^1D^o)$, respectively. Comparison shows low and smooth

Fig. 7.5 Comparison of photoionization cross sections σ_{PI} between 60CC calculations and from the OP (see [31]) for excited singlet levels (no fine structure): (**a, b**) $2p^53p(^1P)$, and (**c, d**) $2p^53d(^1D^o)$. The comparisons demonstrate that without inclusion of $n = 3$ core levels of Fe XVIII, σ_{PI} is considerably underestimated throughout most of the high energy region. These resonances are included as *lines* in the OP work [28, 29], as in other opacities calculations [7, 8]

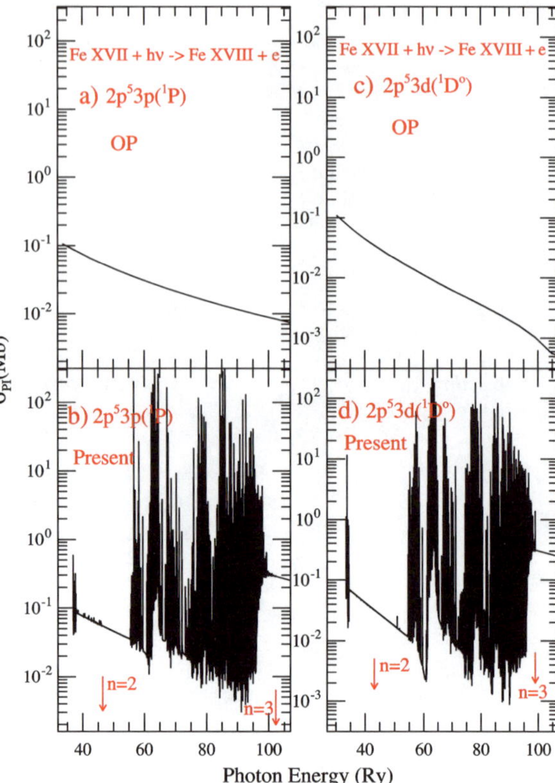

background in OP cross sections in panels (a) and (c) while BPRM σ_{PI} [18] filled with resonant structures. These high-peak resonances in BPRM cross section indicate orders of magnitude higher probability of photoionization. Hence without inclusion of $n = 3$ core states, σ_{PI} is considerably underestimated and give lower opacity as concluded by the observed spectra of the Z-pinch experiment by Bailey et al. [1].

7.5.3 Monochromatic Opacities

Often it is expected that only the ground state and low-lying metastable states in a plasma are significantly populated. The ionization fractions and level populations are computed using an equation-of-state, such as the modified Boltzmann-Saha formulation in the 'chemical picture' [9], based on the assumption that isolated atoms exist, albeit perturbed by the plasma environment [27]. At low densities and temperatures the ion fractions and occupation probabilities of high-lying levels are several orders of magnitude smaller than those for the ground state and metastable levels. But in the high temperature-density regime $N_e > 10^{24}$ cm^{-3}, $T_e > 10^6$ K, similar to that in the solar convection zone, electron-ion recombination rates can be large, and increase rapidly as N_e^2. Even a small population in excited levels would then be susceptible to the resonant enhancements due to PEC resonances, which are currently neglected and the cross sections for excited levels are taken to be nearly hydrogenic, instead of the accurate form exemplified in Figs. 7.3, 7.4 and 7.5.

Monochromatic opacity spectra are sampled at approximately 10,000 points (viz. [28, 29]), although the atomic data are much more finely resolved. It has been verified that the statistical averages of the most important quantity, the Rosseland Mean Opacity (RMO), do not deviate by more than 1–3 % even if the atomic cross sections are 'sampled' at 10^5 or 10^6 points [27] and the energy mesh of $\sim$30,000 points predominantly in the region of covered by the high energy $n = 3$ resonances, should suffice.

Figure 7.6 presents the monochromatic opacity κ (Fe XVII) computed using all of the bound-free, that is, photoionization data of 454 bound levels resolved with fine energy mesh. The calculations are carried out using a newly developed code for high-precision opacities, adapted from the earlier OP work [27]. The new code also employs a frequency mesh of 10^5 points, an order of magnitude finer mesh than OP or OPAL, thus obviating some resolution issues in the monochromatic opacity spectra [31]. Other components of the opacity calculations related to bound-bound transitions are retained as in the OP code: electron-impact and Stark broadening, free-free scattering, and electron-photon interaction in the Rayleigh, Thomson, and Compton scattering limits. However, resonance profiles have not yet been broadened. Details of the computations are given in [20].

Since the resonances would dissolve more readily than lines, it is likely to significantly affect *continuum lowering*, manifest in opacity spectra, are highly sensitive to temperature and density. Lines dissolve and eventually merge into the continuum in high temperature-density regime. Then a precise accounting may not be necessary; they are often treated as "unresolved transition arrays" (UTA) since J-J' transitions are merged together and often subsumed by line broadening [25]. However, our aim is to focus on delineation of atomic features as fully as possible so that their contributions to opacity in different temperature-density regimes can be accurately ascertained.

The BPRM opacity cross sections in Fig. 7.6 are compared to two other sets of results from OP. Although over 20,000 oscillator strengths for bound-bound transitions are also computed, we use the same data for lines as the earlier OP calculations [27]. Thus the differences between the 60CC BPRM results from the present work in the upper panel of Fig. 7.6, and the limited OP without resonances in the middle panel, are entirely due to the differences in the bound-free data sets. The opacity calculations are done at a temperature-density representative of the plasma conditions at the base of the solar convection zone: $\log_{10} T\,(K) = 6.35$ and $\log_{10} N_e = 23$. In addition to resonance contributions, there are some other differences. The OP data include extrapolated cross sections out to very high energies, $\sim$500 Ry. The BPRM data have also been processed to include these high-energy "tails", but the form is slightly different. The background opacity is important to obtain a value for RMOs that spans over 4 decades in temperature, $\log_{10} T\,(K) = 3.5$–7.5.

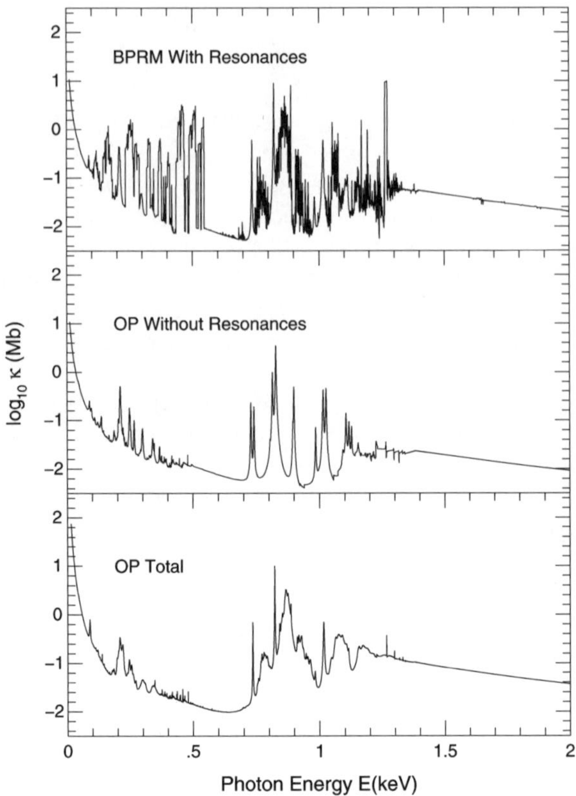

Fig. 7.6 Partial monochromatic opacity of Fe XVII: $\log_{10} \kappa$ (Mb) at temperature $T = 2.24 \times 10^6$ K and electron density $N_e = 10^{23}$ cm^{-3}, corresponding to the base of the solar convection zone where the Fe XVII ion is the largest contributor to opacity. *Top*: Opacities using BPRM photoionization cross sections with resonances computed with 60CC wave function; *Middle*: Without resonances using earlier data from the Opacity Project. Identical dataset for oscillator strengths from the OP is employed, that is, the differences are mainly due to the resonances in the 60CC calculations. *Bottom*: For comparison, total OP opacities [31] that include core-excitation resonances as lines, with autoionization widths considered perturbatively, as well as the high-energy K-shell continuum opacity not yet included in BPRM calculations

The BPRM value of κ_R, including the bound-bound oscillator strengths computed in this work (as opposed to OP) yields a value of 223.8 cm^2/g. Using the same bound-bound data as OP, the BPRM κ-value is still 200.3 cm^2/g, compared to the OP value of 109.7 cm^2/g (Fig. 7.6). The bottom panel in Fig. 7.6 is the total monochromatic opacity spectrum of Fe XVII from OP, including all contributions, with the total RMO $\kappa_R = 306.9$ cm^2/g. Thus the BPRM value using the data computed in this work is 27 % lower. This is primarily because of two factors missing in the BPRM calculations: (I) The high-energy "tails" including the K-ionization thresholds are missing. (II) The bound-bound oscillator strengths are computed up to $n = 10$. A relatively small gap in cross sections in the region between $n = 10$–∞, below the photoionization thresholds of the 454 bound levels are also missing. Following OP work, the tabulation of cross sections at $E = -z^2/v^2$ ($v \sim 10$) below each threshold was carried out resulting in BPRM RMO value from 223.8 cm^2/g to 260.7 cm^2/g, to within 20 % of the total OP RMO. These differences are being investigated.

7.6 Conclusion

The comprehensive calculations were carried out so as to compare with the erstwhile OP calculations that treated resonances as lines and effect of accuracy, correct features and completeness of atomic data on resolving the solar abundances and opacity in the convection zone. A clear distinction is made between pure bound-bound transitions and bound-free transitions into autoionizing levels. The results from this work demonstrate that opacities may be computed using BPRM cross sections and transition probabilities, and it is likely that plasma opacities in general, and those of Fe ions in particular, would be different from earlier ones using atomic cross sections that accurately consider the energy distribution of resonances.

Acknowledgements Partial supports from NSF and DOE are acknowledged. Computations were carried out at the Ohio Supercomputer Center.

References

1. J.E. Bailey et al., in *51st Annual Meeting of the Division of Plasma Physics (DPP) of APS*, Atlanta, Georgia, November 2–6 (2009) TOc.010 (22 authors)
2. K.A. Berrington et al., J. Phys. B **20**, 6379 (1987)
3. K.A. Berrington et al., Comput. Phys. Commun. **92**, 290 (1995)
4. W. Eissner, G. Chen (unpublished)
5. W. Eissner, M. Jones, H. Nussbaumer, Comput. Phys. Commun. **8**, 270 (1974)
6. D.G. Hummer et al., Astron. Astrophys. **279**, 298 (1993)
7. C.A. Iglesias, F.J. Rogers, Astrophys. J. **371**, 40 (1991)
8. C.A. Iglesias, F.J. Rogers, Astrophys. J. **464**, 943 (1996)
9. D. Mihalas, D.G. Hummer, W. Däppen, Astrophys. J. **331**, 815 (1988)
10. S.N. Nahar, Astron. Astrophys. Suppl. Ser. **127**, 253 (2000)
11. S.N. Nahar, Astron. Astrophys. **448**, 779 (2006)
12. S.N. Nahar, Astron. Astrophys. **457**, 721 (2006)
13. S.N. Nahar, J. Quant. Spectrosc. Radiat. Transf. **109**, 2417–2426 (2008)
14. S.N. Nahar, At. Data Nucl. Data Tables **97**, 403 (2011)
15. S.N. Nahar, A.K. Pradhan, Phys. Scr. **61**, 675 (2000)
16. S.N. Nahar, A.K. Pradhan, Astron. Astrophys. **437**, 345 (2005)
17. S.N. Nahar, A.K. Pradhan, H.L. Zhang, Phys. Rev. A **63**, 060701 (2001). Rapid Commun.
18. S.N. Nahar, W. Eissner, G.X. Chen, A.K. Pradhan, Astron. Astrophys. **408**, 789 (2003)
19. S.N. Nahar, M. Montenegro, W. Eissner, A.K. Pradhan, Phys. Rev. A **82**, 065401 (2010). Brief Report
20. S.N. Nahar, A.K. Pradhan, G.X. Chen, W. Eissner, Phys. Rev. A **83**, 053417 (2011)

21. NORAD-Atomic-Data,   www.astronomy.ohio-state.edu/~nahar/nahar_radiativeatomicdata/ index.html
22. OPServer, http://opacities.osc.edu
23. A.K. Pradhan, S.N. Nahar, in *Atomic Astrophysics and Spectroscopy* (Cambridge University Press, Cambridge, 2011)
24. A.K. Pradhan, S.N. Nahar, H.L. Zhang, Astrophys. J. Lett. **549**, L265–L268 (2001)
25. B.F. Rozsnyai, S.D. Bloom, D.A. Resler, Phys. Rev. A **44**, 6791 (1991), and references therein
26. M.J. Seaton, J. Phys. B **20**, 6363 (1987)
27. M.J. Seaton, Y. Yu, D. Mihalas, A.K. Pradhan, Mon. Not. R. Astron. Soc. **266**, 805 (1994)
28. The Opacity Project Team, *The Opacity Project*, vol. 1 (IOP Publishing, Bristol, 1995)
29. The Opacity Project Team, *The Opacity Project*, vol. 2 (IOP Publishing, Bristol, 1996)
30. TIPbase, http://cdsweb.u-strasbg.fr/tipbase/home.html
31. TOPbase, http://vizier.u-strasbg.fr/topbase/topbase.html
32. Y. Yu, M.J. Seaton, J. Phys. B **20**, 6409 (1987)
33. H.L. Zhang, S.N. Nahar, A.K. Pradhan, Phys. Rev. A **64**, 032719 (2001)

Chapter 8
Hybrid Theory of Electron-Hydrogenic Systems Elastic Scattering

A.K. Bhatia

Abstract Accurate electron-hydrogen and electron-hydrogenic cross sections are required to interpret various experiments. Scattering from such simple systems allows us to test various scattering theories. The incident electron produces a distortion of the target which results into a long-range polarization potential in the Schrödinger equation for the scattering function. There are also short-range correlations. In this article, we show how to take into account these two effects at the same time **variationally**. The polarization is considered to take place whether the electron is outside the target-electron orbit or inside it. Phase shifts have been calculated for the scattering from hydrogen atom, He^+ and Li^{++} positive ions. The S-wave phase shifts calculated here have lower bounds and they increase as the number of terms in the correlation function is increased. It is shown that *only* short expansions are needed to obtain accurate results. The results are compared with the previous calculations. Resonance parameters for the resonances in He and Li^+ have been calculated by calculating phase shifts in the resonance regions. The results are compared with the previous calculations and the merit of the present approach compared to the previous approach is pointed out.

8.1 Introduction

Accurate electron-hydrogen and electron-hydrogenic cross sections are required to interpret fusion experiments, laboratory plasma physics and properties of the solar and astrophysical plasmas.

The advantage of working with hydrogenic targets is that the wave function is known exactly which allows us to test any new theories on scattering of the simplest three-body systems. The incident electron produces a distortion of the target orbital and various methods have been used to take into account this distortion producing a long-range potential. Among them is the method of polarized orbitals [1]. In spite of the fact that this method includes the essential physics by modifying the target

---

A.K. Bhatia (✉)
NASA/Goddard Space Flight Center, Greenbelt, MD 20771, USA
e-mail: Anand.K.Bhatia@nasa.gov

M. Mohan (ed.), *New Trends in Atomic and Molecular Physics*,
Springer Series on Atomic, Optical, and Plasma Physics 76,
DOI 10.1007/978-3-642-38167-6_8, © Springer-Verlag Berlin Heidelberg 2013

wave function suitably in the presence of the incident electron, the results obtained for the phase shifts do not obey any bound principles. In spite of this shortcoming, this method has been used extensively on electron scattering from atoms [1] and molecules [2].

In spite of singularities in his calculations, Schwartz [3] used the Kohn variational principle to obtain accurate phase shifts for electron-hydrogen scattering. More recently, scattering from hydrogenic systems has been carried out by using the Feshbach projection operator formalism [4]. In these calculations [5–7], the short-range correlations could be included explicitly by introducing separate correlation functions and then amalgamating them into the scattering equation via an optical potential, thus replacing the many-particle Schrodinger equation with a single-particle equation. The results obtained for the phase shifts are very accurate and they have an important property namely that they have the rigorous lower bounds to the exact phase shifts. That is the exact phase shifts are always higher than the calculated phase shifts. In this way, we obtained accurate S-wave and P-wave phase shifts for electron-hydrogenic systems. This approach also helps us to get very accurate scattering wave functions for S and P states in He$^+$ ion and H atoms. Very accurate cross sections for photoionization of He and photodetachment cross sections of H-ion were obtained along with radiative attachment cross sections [8]. A detailed account of these methods and calculations has been given in [9].

In these calculations [5–8], the short-range correlations could be included explicitly but not the long-range correlations at the same time. In this article, we present an approach which includes the long-range correlations as well as short-range correlations explicitly at the same time [10]. Also, we do not use the projection operator formalism of Feshbach [4]. Let us write the wave function for an angular momentum L in the form

$$\Psi_L(\vec{r}_1,\vec{r}_2) = \frac{u_L(r_1)}{r_1} Y_{L0}(\Omega_1)\phi_0(r_2) \pm (1 \leftrightarrow 2) + \sum_\lambda C_\lambda \Phi_L^\lambda(\vec{r}_1,\vec{r}_2). \tag{8.1}$$

The summation index λ takes values from 1 to the number of terms N in the wave function Φ, the correlation function. The target function is given by

$$\phi(r_2) = \sqrt{\frac{Z^3}{\pi}}e^{-Zr_2}. \tag{8.2}$$

Z is the nuclear charge of the target. The $(\pm)$ above in Eq. (8.1) refers to singlet (upper sign) or triplet (lower sign) scattering, respectively. The first term represents the static approximation. Beyond the first term are the terms giving rise to the exchange approximation [11]. The function Φ_L is the correlation function which for arbitrary L can be written in terms of symmetric Euler angles [12]:

$$\Phi_L = \sum_\kappa [f_L^{\kappa,+1}(r_1,r_2,r_{12})D_L^{\kappa,+1}(\theta,\varphi,\psi) + f_L^{\kappa,-1}(r_1,r_2,r_{12})D_L^{\kappa,-1}(\phi,\varphi,\psi)].$$

$$\tag{8.3}$$

The D functions have been called rotational harmonics and θ, ϕ and ψ are the Euler angles. The f's above are the generalized 'radial functions' which depend on the three residual coordinates which are required to define the two vectors r_1 and r_2. The radial functions have to be defined for each angular momentum L [12].

The Euler angles under exchange are given by

$$\theta \to \pi - \theta, \qquad \varphi \to \pi + \varphi, \qquad \Psi \to 2\pi - \Psi. \tag{8.4}$$

The Hamiltonian in Rydberg units can be written as

$$H = -\nabla_1^2 - \nabla_2^2 - \frac{2Z}{r_1} - \frac{2Z}{r_2} + \frac{2}{r_{12}}, \tag{8.5}$$

and the total energy is given by $E = k^2 - Z^2$, where k^2 is the kinetic energy of the incident electron.

Taking $N = 1$, for illustration, calculate the functional (arising in the Kohn variational principle)

$$I = \langle \psi_L(\vec{r}_1, \vec{r}_2) | H - E | \psi_L(\vec{r}_1, \vec{r}_2) \rangle, \tag{8.6}$$

which can be written as

$$I = A + C_1 B + C_1^2 D, \tag{8.7}$$

where

$$A = \langle [\phi_0(r_2)u(\vec{r}_1) \pm (1 \leftrightarrow 2)] | H - E | \phi_0(r_2)u(\vec{r}_1) \pm (1 \leftrightarrow 2)] \rangle \tag{8.8}$$

and

$$\begin{aligned} B &= 2\langle \Phi_L^{(1)}(\vec{r}_1, \vec{r}_2) | H - E | \varphi_0(r_2)u(\vec{r}_1) \pm (1 \leftrightarrow 2) \rangle \\ &= 4\langle V_1(\vec{r}_1)u(\vec{r}_1) \rangle. \end{aligned} \tag{8.9}$$

In the above equation,

$$V_1(\vec{r}_1) = \langle \Phi_L^{(1)}(\vec{r}_1, \vec{r}_2) | H - E | \varphi_0(\vec{r}_2) \rangle. \tag{8.10}$$

In Eq. (8.7)

$$D = \langle \Phi_L^{(1)}(\vec{r}_1, \vec{r}_2) | H - E | \Phi_L^{(1)}(\vec{r}_1, \vec{r}_2) \rangle = \varepsilon_1 - E, \tag{8.11}$$

where ε_1 is the expectation value of H

$$\langle \Phi_L^{(1)} | H | \Phi_L^{(1)} \rangle = \varepsilon_1. \tag{8.12}$$

The wave function $\Phi_L^{(1)}$ is normalized to unity. Now we determine the eigenvector C_1. The variation with respect to C_1 in Eq. (8.7) is

$$\frac{\partial I}{\partial C_1} = 0. \tag{8.13}$$

Implies

$$B + 2C_1 D = 0. \tag{8.14}$$

This gives

$$C_1 = -\frac{B}{2D} = \frac{2\langle \Phi_L^{(1)}(\vec{r}_1, \vec{r}_2)|H - E|\varphi_0(\vec{r}_2)u_L(\vec{r}_1)\rangle}{E - \varepsilon_1}$$
$$= \frac{2\langle V_1(\vec{r}_1)u_L(\vec{r}_1)\rangle}{E - \varepsilon_1}. \tag{8.15}$$

Now the correlation term in Eq. (8.1) is known. Thus the variational principle reduces to

$$\langle \phi_0(r_2)|H - E|\psi L\rangle = 0, \tag{8.16}$$

which can be simplified to

$$\langle \varphi_0(\vec{r}_2)|H - E|[\varphi_0(\vec{r}_2)u(\vec{r}_1) \pm (1 \leftrightarrow 2)]\rangle + C_1 V_1(\vec{r}_1) = 0. \tag{8.17}$$

Where $V_1(r_1)$ is given in Eq. (8.10). Substitution of C_1 from Eq. (8.15) into Eq. (8.17) gives the equation

$$\langle \varphi_0(\vec{r}_2)|H - E|[\varphi_0(\vec{r}_2)u(\vec{r}_1) \pm (1 \leftrightarrow 2)]\rangle + \frac{2V_1(\vec{r}_1)\langle V_1(\vec{r})u(\vec{r})\rangle}{E - \varepsilon_1} = 0. \tag{8.18}$$

Now the resulting equation for $u_L(r)$, letting $r_1 = r$, can be written in the form

$$\left[\frac{d^2}{dr^2} - \frac{L(L+1)}{r^2} + V_d \pm V_{ex} - V_{op} + k^2\right]u_L = 0, \tag{8.19}$$

where

$$V_{op}u_L = 2r \sum_{\lambda=1}^{N} \frac{V_\lambda(\vec{r})\langle V_\lambda(\vec{r}_1)u_L(\vec{r}_1)\rangle}{E - \varepsilon_1}. \tag{8.20}$$

The above equation also holds when λ ranges from 1 to N, the number of terms in the correlation function (8.3). The potentials V_d is the direct potential and V_{ex} is the exchange potential. The optical potential has been obtained without the use of the Feshbach projection operator formalism and hence is independent of the projection operators P and Q [4]. The phase shifts obtained by two approaches are the same within the numerical accuracy.

8.2 Optical Potential with Polarization

To take into account the long range correlations, we replace $\phi_0(r_2)$ in Eq. (8.1) by

$$\Phi^{pol}(r_1, r_2) = \phi(r_2) - \frac{\varepsilon(r_1, r_2)}{r_1^2} \frac{u_{1s\to p}(r_2)}{r_2} \frac{\cos(\theta_{12})}{\sqrt{Z\pi}}, \tag{8.21}$$

where

$$u_{1s\to p}(r_2) = e^{-Zr_2}\left(\frac{Zr_2^3}{2} + r_2^2\right), \tag{8.22}$$

which is the dipole part of the polarized orbital and θ_{12} is the angle between r_1 and r_2. Temkin [1] introduces the function

$$\varepsilon(r_1, r_2) = \begin{cases} 1, & r_1 > r_2 \\ 0, & r_1 < r_2, \end{cases} \qquad (8.23)$$

which ensures that the polarization takes place when the scattered electron is outside the orbital electron. Now we want the polarization function in Eq. (8.21) to be valid throughout the range, rather than only just for $r_1 > r_2$. We, therefore, replace the function $\varepsilon(r_1, r_2)$ by a cutoff function and in order to avoid discontinuity at $r_1 = r_2$ we choose it of the form

$$\chi_\beta(r_1) = \left(1 - e^{-\beta r_1}\right)^n, \qquad (8.24)$$

where the exponent $n \geq 3$. The nonlinear parameter β, which is function of k, can be used to optimize the results. This function vanishes at large values of r_1, it also contributes to the short-range correlations in addition to those obtained from the correlation function Φ_L. Another form of the cutoff has been proposed by Shertzer and Temkin [13]

$$\chi_{ST} = \frac{1}{N} \int d^3 r_2 \phi_0(\vec{r}_2) \frac{u_{1s \to p}(r_2)}{r_2} \frac{\varepsilon(r_1, r_2)}{r_1^2}, \qquad (8.25)$$

where N denotes the normalization and is given by

$$N = \int d^3 r_2 \phi_0(r_2) \frac{u_{1s \to p}(r_2)}{r_2} \frac{1}{r_1^2}. \qquad (8.26)$$

In the above equations the integrations are from 0 to infinity. This gives

$$\chi_{ST}(r_1) = 1 - e^{-2Zr_1}\left[\frac{1}{3}(Zr_1)^4 + \frac{4}{3}(Zr_1)^3 + 2(Zr_1)^2 + 2Zr_1 + 1\right], \qquad (8.27)$$

which guarantees that $\chi(r_1)/r_1^2$ approaches 0 when r_1 approaches 0. Now the scattering wave function can be written as

$$\Psi_L(\vec{r}_1, \vec{r}_2) = \frac{u_L(r_1)}{r_1} Y_{L0}(\Omega_1) \Phi^{pol}(r_1, r_2) \pm (1 \leftrightarrow 2) + \sum_\lambda C_\lambda \Phi_L^\lambda(\vec{r}_1, \vec{r}_2), \qquad (8.28)$$

where the target function with polarization is given by

$$\Phi^{pol}(r_1, r_2) = \phi(r_2) - \frac{\chi(r_1)}{r_1^2} \frac{u_{1s \to p}(r_2)}{r_2} \frac{\cos(\theta_{12})}{\sqrt{Z\pi}}. \qquad (8.29)$$

The cutoff function χ can be of either form given in Eq. (8.24) or Eq. (8.27). We arrive at the same form of Eq. (8.18) by replacing ϕ_0 by $\Phi^{pol}(\vec{r}_1, \vec{r}_2)$ in Eqs. (8.6)–(8.20). We restrict ourselves to $L = 0$ and we can write the final equation in the form

$$\left[D(r)\frac{d^2}{dr^2} + k^2 + V_d + V^{pol} \pm \left(V_{ex} + V_{ex}^{pol}\right) - V_{op}^{pol}\right] u(r) = 0. \qquad (8.30)$$

The various quantities are given below:

$$D(r) = 1 + \frac{43}{8Z^6}\left(\frac{\chi(r)}{r^2}\right)^2. \qquad (8.31)$$

The direct potential is given by

$$V_d = \frac{2(Z-1)}{r} + 2e^{-2Zr}\left(Z + \frac{1}{r}\right) \tag{8.32}$$

and

$$V_d^{pol} = (x_1 + x_3) + x_2\frac{d}{dr}, \tag{8.33}$$

$$x_1 = \frac{43}{8Z^6}\left(\frac{2Z}{r} - \frac{2}{r^2} + k^2\right)\left(\frac{\chi(r)}{r^2}\right)^2, \tag{8.34}$$

$$x_2 = B_1(r)\frac{43}{8Z^6}\left(\frac{\chi(r)}{r^2}\right), \tag{8.35}$$

$$x_3 = B_2(r)\frac{43}{8Z^6}\left(\frac{\chi(r)}{r^2}\right) + \frac{2\alpha(r)}{(Zr)^4}\chi(r) - \frac{4.5\chi(r)^2}{(Zr)^4} - \frac{8}{3Z}d(r)\left(\frac{\chi(r)}{r^2}\right)^2, \tag{8.36}$$

$$B_1(r) = \frac{1}{r^2}\left[-2Z\chi(r) + 2Z - \frac{2}{r} + e^{-2Zr}\left(-\frac{2}{3}Z^4r^3 - \frac{4}{3}Z^3r^2 + 2Z + \frac{2}{r}\right)\right], \tag{8.37}$$

$$B_2(r) = -2ZB_1(r) - \frac{4Z}{r^3} + \frac{6}{r^4} + e^{-2Zr}\left(\frac{4}{3}Z^5r + 2Z^4 - \frac{4Z^2}{r^2} - \frac{8Z}{r^3} - \frac{6}{r^4}\right). \tag{8.38}$$

The polarizability of the target as a function of r is given by

$$\alpha(r)/Z^4, \tag{8.39}$$

where

$$\alpha(r) = 4.5 - \frac{2}{3}e^{-2Zr}\left(\frac{3}{2}(Zr)^4 + 7.5(Zr)^3 + 13.5(Zr)^2 + 13.5(Zr) + 6.75\right), \tag{8.40}$$

and

$$d(r) = \frac{129}{32Z^5}\frac{1}{r} + \frac{18}{Z^7}\frac{1}{r^3} - e^{-2zr}\left(\frac{3}{16}r^4 + \frac{27}{16Z}r^3 + \frac{54}{8Z^2}r^2\right.$$
$$\left. + \frac{135}{8Z^3}r + \frac{975}{32Z^4} + \frac{1281}{32Z^5}\frac{1}{r} + \frac{36}{Z^6}\frac{1}{r^2} + \frac{18}{Z^7}\frac{1}{r^3}\right), \tag{8.41}$$

$x_3 \to \frac{9}{2Z^4r^4}$, $r \to$ infinity.

In the above expression $9/(2Z^4)$ is the dipole polarizability of the target with nuclear charge Z. The direct polarization potential has also terms proportional to $1/r^2$ and $1/r^3$ which go to zero for $r \to 0$.

For completeness, we give other terms in Eq. (8.30). The exchange terms are given by

$$V_{ex}u(r) = 4Z^3 e^{-Zr}\left[(k^2+Z^2)r\int_0^\infty dx e^{-Zx}xu(x) - 2\int_0^r dx e^{-Zx}u(x)\right.$$
$$\left. - 2r\int_r^\infty dx e^{-Zx}u(x)\right], \tag{8.42}$$

$$V_{ex}^{pol}u(r) = J_2u(r) + J_3u(r) + J_4u(r), \tag{8.43}$$

where

$$J_2u(r) = \frac{8Z}{3}\frac{u_{1s\to p}(r)}{r}\left[\frac{1}{r}\int_0^r dx e^{-Zx}\chi(x)u(x) + r^2\int_r^\infty dx e^{-Zx}\chi(x)\frac{u(x)}{x^3}\right], \tag{8.44}$$

$$J_3u(r) = \frac{8Ze^{-Zr}\chi(r)}{3r^2}\left[\frac{1}{r}\int_0^r dx u_{1s\to p}(x)xu(x) + r^2\int_r^\infty dx u_{1s\to p}(x)xu(x)\right.$$
$$\left. + r^2\int_r^\infty dx u_{1s\to p}(x)\frac{u(x)}{x^2}\right], \tag{8.45}$$

and

$$J_4u(r) = -\frac{8Ze^{-Zr}\chi(r)}{3}\left[\int_0^\infty dx u_{1s\to p}(x)\frac{u(x)}{x^2}\chi(x)\right] + G_2u(r) + G_3u(r). \tag{8.46}$$

Now we will write below the expressions for G_2u and G_3u.

$$G_2u(r) = \frac{4\chi(r)u_{1s\to p}(r)}{3Zr^2}$$
$$\times \int_0^\infty dx e^{-Zx}\left[\frac{Z^3x^3}{2} - 2Z^2x^2 - Zx + 2\right.$$
$$\left. + \left(\frac{Zx^3}{2} + x^2\right)\left(\frac{2Z}{x} - \frac{2}{x^2} + k^2\right)\right]\frac{\chi(x)u(x)}{x^2}, \tag{8.47}$$

$$G_3u(r) = -\frac{8\chi(r)u_{1s\to p}(r)}{8Zr^2}\left[\frac{1}{r}\int_0^r dx \frac{u_{1s\to p}(x)}{x^2}\chi(x)u(x)\right.$$
$$\left. + \int_r^\infty dx \frac{u_{1s\to p}(x)}{x^3}\chi(x)u(x) + \frac{2}{5}Gu(r)\right], \tag{8.48}$$

$$Gu(r) = \frac{1}{r^3}\int_0^r dx u_{1s\to p}(x)\chi(x)u(x) + r^2\int_r^\infty dx \frac{u_{1s\to p}(x)}{x^5}\chi(x)u(x). \tag{8.48a}$$

The polarization term in Eq. (8.21) gives rise to terms in Eqs. (8.33–8.41) and Eqs. (8.43–8.48). These terms make the calculations much more intricate compared to the calculations without these term.

The optical term appearing in Eq. (8.30) is given by

$$V_{op}^{pol}u(r_1) = r_1\sum_s^{N\omega} \frac{\langle Y_{00}^*(\Omega_1)\Phi^{pol}(\vec{r}_1,\vec{r}_2)|H-E|\Phi_0^{(s)}\rangle\langle\Phi_0^{(s)}|H-E|\Psi_0'\rangle}{E-\varepsilon_s}. \tag{8.49}$$

Table 8.1 Phase shifts (radians) of ^1S state for various k

k	Ref. [5]		C	C	Ref. [3]	Ref. [14]
	A	B	χ_{ST}	χ_β	$\eta_{Schwartz}$	η_{SSB}
0.0^a		6.05327	6.00092	5.99567	5.965	
0.1	2.55358	2.55158	2.55372	2.55370	2.553	2.550
0.2	2.06678	2.06644	2.06699	2.06717	2.0673	2.062
0.3	1.69816	1.69640	1.69853	1.69684	1.6964	1.691
0.4	1.41540	1.41783	1.41561	1.41554	1.4146	1.410
0.5	1.20094	1.20084	1.20112	1.20195	1.202	1.196
0.6	1.04083	1.04074	1.04110	1.04191	1.041	1.035
0.7	0.93111	0.93105	0.93094	0.93084	0.930	0.925
0.8	0.88718	0.88717	0.88768	0.88802	0.886	

$^a k = 0$ results represent scattering lengths

It should be noted that in the above expression, we have Ψ_0' instead of Ψ_0 given in Eq. (8.1). The function Ψ_0' is the wave function without the correlation terms in Eq. (8.1). This is very clear in the projection operator-formalism of Feshbach [4] used in [5–8] to obtain the scattering equations. There the function Ψ_0 is written as $P\Psi_0 + Q\Psi_0$ and the function $P\Psi_0$ only occurs on the right-hand side of the formulation. It should be pointed out that $P + Q = 1$. Now, we have given most of the quantities occurring in Eq. (8.30).

For S-waves (i.e., $L = 0$), $D_L =$ constant in Eq. (8.3) and the correlation function is a function of "radial" coordinates. We take Φ_0 of Hylleraas form

$$\Phi_{L=0}(\vec{r}_1, \vec{r}_2) = \sum_{lmn}^{N_\omega} C_{lmn}\left[e^{-\gamma r_1 - \delta r_2} r_1^l r_2^m r_{12}^n \pm (1 \leftarrow 2)\right]. \tag{8.50}$$

The sum includes all values of l, m and n such that $l + m + n = \omega$ and ω is equal to $0, 1, 2, 3, \ldots, 8$. The total number of terms N_ω depends on spin and whether $\gamma = \delta$ or not. There are 95 terms for ^1S state and 84 terms for ^3S state for $N_\omega = 8$ in these calculations. Solutions of the integrodifferential equations have been carried by the noniterative method [see Appendix] and by using quadruple precision. We present results for the ^1S and ^3S states in Tables 8.1 and 8.2 in three approximations:

A. Phase shifts calculated using optical potential obtained by using the projection operator formalism [4].
B. Phase shifts calculated using optical potential obtained without using the projection operator formalism [4], Eq. (8.19).
C. Phase shifts calculated using optical potential without using the projection operator formalism but including the effects of polarization, Eq. (8.30).

The scattering length a is defined by

$$\lim_{k \to 0} k \cot \eta = -1/a. \tag{8.51}$$

Table 8.2 Phase shifts (radians) of ^3S state for various k

k	Ref. [5]		C	C	Ref. [3]	Ref. [14]
	A	B	χ_{ST}	χ_β	$\eta_{Schwartz}$	η_{SSB}
0.0^a		1.81644	1.78467	1.78154	1.7686	
0.1	2.93853	2.93850	2.93856	2.93856	2.9388	2.939
0.2	2.71741	2.71740	2.71751	2.71751	2.7171	2.717
0.3	2.49975	2.49956	2.49987	2.49987	2.4996	2.500
0.4	2.29408	2.29394	2.29465	2.29457	2.2938	2.294
0.5	2.10454	2.10414	2.10544	2.10574	2.1046	2.105
0.6	1.93272	1.93280	1.93322	1.93336	1.9329	1.933
0.7	1.77950	1.78010	1.77998	1.77998	1.7797	1.780
0.8	1.64379	1.64408	1.64425	1.64444	1.643	

$^a k = 0$ results represent scattering lengths

It should be noted that the phase shifts given above have the rigorous lower bounds to the exact phase shifts while the scattering lengths have the rigorous upper bounds to the exact scattering lengths.

Temkin [15] has shown that the scattering length is significantly affected by the long-range polarization. The scattering length obtained at a short distance R can be corrected by the following formula

$$a = a(R) - \alpha\left(\frac{1}{R} - \frac{\alpha}{R^2} + \frac{1}{3}\frac{\alpha^2}{R^3}\right), \tag{8.52}$$

where a is the true scattering length. The scattering length $a(R)$ for ^1S state obtained using χ_β is 5.99567 at $R = 146.0782$ for $N = 95$ terms. Using the formula (8.52), we get $a = 5.96611$ which agrees well with the value $a = 5.965 \pm 0.0003$ given by Schwartz [3]. He also included long-range polarization in his zero-energy calculation. The scattering lengths are rigorous upper bounds to the exact scattering results.

We find that the phase shifts obtained with the Feshbach formalism [4] are fairly close to the present results. That indicates that the short-range correlations do try to assimilate the effect of the long-range correlations to some extent. The present phase shifts have rigorous lower bounds to the exact results. We compare our results with those of Schwartz [3]. His calculation was carried out using the Kohn variational principle. However, his calculations do not provide any bounds on the phase shifts though the Kohn variational principle does provide a bound on the scattering length.

In Table 8.2, we give the results for the ^3S state. The ^3S scattering length for $N = 84$ with χ_β, $\beta = 0.6$ and $n = 4$ is $a(R) = 1.7681542$ at $R = 349.0831$. The corrected value, using formula (8.52), is 1.76815 is a little lower compared to the Schwartz value 1.7686 ± 0.0002. The formula (8.52) certainly gives a good agreement with the Schwartz value. Lower scattering lengths are consistent with the rigorous upper bounds, as stated earlier.

Table 8.3 Comparison of the presently obtained ^1S phase shifts in e$^-$He$^+$ scattering with other calculations: (OP) Ref. [6], (CC) Ref. [16] and Harris-Kohn Ref. [17]

k	Hybrid	(OP)	(CC)	Harris-Kohn
0.1	0.43808			
0.2	0.43550			
0.3	0.43142			0.4300
0.4	0.42608	0.42601		0.4228
0.5	0.41974	0.41964		0.4078
0.6	0.41265	0.41278	0.4111	0.4086
0.7	0.40568	0.40561	0.4046	0.4024
0.8	0.39865	0.39857	0.3974	0.3968
0.9	0.39213	0.39202	0.3906	0.3893
1.0	0.38644	0.38634	0.3850	0.3836
1.1	0.38200	0.38187	0.3805	0.3794
1.2	0.37914	0.37899	0.3780	0.3741
1.3	0.37846	0.37832	0.3744	0.3721
1.4	0.38158	0.38560		0.3786
1.5	0.39802			0.4014
1.6	0.34480			

In Tables 8.1 and 8.2, we have given η_{SSB} obtained by Scholz, Scott and Burke [14] using the well known R-matrix formulation. In these calculations, all the correlations are included in the inner region of a certain radius R. The functions in the two regions are then matched at this radius R. Their results are fairly close to the presently calculated phase shifts.

We see that the hybrid theory developed here includes both the short-range and long-range forces. The present development is significant even when the present results do not differ significantly from those obtained using the Feshbach formalism.

8.3 Electron-He$^+$ Scattering

The above approach has been applied to the scattering of electrons from helium ions. Now the phase shifts are given by

$$\lim_{r \to \infty} u(r) \propto \sin\left[kr + \frac{Z-1}{k} \ln(2kr) + \arg \Gamma\left(1 - \frac{i(Z-1)}{k} \right) + \eta \right]. \quad (8.53)$$

Phase shifts obtained for e$^-$He$^+$ in ^1S and ^3S states are given in Tables 8.3 and 8.4.

They are given at low energies as well and have been compared with the results obtained using the Feshbach projection operator formalism [6]. They are also compared with those obtained in the close-coupling approximation [16] and also obtained using the Harris-Kohn method [17].

Table 8.4 Comparison of the presently obtained ^3S phase shifts in e$^-$He$^+$ scattering with other calculations: (OP) Ref. [6], (CC) Ref. [16] and Harris-Kohn Ref. [17]

k	Hybrid	(OP)	(CC)	Harris-Kohn
0.1	0.93065			
0.2	0.92704			0.9270
0.3	0.92114			0.9210
0.4	0.91302	0.91300		0.9128
0.5	0.90282	0.90275	0.9019	0.9025
0.6	0.89057	0.89050	0.8910	0.8902
0.7	0.87645	0.87640	0.8777	0.8762
0.8	0.86066	0.86069	0.8617	0.8605
0.9	0.84366	0.84356	0.8440	0.8435
1.0	0.82536	0.82531	0.8253	0.8251
1.1	0.80636	0.80625	0.8062	0.8062
1.2	0.78677	0.78666	0.7868	0.7865
1.3	0.76696	0.76684	0.7672	0.7665
1.4	0.74708	0.74697		0.7466
1.5	0.72746			0.7274
1.6	0.70815			0.7095

Phase shifts are higher for triplet states and both ^1S and ^3S phase shifts decrease with increasing incident energy. It is seen that phase shifts start rising after $k = 1.4$, indicating the existence of a resonance which we will discuss below.

8.4 Electron-Li^{++} Scattering

A similar calculation has been carried out for e$^-$Li^{++} scattering. As in the case of He$^+$, there is a long-range Coulomb potential in addition to the long-range potential due to polarization of the target.

Gien [18] has carried out calculations by using Harris-Nesbet method and has used various combinations of target states in his calculations. We have given results which are labeled (E4S) in his paper. He used $1s$, $2s$, $2p$ states of the target along with a $2\bar{\bar{p}}$ pseudostate [19] and correlations. The pseudostate is given by

$$2\bar{\bar{p}} = Z^{\sqrt{5/2}}[0.340r^2 e^{-0.5Zr} - 0.966r^2 e^{-Zr}(1 + 0.5Zr)]. \qquad (8.54)$$

This state has been formed as a linear combination of target $2p$ state and $u_{1s \to p}$ with a constant factor such that it is normalized and orthogonal to the $2p$ state. The inclusion of the pseudostate gives the exact polarizability of the target. These results are given in Table 8.5.

Table 8.5 Comparison of the presently calculated ^1S and ^3S phase shifts in e$^-$Li^{++} scattering with those obtained by Gien [18]

k	^1S (Hybrid)	^1S (Gien)	^3S (Hybrid)	^3S (Gien)
0.1	0.23188		0.56084	
0.2	0.23176		0.56020	
0.3	0.23148		0.55869	
0.4	0.23109		0.55678	
0.5	0.23064	0.2273	0.55435	0.5526
0.6	0.23012	0.2264	0.55142	0.5499
0.7	0.22960	0.2265	0.54799	0.5467
0.8	0.22906	0.2272	0.54413	0.5430
0.9	0.22855	0.2277	0.53925	0.5390
1.0	0.22807	0.2275	0.53514	0.5345
1.1	0.22769	0.2262	0.53000	0.5296
1.2	0.22740	0.2250	0.52456	0.5244
1.3	0.22724	0.2258	0.51880	0.5189
1.4	0.22724	0.2328	0.51276	0.5131
1.5	0.22742	0.2521	0.50646	0.5069
1.6	0.22782		0.49997	0.5005

Gien [17, 18] obtained phase shifts at irregular energy points, obtained from the diagonalization of the Hamiltonian. His results could be fitted to

$$\eta = A + BE + CE^2 + DE^3 + F\exp[aE]. \tag{8.55}$$

In the above equation $E = k^2$, k^2 is the incident energy and A, B, C, D, E, F and a are the fitting parameters.

8.5 Resonances

The first time, a resonance was calculated, was in 1962 by Burke and Schey [20]. This was in electron-hydrogen scattering. They used the close-coupling approximation and found that the resonance was centered at 9.61 eV.

Resonances have been known by many names: autoionization states, doubly excited states and Feshbach resonances. It is observed that there is a rapid change of phase shift in the resonance region. This phase shift can be fitted to the Breit-Wigner form to obtain the resonance parameters,

$$\eta_{calc.}(E) = \eta_0 + AE + \tan^{-1}\frac{0.5\Gamma}{(E_R - E)}, \tag{8.56}$$

where $E = k^2$, k^2 is the incident energy, $\eta_{calc.}$ are the calculated phase shifts, η_0, A, Γ, and E_R are the fitting parameters. E_R and Γ are the resonance position

Table 8.6 Phase shifts in the resonance region for the lowest ^1S resonance in He and Li$^+$

Target	k	η	Target	k	η
He$^+$	1.555	0.5483195	Li^{++}	2.282	-0.090487
	1.558	0.6398967		2.2825	-0.058190
	1.560	0.7803777		2.283	-0.031793
	1.562	1.177147		2.2835	-0.009842
	1.5634	1.938293		2.284	0.008626
	1.56345	1.972649		2.2845	0.002451
	1.565	2.776815		2.285	0.038199
	1.5655	-3.372919			
	1.566	-3.276240			
	1.567	-3.148991			

and resonance width. We would like to calculate the resonance position and width of the lowest ^1S resonance in He (below the $n = 2$ threshold of He$^+$). In Table 8.6, we give phase shifts for 70 terms in the correlation function in the resonance region.

We calculate the square of the difference between the left and right side of Eq. (8.57). The difference is being calculated at the energies given in Table 8.6. Minimize the sum of the squares for various fitting parameters. We find $\eta_0 = 0.3761$, $A = 0$, $E_R = 2.4426$ Ry with respect to He$^+$ and $\Gamma = 0.00906$ Ry when the minimized sum is 1.57×10^{-7}. This gives resonance position $E_R = 57.8481$ eV with respect to the ground state of He and width $\Gamma = 0.1233$ eV. This agrees very well with $E_R = 57.8435$ eV and $\Gamma = 0.125$ eV, obtained using the Feshbach formalism [21]. These resonance parameters, calculated now, include the contribution from the long-range and short-range correlations. But these parameters do not have any rigorous bonds although the calculated phase shifts do have rigorous lower bonds. This is the disadvantage of fitting to the Breit-Wigner form Eq. (8.56).

A similar calculation has been carried out for the lowest ^1S state in Li$^+$. The phase shits in the resonance region for 70 terms in the correlation function are given in Table 8.6. By fitting, wc obtain $E_R = 70.5904$ eV with respect to ground state of Li^{++} and $\Gamma = 0.1657$ eV which compares very well with $E_R = 70.5837$ eV and $\Gamma = 0.157$ eV [22]. In the Feshbach formalism, the resonance parameters, position and width, are defined as

$$E_R = \langle \Phi | QHQ | \Phi \rangle + \Delta = \varepsilon_R + \Delta, \tag{8.57}$$

$$\Gamma = 2k \langle \psi | PHQ | \Phi \rangle. \tag{8.58}$$

It should be noted in the projection operator formalism of Feshbach, the contribution Δ to the position of the resonance from the continuum and nearby resonances has to be calculated separately. This contribution is already included in the present calculation which is a great advantage.

8.6 Conclusions

We have developed a method in which the short-range and long-range correlations can be included at the same time in the scattering equations. The phase shifts have rigorous lower bounds and the scattering lengths have rigorous upper bounds. The phase shifts in the resonance region can be used to calculate very accurately the resonance parameters.

Appendix

We will briefly describe the non-iterative method for solving integrodifferential equations [23]. Consider the equation for the scattering function $u(r)$ given by

$$\left[\frac{d^2}{dr^2} + V(r) + k^2\right]u(r) = g(r)\int_0^\infty f(x)u(x)dx. \tag{A.1}$$

Let

$$u(r) = u_0(r) + Cu_1(r). \tag{A.2}$$

The constant C represents the definite integral in Eq. (A.1).

Substituting $u(r)$ in Eq. (A.1), we get two equations

$$\left[\frac{d^2}{dr^2} + V(r) + k^2\right]u_0(r) = 0, \tag{A.3}$$

$$\left[\frac{d^2}{dr^2} + V(r) + k^2\right]u_1(r) = g(r). \tag{A.4}$$

These two equations can be solved easily. The substitution of (A.2) in C gives

$$C = \int_0^\infty f(x)u_0(x)dx + C\int_0^\infty f(x)u_1(x)dx = I_0 + CI_1. \tag{A.5}$$

Having calculated u_0 and u_1, I_0 and I_1 can be calculated. From the above equation, we can now solve for C. We get

$$C = \frac{I_0}{(1 - I_1)}. \tag{A.6}$$

Equation (A.1) can now be written as

$$\left[\frac{d^2}{dr^2} + V(r) + k^2\right]u(r) = Cg(r). \tag{A.7}$$

Now the right-hand side is known and this equation can be solved for $u(r)$. This method can be generalized to any number of definite integrals.

References

1. A. Temkin, Phys. Rev. **107**, 1004 (1957)
2. K.V. Vasavada, A. Temkin, Phys. Rev. **160**, 109 (1967)
3. C. Schwartz, Phys. Rev. **124**, 1468 (1968)
4. H. Feshbach, Ann. Phys. **19**, 287 (1962)
5. A.K. Bhatia, A. Temkin, Phys. Rev. **64**, 032709 (2001)
6. A.K. Bhatia, Phys. Rev. A **66**, 064702 (2002)
7. A.K. Bhatia, Phys. Rev. A **69**, 032714 (2004)
8. A.K. Bhatia, Phys. Rev. A **73**, 012705 (2006)
9. A.K. Bhatia, in *Atomic Structure and Collision Processes*, edited by M. Mohan (Narosa Publishing House, New Delhi, 2010), pp. 87–103
10. A.K. Bhatia, Phys. Rev. A **75**, 032713 (2007)
11. P.M. Morse, W.P. Allis, Phys. Rev. **44**, 269 (1933)
12. A.K. Bhatia, A. Temkin, Rev. Mod. Phys. **36**, 1050 (1964)
13. J. Shertzer, A. Temkin, Phys. Rev. A **74**, 052701 (2006)
14. T. Scholz, P. Scott, P.G. Burke, J. Phys. B **21**, L139 (1988)
15. A. Temkin, Phys. Rev. Lett. **6**, 354 (1961)
16. P.G. Burke, A.J. Taylor, J. Phys. B **2**, 44 (1969)
17. T.T. Gien, J. Phys. B **35**, 4475 (2002)
18. T.T. Gien, J. Phys. B **36**, 2291 (2003)
19. P.G. Burke, D.F. Gallaher, S. Geltman, J. Phys. B **2**, 1142 (1969)
20. P.G. Burke, H.M. Schey, Phys. Rev. **126**, 147 (1962)
21. A.K. Bhatia, A. Temkin, Phys. Rev. A **11**, 2018 (1975)
22. A.K. Bhatia, Phys. Rev. A **15**, 1315 (1977)
23. K. Omidvar, Phys. Rev. **133**, A970 (1964)

Chapter 9
Relativistic Electron-Atom Collisions: Recent Progress and Applications

Rajesh Srivastava and Lalita Sharma

Abstract A review of the present relativistic methods to study electron impact excitation of atoms (ions) is given. Application of relativistic distorted wave theory has been described in detail. The importance of implementing fine-structure resolved cross sections to a collisional radiative model for Ar-plasma has been discussed.

9.1 Introduction

Electron impact excitation of atoms has been actively growing field of Atomic and Molecular Collision Physics due to its significant contributions to fundamentals of physics as well as practical applications in many fields of science which include providing of cross section data for the modeling of fusion plasma, astrophysics, laser physics and the physics of planetary atmosphere. In an electron-atom collision process where the incident beam of electrons interacts with the target atoms and is scattered after the collision, the experimentalists measure the differential cross section (DCS) or the integrated cross section (ICS). The DCS is a measure of the number of electrons scattered per unit solid angle per unit time. The ICS can be obtained by angle integration of DCS over the scattering angles of the projectile electrons. It is found that in comparison to ICS, the DCS is more sensitive scattering parameter as the effect of the choice of target wave functions and the merit of the theoretical method employed for the calculation are better reflected. Earlier measurements and the theoretical reports on cross sections have focused mostly on the excitation of the atoms from their ground states. However, with the advancement in the experimental technology the collision targets can be prepared in their excited states and cross section measurements are being made for electron induced excitation processes of atoms from their initially excited states *e.g.* metastable and other higher states [1–3]. These measurements cast light on the sensitivity of cross sections to both the atomic structure of the target and the method employed to study the collision process. Moreover, such studies are of great interest for plasma diagnostics

R. Srivastava (✉) · L. Sharma
Indian Institute of Technology Roorkee, Roorkee 247667, India
e-mail: rajsrfph@gmail.com

M. Mohan (ed.), *New Trends in Atomic and Molecular Physics*,
Springer Series on Atomic, Optical, and Plasma Physics 76,
DOI 10.1007/978-3-642-38167-6_9, © Springer-Verlag Berlin Heidelberg 2013

owing to the fact that in comparison to excitation from the ground state, the cross sections corresponding to excitations from metastable states or between the excited states of atomic targets are order of magnitude larger [3] due to their less excitation energies involved.

Further, as a result of improved experimental techniques, more detailed insight of an electron-atom collision process can be acquired by using spin-polarized projectile electrons and spin-polarized atomic targets and to perform experiments with the detection of their spins. One can also analyze the polarization of the characteristic emission following the decay of the collisionally excited atoms. Such experiments are known as "scattered electron-emitted photon coincidence experiment", in which the polarization of photons emitted from the atom is analyzed in coincidence with the scattered electrons. These experiments are also part of the so called perfect or complete experiment as proposed by Bederson [4–6]. The polarization is usually measured in terms of differential Stokes parameters [7] which describe the shape and orientation of the charge cloud of the excited state of the atom and provide information of the excited magnetic substates. Consequently, these measurements provide a better understanding and insight of the collisional dynamics of the electron-atom collision process and also serve as a testing ground for various theoretical approaches. The constant development in the quality of experimental technology and the increasing power of modern computers have moved the level of comparison between experiment and theory to a very high degree of sophistication.

In spite of the advance experimental techniques, the growing demand of electron-atom collision data can not be fulfilled solely through the experimental measurements. Therefore, to meet the requirement of atomic data the theoretical studies have dominated the research in this area. In a theoretical study of electron-atom collision process, all the scattering amplitudes are to be determined primarily to further obtain various scattering parameters. However, the evaluation of the scattering amplitudes can not be achieved exactly being a three body interaction and thus the theoretical calculations of scattering parameters depend on various approximation methods. It is presently understood that a more suitable method based on solving the Dirac equations is required for the description of scattering processes involving heavy targets. Thus, in recent years, much progress has been achieved in calculating reliable cross section data for electron scattering from atoms and ions, by using relativistic theoretical methods. A number of theoretical methods are available at present that follow a fully relativistic formulation based on suitable non-perturbative and perturbative approaches.

The non-perturbative approaches are the Dirac R-matrix method [8], the recently developed Dirac B-spline R-matrix method (DBRM) [9, 10], Dirac R-matrix with pseudo states method (DRMPS) [11], and relativistic convergent close coupling (RCCC) method [12–15]. The relativistic B-spline R-matrix method is based on the solution of the many-electron Dirac-Fock equation and allows to employ non-orthogonal sets of atomic orbitals. A B-spline basis is used to generate both the target description and the R-matrix basis functions in the inner region. Employing B-splines of different orders for the large and small components prevents the appearance of spurious states in the spectrum of the Dirac equation. Using term-dependent and thus non-orthogonal sets of one-electron functions enables one to generate accurate and flexible representations of the target states and the scattering function. These R-matrix calculations are most computationally effective at low energies where the size of the close-coupling expansion does not have to be large to obtain converged results. However, at intermediate energies, a large close-coupling expansion is often required. In a relativistic formulation, the size of the close-coupling expansion is about twice larger than in the corresponding non-relativistic method which makes the relativistic R-matrix calculations computationally difficult to perform [16].

In the Dirac R-matrix with pseudo-states (DRMPS) method an N-electron atom is represented by an anti-symmetrized product of single-particle spinors comprising the usual four-component Dirac spinors as well as paired two component Laguerre spinors (L-spinors). An $(N + 1)$th 'scattering' electron is represented by the exact same L-spinor basis plus the usual R-matrix box-state spinors which are chosen so as to form a combined complete finite linearly independent orthogonal basis. The (non-diagonal) Buttle correction is determined consistently. This representation has been implemented quite generally within the Dirac atomic R-matrix code (DARC) [8].

The RCCC method is based on the solution of the Dirac equation describing electron scattering from quasi-one-electron atoms. A square-integrable Dirac L-spinor basis has been used to obtain a set of target states representing both the bound and continuum spectra of the target. A set of momentum-space coupled Lippmann-Schwinger equations for the T matrix is then formulated and solved.

It is only very recently that the relativistic effects have been incorporated in the non-perturbative theoretical methods *viz.* the R-matrix and the CCC methods in a complete manner resulting in their relativistic version as already discussed above. However, these methods have been applied so far to study scattering of electrons from a limited number of targets [9, 10, 12–16].

Among the perturbative methods, the relativistic distorted wave (RDW) theory has been well known and extensively applied from nearly two decades and proved to be quite successful in explaining the experiments especially, at intermediate and high energies [17, 18]. The main advantage of using the distorted wave approximation is its flexibility which allows inclusion of more of the physical effects in the leading terms of the perturbation series expansion. This leads distorted wave series to converge faster and therefore, even the first order term of distorted wave series has been found to produces reasonably accurate results [19]. In the method, the target atom is represented by multi-configuration Dirac-Fock wavefunctions while the continuum wavefunction for the scattered electron is calculated via Dirac equations, consequently, the relativistic effects are included to all orders. Such a RDW theory has been able to describe not only differential and integrated cross sections but also the parameters for complete experiments in heavier atoms like Hg [20].

The present chapter gives a review of the recent progress and applications of relativistic theoretical approaches to study electron impact excitation of atoms (ions). Since RDW approach has been used most widely, in Sect. 9.2 we briefly describe RDW theory for electron-atom (ion) excitation. Further, in Sect. 9.3 we give account of the recent applications of RDW method to consider excitations of various open and closed shell atoms from ground and excited states as these relativistic calculations were needed in order to compare with the experimental data. Several other applications of RDW theory to excitation of singly and multiply charged ions are given in Sect. 9.4. Finally, in Sect. 9.5, a collisional radiative (CR) model where RDW cross sections have been implemented for plasma modeling is described. We would like to mention that the present chapter is not a complete review of the work done for electron-atom collision but largely takes account of the work done using relativistic theories.

9.2 Relativistic Distorted Wave Theory

9.2.1 Scattering Amplitude

All the collisional parameter of an excitation process can be expressed in terms of the scattering amplitude or through a transition matrix. It can be written within the first order perturbation theory using RDW approximation for a projectile electron exciting an atom (or ion) having N electrons as (atomic units will be used throughout),

$$f(J_i, M_i, \mu_i; J_f, M_f, \mu_f, \theta)$$

$$= (2\pi)^2 \sqrt{\frac{k_f}{k_i}} \langle \phi_f(\mathbf{1}, \mathbf{2}, \ldots, \mathbf{N}) F_f^{DW-}(\mathbf{k}_f, \mathbf{N}+\mathbf{1}) | V_{in}$$

$$- U_d(N+1) | \mathcal{A}\phi_i(\mathbf{1}, \mathbf{2}, \ldots, \mathbf{N}) F_i^{DW+}(\mathbf{k}_i, \mathbf{N}+\mathbf{1}) \rangle. \tag{9.1}$$

In this amplitude the target atom (or ion) before and after excitation is supposed to be in the states $|\alpha_i, J_i, M_i\rangle$ and $|\alpha_f, J_f, M_f\rangle$ with their well-defined total angular momenta J and magnetic components M and α denote all the additional quantum numbers as needed for a unique specification of the states. $(\mathbf{1}, \mathbf{2}, \ldots, \mathbf{N})$ refers to the position coordinates of the target electrons with respect to the nucleus. In the above expression, V_{in} is the interaction between incident electron and the target atom (ion). U_d is the distortion potential which is chosen to be the function of the position coordinate of the scattered electron only i.e., r_{N+1}. In Eq. (9.1) $\phi_{i,f}$ represent N-electron target wave functions while $F_{i,f}^{DW+(-)}$ denote the projectile electron distorted-wave functions which are calculated in the presence of distortion potential U_d. In this notation, '+' sign refers to an outgoing wave, while '−' sign denotes an incoming wave. $\mathcal{A}$ is the antisymmetrization operator to take into account the exchange of the projectile electron with the target electrons. $\mu_{i,f}$ are the spin projections of electron in initial and final channel. $\mathbf{k}_{i,f}$ are the relativistic wave vectors whose magnitude can be given by,

$$k_{i,f} = \frac{\sqrt{E_{i,f}(E_{i,f} + 2c^2)}}{c} \tag{9.2}$$

where $E_{i,f}$ are the relativistic energies (including the rest mass of the electron). The coordinate system is chosen in such a way that the incident electron defines the z axis, and the electrons are scattered in the xz plane with the scattering angle θ as the angle between the wave vectors $\mathbf{k}_i$ and $\mathbf{k}_f$ of the incident and scattered electron. In order to evaluate the amplitude (Eq. (9.1)) we need to know the explicit form of both, interaction operators and electronic wavefunctions of the projectile and target electrons. These are described in the following subsections in more details.

9.2.2 Interaction Potential

The target-projectile interaction in Eq. (9.1) contains both, the Coulomb V^C as well as the Breit V^B interactions i.e.,

$$
\begin{aligned}
V_{in}(\mathbf{r}_i, \mathbf{r}_j) &= V^C(\mathbf{r}_i, \mathbf{r}_j) + V^B(\mathbf{r}_i, \mathbf{r}_j) \\
&= \sum_{i<j}\left(\frac{1}{r_{ij}} - \boldsymbol{\alpha}_i \cdot \boldsymbol{\alpha}_j \frac{1}{r_{ij}} + \frac{1}{2}(\boldsymbol{\alpha}_i \cdot \nabla_i)(\boldsymbol{\alpha}_j \cdot \nabla_j) r_{ij}\right).
\end{aligned}
\tag{9.3}
$$

The first term is the usual Coulomb interaction between the projectile and target electrons. The second and third terms represent the contributions from the magnetic interaction between electrons and classical retardation due to the finite propagation time of the interaction between the electrons. In the above expression, $\boldsymbol{\alpha}_i$ are the Dirac matrices. The reduction of scattering amplitude in terms of angular coefficients and radial integrals for Coulomb interaction has been discussed in more details in [21]. In order to include the Breit interaction to reduce the scattering amplitude as product of radial integrals and angular coefficients, one can follow the paper of Grant and Pyper [22].

9.2.3 Bound State Wavefunction

The bound state wavefunction of N-electron atom is taken as a Slater determinant of single electron Dirac-Fock orbitals as given by,

$$
\phi_{n\kappa m}(\mathbf{r}, \sigma) = \frac{1}{r}\begin{pmatrix} P_{n\kappa}(r)\xi_{\kappa m}(\hat{\mathbf{r}}, \sigma) \\ i Q_{n\kappa}(r)\xi_{-\kappa m}(\hat{\mathbf{r}}, \sigma) \end{pmatrix},
\tag{9.4}
$$

where $P_{n\kappa}$ and $Q_{n\kappa}$ are the large and small components of the radial wavefunctions and the spin-angular functions ξ are expressed as

$$
\xi_{\kappa m}(\hat{\mathbf{r}}, \sigma) = \sum_{\mu\nu}\left(l\mu\,\frac{1}{2}\nu\,\Big|\,jm\right) Y_{l\mu}(\hat{\mathbf{r}})\psi_{\frac{1}{2}\nu}(\sigma),
\tag{9.5}
$$

$$\xi_{-\kappa m}(\hat{\mathbf{r}}, \boldsymbol{\sigma}) = \sum_{\mu \nu} \left(\tilde{l}\mu \frac{1}{2} \nu \Big| jm \right) Y_{\tilde{l}\mu}(\hat{\mathbf{r}}) \psi_{\frac{1}{2}\nu}(\boldsymbol{\sigma}). \tag{9.6}$$

Here j is the total angular momentum of the electron, m is the z-component of j, $\tilde{l} = 2j - l$, $(l_1 m_1 l_2 m_2 | l_3 m_3)$ is a Clebsch-Gordan coefficient, the $Y_{lm}(\hat{\mathbf{r}})$ is spherical harmonic and $\psi_{\frac{1}{2}\nu}(\boldsymbol{\sigma})$ is a spinor basis function.

The bound state orbitals satisfy the orthogonal conditions

$$\int_0^\infty \left(P_{n\kappa}(r) P_{n'\kappa}(r) + Q_{n\kappa}(r) Q_{n'\kappa}(r) \right) dr = \delta_{nn'}, \tag{9.7}$$

and

$$\left\langle \xi_{\kappa m}(\hat{\mathbf{r}}, \boldsymbol{\sigma}) \Big| \xi_{\kappa'm'}(\hat{\mathbf{r}}, \boldsymbol{\sigma}) \right\rangle = \delta_{\kappa\kappa'} \delta_{mm'}. \tag{9.8}$$

The wavefunctions for target atom (ion) are obtained form the solution of Dirac equations using Dirac-Fock potential through GRASP92 code of Parpia *et al.* [23] or GRASP2K [24].

9.2.4 Distortion Potential

In order to obtain distorted waves, distortion potential U_d can be taken as the spherically symmetric static potential $V_{stat}(r)$ which is obtained from the charge distribution of initial and final states calculated from the atomic wavefunctions. In calculating the static potential the atomic wavefunctions are assumed not to be distorted by the projectile electron. In such calculations one can choose $V_{stat}(r)$ as the final excited-state potential of the atom or ion with nuclear charge Z and number of electrons N,

$$V_{stat}(r) = -\frac{Z-N}{r} + \sum_{j \in \text{all subshells}} \mathcal{N}_j \int_0^\infty [P_{n_j\kappa_j}^2(r_j) + Q_{n_j\kappa_j}^2(r_j)] \frac{1}{r_>} dr. \tag{9.9}$$

Here $\mathcal{N}_j$ is the occupation number of the jth subshells and the electron in it is represented by quantum number $n_j\kappa_j$. P and Q are the larger and smaller components of the radial wavefunctions of atomic (ionic) orbitals as discussed above.

9.2.5 Continuum Distorted Wavefunction

The projectile electron distorted-wave functions $F_{ch,\mu_{ch}}^{DW\pm}$ are expanded into relativistic partial waves as:

$$F_{ch,\mu_{ch}}^{DW\pm}(\mathbf{k}_{ch},\mathbf{r}) = \frac{1}{(2\pi)^{3/2}} \sum_{\kappa m} e^{\pm i\Delta_\kappa} a_{ch,\kappa m}^{\mu_{ch}}(\hat{\mathbf{k}}_{ch}) \frac{1}{r} \left(\begin{array}{c} f_\kappa(r)\chi_{\kappa m}(\hat{\mathbf{r}}) \\ i g_\kappa(r)\chi_{-\kappa m}(\hat{\mathbf{r}}) \end{array} \right), \quad (9.10)$$

with

$$a_{ch,\kappa m}^{\mu_{ch}}(\hat{\mathbf{k}}_{ch}) = 4\pi i^l \left[\frac{E_{ch}+c^2}{2E_{ch}} \right]^{\frac{1}{2}} \sum_{m_l} (l m_l 1/2\mu_{ch}|jm) Y_{l m_l}^*(\hat{\mathbf{k}}_{ch}). \quad (9.11)$$

For sake of convenience here we write 'ch' instead of '$i(f)$' to refer to the two channels i.e., initial 'i' and final 'f'. Δ_κ is the phase (Coulomb + scattering) shift. f_κ and g_κ are the large and small components of the radial wavefunctions and the $\chi_{\pm\kappa m}$ are the spinor spherical harmonics as given by Eqs. (9.5) and (9.6).

The large and small components f_κ and g_κ of the continuum wave functions satisfy the following coupled Dirac equations,

$$\left(\frac{d}{dr} + \frac{k}{r} \right) f_\kappa(r) - \frac{1}{c}(c^2 - U_d + E_{ch})g_\kappa(r) - \frac{1}{cr}W_Q(\kappa;r) = 0, \quad (9.12)$$

$$\left(\frac{d}{dr} - \frac{k}{r} \right) g_\kappa(r) + \frac{1}{c}(-c^2 - U_d + E_{ch})f_\kappa(r) + \frac{1}{cr}W_P(\kappa;r) = 0. \quad (9.13)$$

Here W_P and W_Q are non local exchange kernels which result due to exchange of projectile electron with bound electron. These are described in details by Zuo [25]. However, by dropping the exchange terms W_P and W_Q the solutions of the two equations (9.12) and (9.13) can be simplified and the effect of exchange of electrons can be incorporated by adding a local exchange potential [26] to the distortion potential (Eq. (9.9)). Such simplification has been widely used in the literature for studying electron impact excitation of ions.

These coupled differential equations (9.12) and (9.13) are solved numerically by subjecting to the following boundary conditions;

$$f_\kappa(r) \xrightarrow{r\to\infty} \frac{1}{k} \sin\left(k_{ch}r - \frac{l\pi}{2} - \eta \ln 2k_{ch}r + \Delta_\kappa \right), \quad (9.14)$$

$$g_\kappa(r) \xrightarrow{r\to\infty} \frac{c}{c^2 + E_{ch}} \cos\left(k_{ch}r - \frac{l\pi}{2} - \eta \ln 2k_{ch}r + \Delta_\kappa \right), \quad (9.15)$$

where $\eta = Z/v$ is the Sommerfeld's parameter with v as the velocity of the electron.

After one has obtained the wavefunctions for the target atom (ion) as well as projectile electrons, the scattering amplitudes (9.1) can be evaluated [21]. These amplitudes can be utilized to obtain various scattering parameters *viz.* cross sections (DCS and ICS), spin S, T and U parameters, polarization of the characteristic emission in terms of differential and integrated Stokes parameters [7].

9.3 Electron Excitation of Neutral Atoms

In recent years, the attention has been paid to the study of electron-impact excitation of inert gas atoms because their excitation cross sections are important for optical emission based diagnostics of rare gases plasmas. These are widely used in laboratory and astrophysical plasma studies and industrial applications. There have been many theoretical and experimental investigations of electron collisions with the rare gases that have reported results for both differential cross sections (DCS) and integrated cross sections (ICS) [3, 27]. Most of these studies have only considered excitation from the ground (np^6) state to the excited $np^5(n+1)s$ and $np^5(n+1)p$ states in Ne $(n=2)$, Ar $(n=3)$, Kr $(n=4)$ and Xe $(n=5)$. However, metastable states of the noble gases are significantly populated in low temperature plasmas and consequently, transitions out of these levels to higher states play an important role. Two factors contribute to the role of metastables in such plasma. First, the peak value of the excitation cross sections from the metastable levels which can be up to three orders of magnitude larger than the peak excitation cross sections from the ground state. Second, metastable atoms are already in an excited state, so it takes only few eV energy to excite a metastable into a higher level or to ionize them. Hence, the success of many forms of plasma modeling and plasma diagnostics requires starting with accurate electron-atom cross section data [3, 27, 28]. Thus, a thorough understanding of the fundamental physical nature of excitation from metastable states in noble gases is important even if their number density is small relative to the ground state atoms. In this connection, RDW calculations have been performed to obtain the integral cross sections for the fine-structure transitions in noble gases from their two lowest $J=0$ and $J=2$ metastable states with configuration $np^5(n+1)s$ to the higher lying states of $np^5(n+1)p$ and $np^5(n+2)p$ configurations [29–31]. The other calculations which are available for these transitions are either non-relativistic or semi-relativistic.

There are several more fine-structure transitions of higher lying levels for which excitation cross sections are needed for plasma modeling. These cross sections have been calculated for the excitation of the np^5nd, $np^5(n+2)s$ and $np^5(n+2)p$ manifolds from the ground level of the Ar and Kr atoms [32, 33]. In these calculations of the ICS as well as DCS results for all the earlier said excitations are reported and compared with the previous non-relativistic and semi-relativistic calculations as well as measurements where available. As an illustration Fig. 9.1 displays the ICS results for the excitation of the $3p^55p$ manifold of Ar from the ground state. These cross sections show the expected energy dependence associated with the total angular momentum J_f values of the excited levels. Since the initial state is the closed shell ground state of Ar, we have $J_i=0$. According to the angular momentum coupling of the final state, the direct term of the T-matrix is non-zero only if J_f is odd for the odd parity ($3d$ and $5s$) states while J_f must be even for the even parity ($5p$) state. In the other cases, only the exchange terms of the T-matrix contribute to the cross sections. For exchange transitions, the ICS drop off more rapidly with energy than those for which the direct term is non-zero. Generally, the differential cross section for exchange transitions are relatively flat in the forward scattering region

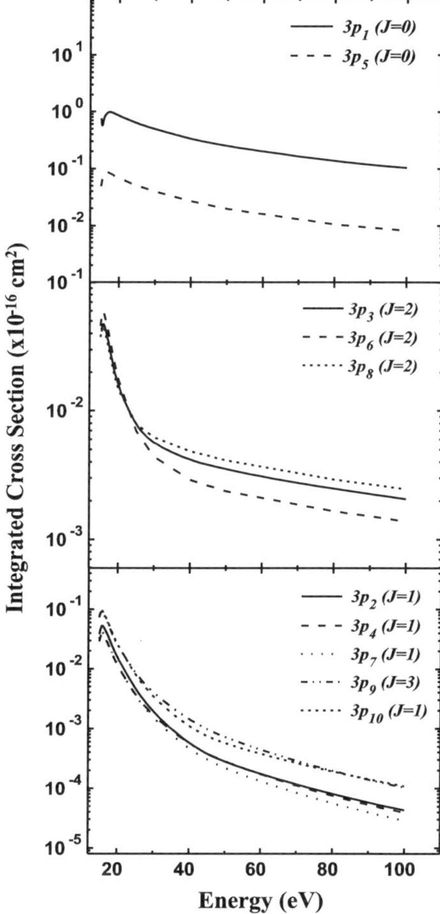

Fig. 9.1 Integrated cross sections in units of 10^{-16} cm^2 for the $3p^5 5p$ excitations from the ground state of Ar atom

whereas the direct transitions show a highly peaked DCS in this region. The transitions for which the direct term is non-zero, the J_f value indicates the multipole moment of the interelectronic interaction which gives rise to this term. For the odd parity levels, the transitions to the ones with total angular momentum $J_f = 1$ are (optically allowed) dipole transitions and are expected to fall off more slowly with increasing impact energy than the other transitions. In the other cases for both even and odd parity levels, the larger the J_f value, the faster the expected decrease in cross sections as the energy increases. Thus the cross sections for excitation to the allowed $J_f = 0$ levels decrease more slowly than the ones for the $J_f = 2$ levels while the cross sections for the forbidden transitions to levels with odd values of J_f fall off most rapidly.

In Figs. 9.2(a)–9.2(f) RDW calculations [32] for ICS results in the range of 20 to 300 eV for all the transitions to the $4p^5 6s$, $4p^5 5p$ and $4p^5 4d$ manifolds of krypton from the ground state have been shown. The general behavior as a function of the

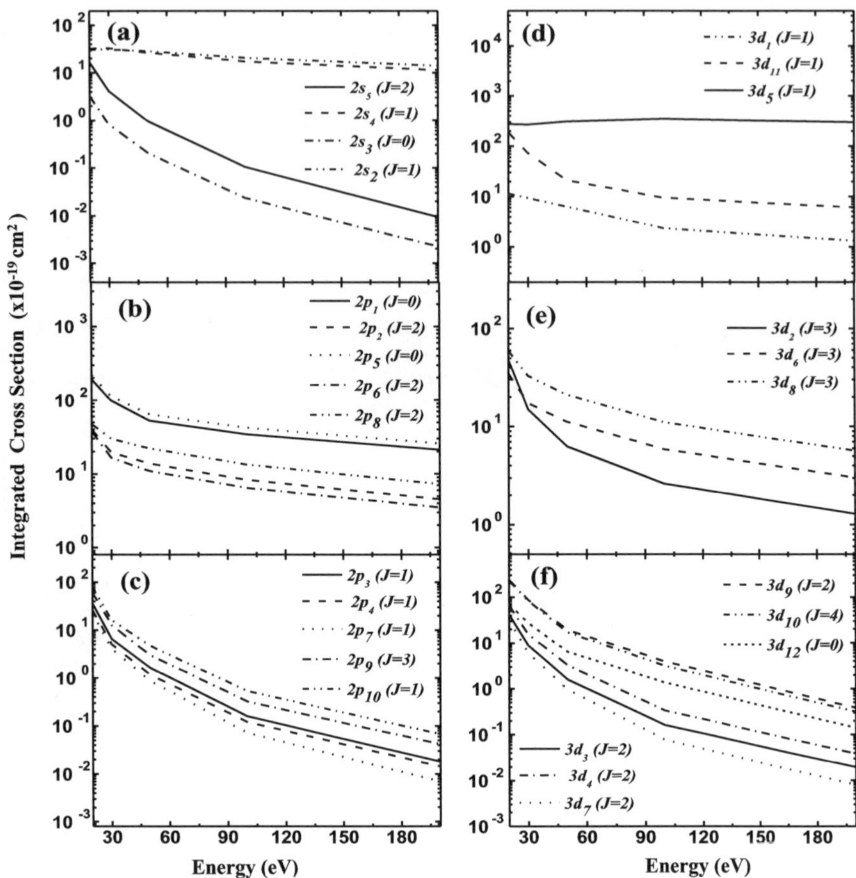

Fig. 9.2 Integrated cross sections in units of 10^{-19} cm^2 for excitation of the $4p^56s$, $4p^55p$ and $4p^54d$ manifolds from the ground state of Kr atom

parity and J_f value which was observed in Ar in Fig. 9.1 is also present in the calculated ICS results of Kr. For example, the ICS for pure exchange transitions drop off more rapidly with energy than those for which the direct T-matrix is non-zero. For forbidden transitions which have a non-zero direct contribution to the T-matrix (J_f odd and not equal to unity for the $4p^54d$ and $4p^56s$ manifolds or J_f even for the $4p^55p$ manifold) the cross sections decrease as $1/E$ with increasing energy E of the incident electron while for the purely exchange transitions (J_f even for the $4p^54d$ and $4p^56s$ manifolds or J_f odd for the $4p^55p$ manifold) the cross sections fall off like E^{-3}. The largest deviations from this pattern occur for the $4p^55p$ levels with $J_f = 0$.

The applicability of the RDW method has also been extended to the excitation of open-shell atoms [34, 35]. Calculations for the electron impact fine-structure transitions of atomic oxygen from its ground $2p^4$ ^{3}P$_{0,1,2}$ state to the excited $2p^33s$

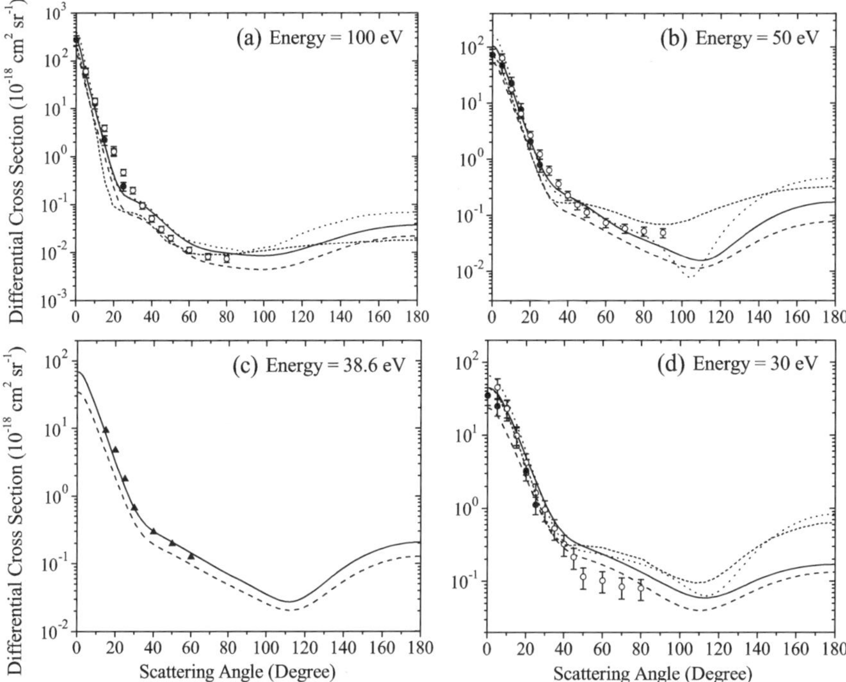

Fig. 9.3 DCS for the $2p^4\ ^3$P–$2p^3 3s\ ^3$S excitation at a projectile energy of: (**a**) 100 eV, (**b**) 50 eV, (**c**) 38.6 eV, (**d**) 30 eV. The *solid line*, RDW-MCGS; *dashed line*, RDW-SCGS; *dotted line*, *R*-matrix calculations of Kanik *et al.* [36]; *short dashed line*: CCO calculations of Wang and Zhou [39]; *solid circles with error bars* from Kanik *et al.* [36]; *solid triangles* from Doering and Yang [37]; *open circles with error bars* from Vaughan and Doering [38]

^{3}S$_1$, ^{3}P$_{0,1,2}$ and ^{3}D$_{1,2,3}$ states and to the $2p^3 3d\ ^3$D$_{1,2,3}$ state have been reported in the light of experimental cross section data published by Jet Propulsion Laboratory, USA [34]. Differential cross section results have been obtained in the energy range from 15 to 100 eV. In order to examine the effect of multi-configuration interaction in the representation of the atomic target states, two types of calculations, single-configuration ground state (SCGS) and multi-configuration ground state (MCGS) calculations corresponding to two different choices of the bound state wavefunctions have been reported [34]. Results for the differential cross sections are compared with both experimental measurements and close coupling and *R*-matrix calculations available for fine-structure unresolved transitions. As an example, Fig. 9.3 shows the DCS for $2p^4\ ^3$P–$2p^3 3s\ ^3$S excitation in the energy range from 30 to 100 eV. One can see from this figure that the MCGS results agree very well with the experimental data sets [36–38] as well as theoretical results from *R*-matrix [36] and momentum-space coupled-channels-optical (CCO) methods [39] at all impact energies.

Further, RDW calculations have been carried out for the electron impact excitation of the $6s\ ^2$S$_{1/2}$ state in open shell indium from the $5p\ ^2$P$_{1/2}$ ground state [35].

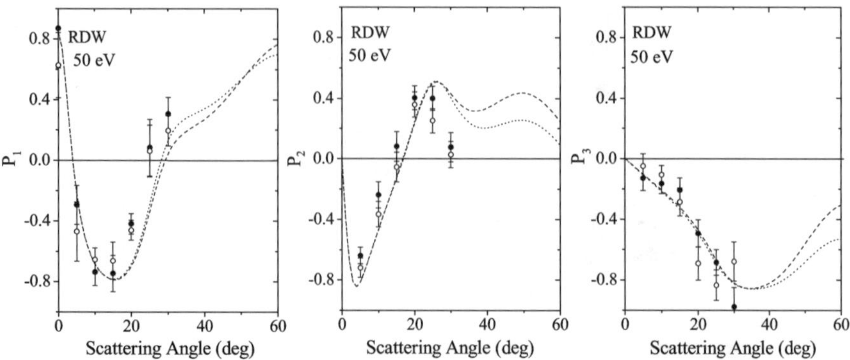

Fig. 9.4 Spin-resolved Stokes parameters P_1, P_2, and P_3, for the transition $6s6p$ ^{1}P$_1$ to $6s^2$ ^{1}S$_0$ in mercury after spin-polarized electron impact excitation at 50 eV. The diagrams show RDW calculations (*dashed lines*: spin-up, *dotted lines*: spin-down) and the experimental results for spin-up (•) and spin-down (o) [41]

Differential cross sections (DCS) are obtained at incident electron energies from 10 to 100 eV and scattering angles up to 10^0. The results are compared with recently reported experimental data [40]. The comparison suggests that the available relative measurements [40] at the smallest scattering angles are too low which resulted in an incorrect normalization.

The electron-photon $(e, e\gamma)$ coincidence studies for the transition $6s6p$ ^{1}P$_1$ and $6s6p$ ^{3}P$_1$ to $6s^2$ ^{1}S$_0$ in mercury atoms with spin polarized electrons have also been considered. This work has been done in the light of recent measurements for Stokes parameters P_i $(i = 1, 4)$ by Münster group, Germany [10, 41–43]. For any process, which involves spin-polarized electrons and where the spin-orbit interaction is important, the Dirac equations are the natural way to describe the system. The Dirac equations explicitly involve the spin of the electrons of the system and include all one-particle spin-orbit interactions. The RDW calculations have been carried out to obtain differential Stokes parameters P_i $(i = 1, 4)$ and total degree of polarization P_{tot} at electron impact energies up to 100 eV. The results have been compared with the experimental data and theoretical calculations based on a five-state Breit-Pauli R-matrix approach and convergent close-coupling model. For example, Fig. 9.4 shows the comparison of RDW results with experiment for spin-resolved differential Stokes parameters P_1, P_2, and P_3 for the transition $6s6p$ ^{1}P$_1$ to $6s^2$ ^{1}S$_0$ in mercury due to spin-polarized electron impact excitation at 50 eV. The RDW calculations show good agreement with the measurements [41]. Integrated Stokes parameters for both the $6s6p$ ^{1}P$_1$ and $6s6p$ ^{3}P$_1$ excitations in Hg have been performed and compared with the recent measurements from Münster group and Dirac B-spline R-matrix method [10]. The agreement has been found to be excellent.

There are further RDW calculations for other heavier atoms *viz.* Yb where partial differential cross sections for electron-impact excitation of Yb atoms from the laser-excited $6s6p$ ^{3}P$_1$ level to the $6s7s$ ^{3}S$_1$, $6s6p$ ^{1}P$_1$ and $6s5d$ ^{3}D$_{1,2,3}$ levels have been reported. P_3 collision Stokes parameter has also been calculated for the collisional de-excitation from the higher lying $^1P_1$, $^3D_{1,2,3}$ and $^3S_1$ levels to the $^3P_1$ level for 20 eV electron-impact energy [44, 45].

9.4 Electron Excitation of Ions

Besides the applications of relativistic methods to excitation of neutral atoms, a number of relativistic calculations have been performed to consider electron impact excitation of ions. Recently, for instance, Sampson *et al.* [17] presented a review of their fully relativistic approach to calculate atomic cross section data for highly charged ions, covering research work of twenty years. They also discussed the importance of generalized Breit interaction in excitation between magnetic sublevels of highly charged ions and hence, on the polarization of the subsequently emitted photons. Dong and co-workers applied RDW theory to study linear polarization of the characteristic emission [46, 47]. The cross sections for inner-shell electron-impact excitation to the magnetic sublevels $M_f = 0$ and $M_f = \pm 1$ of the excited state $1s2s^2 2p_{1/2}\ J = 1$ for highly charged beryllium-like Mo^{38+}, Nd^{56+} and Bi^{79+} ions are calculated by using a fully relativistic distorted-wave (RDW) method. These cross sections are employed in calculating the degree of linear polarization for the corresponding X-ray emission. The influence of Breit interactions on the cross sections and the degree of linear polarizations have been analyzed. Here we take one such example of our RDW calculations for the polarization of Lyman-α_1 emission line of Ar^{17+}, Ti^{21+} and Fe^{25+} ions and presented these results in Table 9.1. We performed two sets of calculations with including and without including Breit potential terms in the interaction potential equation (9.3). Our calculations for polarization with Coulomb and Breit interactions are in good agreement with previous relativistic convergent close coupling calculations of Bostock *et al.* [12] and the experimental results reported by Robbins *et al.* [48]. Thus, our results also confirm that account of Breit relativistic corrections is necessary in the calculation to provide good agreement with experiment.

The relativistic results for cross sections as well as for the degree of linear polarization as mentioned above have been obtained when the projectile electron remains unobserved after scattering. However, it would be interesting to see the behavior of the cross sections with respect to the angle of the scattered electron. A number of measurements have been performed on excitation of singly charged metal ions *viz.* Mg^+, Ca^+, Zn^+, Cd^+, Ba^+ for DCS, ICS and polarization parameters. For these metal ions, the available calculations are based on simple non-relativistic theories only [49–51]. Therefore, to analyze the quality of these data, improved relativistic calculations for the excitation cross sections are needed for singly-ionized metal ions which play an important role in understanding heating and radiation mechanisms in a wide range of high electron-temperature plasmas. In this connection, RDW calculations have been performed for differential and integrated cross sections for the resonance transitions $ns_{1/2} - np_{1/2}$ and $ns_{1/2} - np_{3/2}$ in Mg^+ ($n = 3$), Ca^+ ($n = 4$), Zn^+ ($n = 4$), Cd^+ ($n = 5$) and Ba^+ ($n = 6$) in the electron energy range

Table 9.1 Polarization for Lyman-α_1 emission line of Ar^{17+}, Ti^{21+} and Fe^{25+} ions

Element	Energy (keV)		RDW	RCCC [12]	Experiment [48]
Ar^{17+}	30	Coulomb	0.070	0.070	-0.019 ± 0.025
		Breit	0.034	0.031	
	84	Coulomb	-0.037	-0.056	-0.099 ± 0.045
		Breit	-0.100	-0.144	
Ti^{21+}	10.6	Coulomb	0.290	0.296	0.214 ± 0.066
		Breit	0.273	0.283	
	24.7	Coulomb	0.159	0.161	0.085 ± 0.081
		Breit	0.124	0.127	
	49.6	Coulomb	0.062	0.062	0.050 ± 0.082
		Breit	0.007	-0.0003	
Fe^{25+}	30	Coulomb	0.189	0.186	0.071 ± 0.034^1
		Breit	0.141	0.145	0.051 ± 0.011^2
	120	Coulomb	0.010	0.0095	-0.236 ± 0.109^1
		Breit	-0.083	-0.114	-0.217 ± 0.045^2

[1]Polarization measurements were obtained using the two crystal spectrometer method

[2]Polarization measurements were obtained using the one-crystal method

up to 300 eV [52]. In addition to the cross sections, calculations are performed also on the polarization of the photon emission following the decay of the $np_{3/2}$ state back into the $ns_{1/2}$ ground state in these ions. The results are also compared with the available experimental and other non-relativistic theoretical data. For example, Fig. 9.5 shows our DCS results for fine-structure resolved as well as unresolved $4s-4p$ transition in Zn^+ at 50 eV and 75 eV. Since Zn^+ can be said to be relatively heavier ion we see that relativistic effects play significant role. Consequently, ratio of DCS for $4s_{1/2}-4p_{1/2}$ and $4s_{1/2}-4p_{3/2}$ is not nearly two for Zn^+ as has been seen for Mg^+ and Ca^+ [52]. In Fig. 9.5(a), we also compare our DCS results for $4s-4p$ transition at 50 eV with the relative measurements of Williams *et al.* [53]. These experimental data were normalized to five-state close coupling (CC) calculations [51] at $\theta = 12°$. The results are in good agreement with the experimental values in the small scattering angle range. Figure 9.5(b) shows comparison of the relativistic calculations at 75 eV with the experimental data of Chutjian *et al.* [54] and a good agreement with the experimental data near forward angles can be observed.

9.5 Application of RDW Cross Sections to CR Plasma Model

Noble gases are important constituents of plasma. Since in low temperature plasmas electron impact excitation of atoms is one the most dominant processes, success of plasma diagnostics rely on the excitation cross sections. Hence, there is

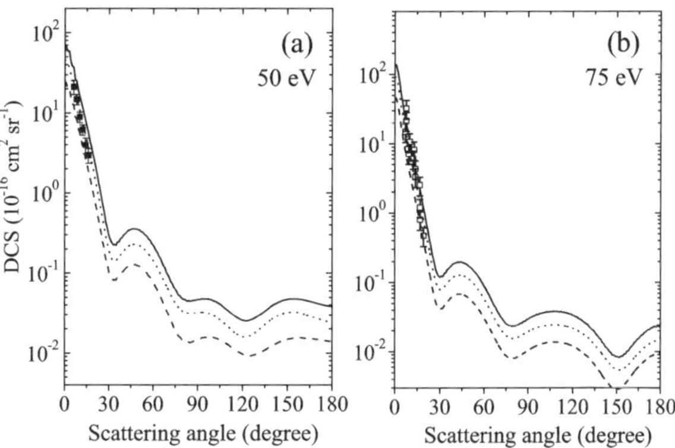

Fig. 9.5 RDW DCS results for electron impact excitation of Zn$^+$ at: (**a**) 50 eV and (**b**) 75 eV for $4s$–$4p$ transition. *Solid curve*, summed DCS for $4s$–$4p$; *dashed curves*, DCS for $4s_{1/2}$–$4p_{1/2}$; *dotted curve*, RDW DCS for $4s_{1/2}$–$4p_{3/2}$; *squares with error bars*, Williams *et al.* [53]; *open squares with error bars*, Chutjian *et al.* [54]

a great demand of accurate cross sections for electron-inert gas excitations. These cross sections can be implemented in collisional radiative (CR) model which further combined with optical emission spectroscopy measurements yield various plasma parameters.

There are several recently published CR models for low temperature Ar discharges [55–57]. These models have used cross section data either from the few available experimental measurements or theoretical data calculated by empirical formula from Drawin [58, 59] or simple classical methods [60]. The role and effect of the available calculated RDW cross sections for the excitations of the $5p^5 6p$ ($2p$ in Paschen notation) fine-structure levels in Xe atom from its metastable $5p^5 6s$ ($1s_5$ and $1s_3$ in Paschen notation) levels has been recently tested through a CR model to describe the plasma of Xe fed thrusters successfully [61]. Recently the applicability of the calculated RDW cross sections is further tested for argon by implementing in a simple collisional radiative model of argon for low temperature plasma discharges. For this purpose a collisional radiative model has been developed for low temperature Ar-plasma which incorporates the calculated RDW cross sections [62]. Excitation cross sections from the two $3p^5 4s$ $J = 1$ resonance levels, $1s_2$ and $1s_4$, to the higher lying $2p$ fine-structure manifold as well as for transitions among individual levels of the $1s$ and $2p$ manifolds are also calculated and included in the present model which were not fully considered in any earlier model. The calculations have been performed for the population densities of the $1s$ and $2p$ levels and these are compared with recent optical emission spectroscopic measurements [57]. The variation of population densities of all the $1s$ and $2p$ levels with electron temperature and density are presented. As an example, Fig. 9.6 shows the simulated argon emission spectra for the decay from $2p$ to $1s$ fine-structure levels at the electron temperature 1.5 eV and line width 0.5 Å. Calculations are also performed for the intensities of the 750.38 nm ($2p_1 \rightarrow 1s_2$) and 696.54 nm ($2p_2 \rightarrow 1s_5$) lines [62] and compared with recently reported experimental results Palmero *et al.* [55]. This work suggests that the inclusion of a complete fine-structure description of the electronic processes occurring in the plasma is important for the success of a collisional radiative model.

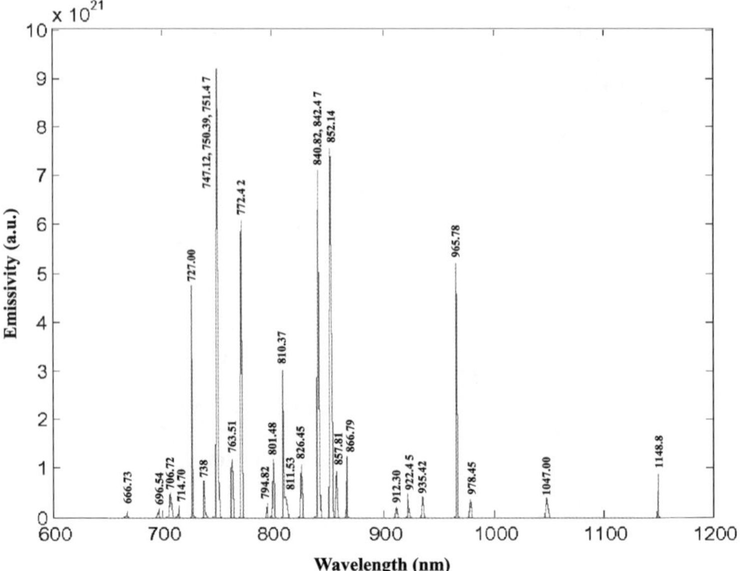

Fig. 9.6 Calculated emission spectra for the decay from $2p$ levels to $1s$ levels of argon using CR model

9.6 Conclusions

In this chapter we have given a review of the recent applications of relativistic methods to study electron impact excitation of atoms and ions. We further discussed implementing of the RDW cross sections of Ar to CR model for Ar-plasma diagnostic and demonstrated the importance of reliable cross sections. In the light of the great demand of excitation cross sections of ions for plasma modeling purposes, it would be worth to apply RDW as well as other relativistic theories for excitation of inert gas ions as a future work.

Acknowledgements We thank Prof. A.D. Stauffer and our other co-workers for their collaborations in using the distorted wave theory presented here which resulted in the form of several joint publications. R.S. acknowledges research grants in support of this work from the Council of Scientific and Industrial Research (CSIR), New Delhi, and IAEA, Vienna.

References

1. R.O. Jung, J.B. Boffard, L.W. Anderson, C.C. Lin, Phys. Rev. A **80**, 062708 (2009)
2. R.O. Jung, J.B. Boffard, L.W. Anderson, C.C. Lin, Phys. Rev. A **75**, 052607 (2007)
3. J.B. Boffard, C.C. Lin, C.A. DeJoseph, J. Phys. D **37**, R143 (2004)
4. B. Bederson, Comments At. Mol. Phys. **2**, 160 (1970)
5. B. Bederson, Comments At. Mol. Phys. **1**, 41 (1969)
6. B. Bederson, Comments At. Mol. Phys. **1**, 65 (1969)

7. N. Andersen, K. Bartschat, *Polarization, Alignment, and Orientation in Atomic Collisions* (Springer, Berlin, 2001)
8. I.P. Grant, J. Phys. B **41**, 055002 (2008)
9. O. Zatsarinny, K. Bartschat, Phys. Rev. A **79**, 042713 (2009)
10. F. Jüttemann, G.F. Hanne, O. Zatsarinny, K. Bartschat, R. Srivastava, R.K. Gangwar, A.D. Stauffer, Phys. Rev. A **81**, 012705 (2010)
11. N.R. Badnell, J. Phys. B **41**, 175202 (2008)
12. C.J. Bostock, D.V. Fursa, I. Bray, Phys. Rev. A **80**, 052708 (2009)
13. D.V. Fursa, C.J. Bostock, I. Bray, Phys. Rev. A **80**, 022717 (2009)
14. C.J. Bostock, D.V. Fursa, I. Bray, Phys. Rev. A **82**, 022713 (2010)
15. C.J. Bostock, J. Phys. B **44**, 083001 (2011)
16. M.J. Berrington, C.J. Bostock, D.V. Fursa, I. Bray, Phys. Rev. A **85**, 042708 (2012)
17. D.H. Sampson, H.L. Zhang, C.J. Fontes, Phys. Rep. **477**, 111 (2009)
18. R. Srivastava, L. Sharma, in *Atomic and Molecular Physics: Introduction to Advanced Topics*, ed. by R. Srivastava, R. Choubisa (Narosa, New Delhi, 2012), p. 41
19. D.H. Madision, K. Bartschat, The distorted-wave method for elastic scattering and atomic excitation, in *Computational Atomic Physics: Electron and Positron Collision with Atoms and Ions*, ed. by K. Bartschat (Springer, Berlin, 1996)
20. K. Muktavat, R. Srivastava, A.D. Stauffer, J. Phys. B **36**, 2341 (2003)
21. L. Sharma Ph.D. Thesis, IIT Roorkee, Roorkee, 2008
22. I.P. Grant, N.C. Pyper, J. Phys. B **9**, 761 (1976)
23. F.A. Parpia, C.F. Fischer, I.P. Grant, Comput. Phys. Commun. **94**, 249 (1996)
24. P. Jönsson, X. He, C. Froese Fischer, I.P. Grant, Comput. Phys. Commun. **177**, 597 (2007)
25. T. Zuo, Ph.D. Thesis, York University, 1991
26. F.A. Gianturco, S. Scialla, J. Phys. B **20**, 3171 (1987)
27. E. Gargioni, B. Grosswendt, Rev. Mod. Phys. **80**, 451 (2008)
28. Y.-H. Chiu, B.L. Austin, S. Williams, R.A. Dressler, G.F. Karabadzhak, J. Appl. Phys. **99**, 113304 (2006)
29. R. Srivastava, A.D. Stauffer, L. Sharma, Phys. Rev. A **74**, 012715 (2006)
30. L. Sharma, R. Srivastava, A.D. Stauffer, Phys. Rev. A **79**, 024701 (2007)
31. L. Sharma, R. Srivastava, A.D. Stauffer, Eur. Phys. J. D **62**, 399 (2011)
32. R.K. Gangwar, L. Sharma, R. Srivastava, A.D. Stauffer, Phys. Rev. A **81**, 052707 (2010)
33. R.K. Gangwar, L. Sharma, R. Srivastava, A.D. Stauffer, Phys. Rev. A **82**, 032710 (2010)
34. L. Sharma, R. Srivastava, A.D. Stauffer, J. Phys. B **40**, 3025 (2007)
35. T. Das, R. Srivastava, A.D. Stauffer, Phys. Lett. A **375**, 568 (2011)
36. I. Kanik, P.V. Johnson, M.B. Das, M.A. Khakoo, S.S. Tayal, J. Phys. B **34**, 2647 (2001)
37. J.P. Doering, J. Yang, J. Geophys. Res. **106**, 203 (2001)
38. S.O. Vaughan, J.P. Doering, J. Geophys. Res. **92**, 7749 (1987)
39. Y. Wang, Y. Zhou, J. Phys. B **39**, 3009 (2006)
40. M.S. Rabasović, S.D. Tosić, D. Šević, V. Pejčev, D.M. Filipović, B.P. Marinković, Nucl. Instrum. Methods B **267**, 279 (2009)
41. G. Außendorf, F. Jüttemann, K. Muktavat, L. Sharma, R. Srivastava, A.D. Stauffer, K. Bartschat, D.V. Fursa, I. Bray, G.F. Hanne, J. Phys. B **39**, 2403 (2006)
42. G. Außendorf, F. Jüttemann, K. Muktavat, L. Sharma, R. Srivastava, A.D. Stauffer, K. Bartschat, D.V. Fursa, I. Bray, G.F. Hanne, J. Phys. B **39**, 4435 (2006)
43. R. Srivastava, R.K. Gangwar, A.D. Stauffer, Phys. Rev. A **80**, 022718 (2009)
44. J.D. Hein, S. Kidwai, P.W. Zetner, C. Bostock, D.V. Fursa, I. Bray, L. Sharma, R. Srivastava, A.D. Stauffer, J. Phys. B **44**, 015202 (2011)
45. J.D. Hein, S. Kidwai, P.W. Zetner, C. Bostock, D.V. Fursa, I. Bray, L. Sharma, R. Srivastava, A.D. Stauffer, J. Phys. B **44**, 075201 (2011)
46. J. Jiang, C.Z. Dong, I.Y. Xie, J.G. Wang, J. Yan, S. Fritzsche, Chin. Phys. Lett. **24**, 691 (2007)
47. Z.W. Wu, J. Jiang, C.Z. Dong, Phys. Rev. A **84**, 032713 (2011)

48. D.L. Robbins, P. Beiersdorfer, A.Ya. Faenov, T.A. Pikuz, D.B. Thorn, H. Chen, K.J. Reed, A.J. Smith, G.V. Brown, R.L. Kelley, C.A. Kilbouren, F.S. Porter, Phys. Rev. A **74**, 022713 (2006)
49. A.W. Pangantiwar, R. Srivastava, J. Phys. B **21**, L219 (1988)
50. M.S. Pindzola, N.R. Badnell, R.J.W. Henry, D.C. Griffin, W.L. van Wyngaarden, Phys. Rev. A **44**, 5628 (1991)
51. I.D. Williams, A. Chutjian, A.Z. Msezane, R.J.W. Henry, Astrophys. J. **299**, 1063 (1985)
52. L. Sharma, A. Surzhykov, R. Srivastava, S. Fritzsche, Phys. Rev. A **83**, 062701 (2011)
53. I.D. Williams, A. Chutjian, R.J. Mawhorter, J. Phys. B **19**, 2189 (1986)
54. A. Chutjian, A.Z. Msezane, R.J.W. Henry, Phys. Rev. Lett. **50**, 1357 (1983)
55. A. Palmero, E.D. Hattum, H. Rudolph, F.H.P.M. Habraken, J. Phys. D, Appl. Phys. **101**, 053306 (2007)
56. Y. Yi-Qing, Y. Xin, N.Z. Yuan, Chin. Phys. B **20**, 015207 (2011)
57. X.-M. Zhu, Y.K. Pu, J. Phys. D, Appl. Phys. **43**, 015204 (2010)
58. J. Vlcek, J. Phys. D, Appl. Phys. **22**, 623 (1989)
59. A. Bultel, B.V. Ootegem, A. Bourdon, P. Vervisch, Phys. Rev. E **65**, 046406 (2002)
60. A. Hartgers, J.V. Dijk, J. Jonkers, J.A.M. van der Mullen, Comput. Phys. Commun. **135**, 199 (2001)
61. R.A. Dressler, Y. Chiu, O. Zatsarinny, K. Bartschat, R. Srivastava, L. Sharma, J. Phys. D **42**, 185203 (2009)
62. R.K. Gangwar, L. Sharma, R. Srivastava, A.D. Stauffer, J. Appl. Phys. **111**, 053307 (2012)

Chapter 10
Formation of a Narrow Group of Intense Lines in the Emission and Photoexcitation Spectra

R. Karazija, S. Kučas, V. Jonauskas, and A. Momkauskaitė

Abstract Formation of a narrow group of intense lines in the emission and photoexcitation spectra, corresponding to the transitions between the electron shells with the same principal quantum number is considered. There are two reasons of such an effect, namely, the strong Coulomb exchange interaction and the configuration interaction with a symmetric exchange of symmetry. The conditions and regularities of their manifestation in the isoelectronic and isonuclear sequences are analyzed.

10.1 Introduction

The excited configurations of atoms, containing two open shells with the same principal quantum number, are distinguished by some particular features. The radial orbitals of such neighboring shells overlap considerably, thus, the Coulomb interaction between these shells remains more important than the spin-orbit interaction even in the highly charged ions of heavy elements (namely, such a case will be considered in our paper). The exchange part of the Coulomb interaction depends more on the overlap of orbitals than the direct interaction, thus, the strong exchange interaction mainly forms the structure of energy level spectrum and determines its rather large width.

Configuration of such a type is obtained by the dipole excitation from the closed shell nl^{4l+2} to the outer open shell $n(l+1)^N$. The two different cases can be distinguished:

R. Karazija (✉) · S. Kučas · V. Jonauskas · A. Momkauskaitė
Institute of Theoretical Physics and Astronomy, Vilnius University, A. Goštauto 12, 01108 Vilnius, Lithuania
e-mail: romualdas.karazija@tfai.vu.lt

S. Kučas
e-mail: sigitas.kucas@tfai.vu.lt

V. Jonauskas
e-mail: valdas.jonauskas@tfai.vu.lt

A. Momkauskaitė
e-mail: alina.momkauskaite@tfai.vu.lt

M. Mohan (ed.), *New Trends in Atomic and Molecular Physics*,
Springer Series on Atomic, Optical, and Plasma Physics 76,
DOI 10.1007/978-3-642-38167-6_10, © Springer-Verlag Berlin Heidelberg 2013

(i) If $l = n - 2$, the other excited configuration $nl^{4l+2}n(l + 1)^{N-1}n(l + 2)$, belonging to the same complex, does not exist. Then usually the main features of transition array

$$nl^{4l+2}n(l + 1)^{N} - nl^{4l+1}n(l + 1)^{N+1} \tag{10.1}$$

can be described by the single-configuration approximation.

(ii) If $l \leq n - 3$, the $nl^{4l+2}n(l + 1)^{N-1}n(l + 2)$ configuration is possible and it strongly interacts with the $nl^{4l+1}n(l + 1)^{N+1}$ configuration. In the second configuration one electron is filling the vacancy in a deeper shell with a smaller orbital quantum number and the other electron is excited to an empty shell with a larger orbital quantum number, thus, this configuration interaction (CI) is named as the interaction with a symmetric exchange of symmetry (SEOS) [1]. Both these excited configurations are related by the electric dipole transitions to the ground configuration; thus, their interaction strongly influences the transition array:

$$nl^{4l+2}n(l + 1)^{N} - \left(nl^{4l+1}n(l + 1)^{N+1} + nl^{4l+2}n(l + 1)^{N-1}n(l + 2) \right). \tag{10.2}$$

The short notation n–n for such transitions is used.

The first striking manifestation of the Coulomb exchange interaction influence on the photoabsorption spectra was obtained in the late sixties. The registered 4d photoabsorption spectra of lanthanide metals are dominated by one broad peak instead of usual for the photoionization from an inner shell saw-tooth shape [2]. These spectra were interpreted in a free-ion model as the giant resonances corresponding to the excitation from the $4d^{10}$ shell to the open $4f^{N}$ shell [3–5]. The resonance is formed by a few intense overlapping lines, concentrated on a short-wavelength side of the spectrum. The 4d spectra of free atoms also contain the similar giant resonances [6], though their interpretation at the beginning of the group is complicated by the problem of the collapse of $4f$ orbital [7, 8]. It was suggested in [3, 9] that the strongest photoexcitation lines correspond to the transitions into the upper group of levels, formed by the Coulomb exchange interaction. Cancellation of many lines on the low-energy side of the spectrum, corresponding to transitions (10.1), was related to the existence of additional selection rule for a number of vacancy-electron pairs in the special wavefunction basis [10]. The giant resonances are also typical of the photoabsorption from the $3d^{10}$ shell in the vapors of elements from the second part of the iron group [11, 12]. However, the single-configuration model becomes inaccurate for the $3p$ photoabsorption spectra of the elements from the first part of this group: due to the near-degeneracy of 3d and 4s orbitals and strong $(s + d)^{N}$ mixing, they contain a large number of sharp resonances also on the high-energy side.

The main feature of the photoexcitation and emission spectra, corresponding to transitions (10.2), is also the formation of a narrow group of intense lines (NGIL) and quenching of many other lines due to CI SEOS. This effect at first was noticed in [13–15] for the spectra corresponding to 3–3 transitions. The strong resonance-like emission of laser-produced plasma involving elements from Cs to Yb was reported in [16, 17] and attributed to 4–4 transitions at ionization degrees $q = 7$–15.

Under the assumption that the intensities of emission lines can be approximated by their strengths the closed formula was derived for the shift of average energy of transition array due to CI [18]. This formula demonstrates that such a shift due to CI SEOS is mainly determined by the dipole term of the interconfigurational Coulomb interaction and is always positive. The concepts of the emissive and receptive zones were introduced as the line-strength-weighted energy level distribution of the upper and lower configurations [19].

NGIL in the spectra of plasma obtains the form of unresolved transition array or quasi-continuum band. Such bands attracted much attention due to their applications. The emission spectra of highly charged ions are considered as a prospective source of EUV and XUV radiation for lithography [20, 21]. For this reason the extensive investigations of band formation in the spectra of Sn^{q+} ions at 13.5 nm [20, 22–25] as well as in the Gd^{q+} and Tb^{q+} ions at 6.5 nm [21, 26, 27] have been performed. On the other hand, such a radiation in the highly charged tungsten ions plays a negative role [28, 29], as it causes large radiation losses in the tokamak plasma. The separate intense lines of less complex spectra of ions, corresponding to 3–3 transitions, are widely used for diagnostics of laboratory and astrophysics plasma [30].

In the low density plasma, obtained in tokamak or EBIT, the ground level of ions is mainly populated. Under these conditions the influence of CI SEOS on the emission and excitation spectra does not necessarily obey the same regularities as in the case of the total transition arrays. The formation of narrow groups of intense lines for such emission and photoexcitation spectra corresponding to 4–4 and 3–3 transitions in the isoelectronic an isonuclear sequences was investigated in [31, 32].

Two possible mechanisms of the formation of NGIL are considered in Sects. 10.2 and 10.3 of this work.

The results of calculations presented in this work were obtained in the single-, two- or many-configuration approximations with quasi-relativistic Hartree-Fock (HFR) [33], relativistic Dirac-Fock-Slater (RDFS) [34] and relativistic Dirac-Fock (RDF) [35] wave functions. All relativistic configurations corresponding to the same nonrelativistic configuration were taken into account.

10.2 Formation of a Narrow Group of Intense Lines by the Coulomb Interaction in the Single-Configuration Approximation

Even in rather heavy atoms and highly charged ions the structure and width of the energy level spectrum of $nl_1^{N_1} nl_2^{N_2}$ configuration mainly depend on the exchange part of the Coulomb interaction. In the second quantization representation the operator of this interaction between two open shells can be presented in the form [9]:

$$H_{\mathrm{ex}}^{C} = \left(\sum_{k} (C^{(k)}(nl_1, nl_2) C^{(k)}(nl_2, nl_1)) - \frac{\hat{N}(nl_2)}{2l_2 + 1} \right)$$

$$\times \langle l_1 \| C^{(k)} \| l_2 \rangle^2 G^k(nl_1, nl_2). \tag{10.3}$$

Here, $C^{(k)}(nl_1, nl_2)$ is the spherical tensor of rank k producing the exchange of electron between shells:

$$C_q^{(k)}(nl_1, nl_2) = \sum_{\nu\xi} a_\nu^+ \langle nl_1\nu | C_q^{(k)} | nl_2\xi \rangle b_\xi, \tag{10.4}$$

a_ν^+ is the operator of electron creation in the state $nl_1\nu$ and b_ξ is the operator of electron annihilation in the state $nl_2\xi$, where ν and ξ denote the sets of orbital and spin momenta projections. In Eq. (10.3), $\hat{N}(nl_2)$ is the operator of electron number in the shell $nl_2^{N_2}$ and its eigenvalue is equal to N_2. $G^k(nl_1, nl_2)$ is the exchange radial integral, and the multiplier before it is the reduced matrix element of spherical tensor of rank k.

The main contribution to the matrix element of the H_{ex}^C operator is given by the smallest order term in its expansion. When $l_2 = l_1 + 1$, such a term is the dipole one with $k = 1$. The $C^{(1)}(nl_1, nl_2)$ operator represents the electric dipole transition operator, but acting only in the spin angular space. Consequently, the matrix element of the H_{ex}^C operator can be expressed in terms of dipole transition amplitudes (without the radial integral). In the case of two considered excited configurations the g_1 coefficient at the Slater integral $G^1(nl_1, nl_2)$ obtains the following expressions [9]:

$$g_1\left(l^{4l+1}(l+1)^{N+1}\gamma_2 L_2 S_2, \gamma_2' L_2' S_2' L S\right)$$

$$= (2J+1)^{-1} \sum_{\gamma_2'' L_2'' S_2'' L' S' J'} \langle l^{4l+1}(l+1)^{N+1}(\gamma_2 L_2 S_2) L S J$$

$$\times \| C^{(1)} \| l^{4l+2}(l+1)^{N}(\gamma_2'' L_2'' S_2'') L' S' J' \rangle$$

$$\times \langle l^{4l+1}(l+1)^{N+1}(\gamma_2' L_2' S_2') L S J \| C^{(1)} \| l^{4l+2}(l+1)^{N}(\gamma_2'' L_2'' S_2'') L' S' J' \rangle$$

$$- \delta(\gamma_2 L_2 S_2, \gamma_2' L_2' S_2') \frac{N+1}{2l+3} \langle l \| C^{(1)} \| l+1 \rangle^2; \tag{10.5}$$

$$g_1\left((l+1)^{N-1}\gamma_1 L_1 S_1, \gamma_1' L_1' S_1' (l+2) L S\right)$$

$$= (2J+1)^{-1} \sum_{\gamma_1'' L_1'' S_1'' L' S' J'} \langle (l+1)^{N-1}\gamma_1 L_1 S_1 (l+2) L S J$$

$$\times \| C^{(1)} \| (l+1)^{N}(\gamma_1'' L_1'' S_1'') L' S' J' \rangle$$

$$\times \langle (l+1)^{N-1}\gamma_1' L_1' S_1' (l+2) L S J \| C^{(1)} \| (l+1)^{N}(\gamma_1'' L_1'' S_1'') L' S' J' \rangle$$

$$- \delta(\gamma_2 L_2 S_2, \gamma_2' L_2' S_2') \frac{1}{2l+5} \langle l+1 \| C^{(1)} \| l+2 \rangle^2. \tag{10.6}$$

The dependence of the right-hand side of Eqs. (10.5) and (10.6) on the total quantum number J disappears when the summation over J' is performed.

Table 10.1 Number of terms in the upper and lower groups of some configurations with two open shells

Configuration		p^5d	p^4d	p^3d	p^2d	pd	d^9f	d^8f	d^7f	d^6f	d^9f^2	d^9f^3
All terms		6	12	18	12	6	10	33	88	118	42	158
Upper terms after diagonalization		1	3	8	7	5	1	3	13	24	3	20
Lower terms	before diagonalization	5	6	6	3	1	9	21	33	21	27	49
	after diagonalization	5	9	10	5	1	9	30	75	94	39	138

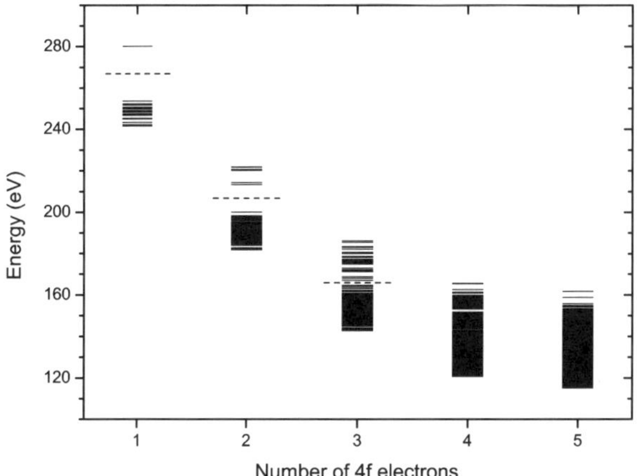

Fig. 10.1 Two groups of levels in the energy level spectra for the isonuclear sequence of praseodymium $4d^9 4f^N$ ($N = 1$–5). Both groups are energetically separated only for $N = 1$–3, this is indicated by a *broken line*. Results of calculations in the single-configuration HFR approximation

Thus, the g_1 coefficient consists of the term-dependent and term-independent parts. For the diagonal matrix element the term-dependent part has a positive value. It gives a contribution to the energy only for those terms of $nl^{4l+1}n(l + 1)^{N+1}$ or $n(l + 1)^{N-1}n(l + 2)$ configuration, which are related by the electric dipole transitions to the terms of $nl^{4l+2}n(l + 1)^N$ configuration. Consequently, the terms of excited configuration have a tendency to separate into two groups: the lower group with a negative and constant g_1 value and the upper group, corresponding to a positive term-dependent g_1 coefficient.

The nondiagonal g_1 coefficients can also obtain rather large values of the same order as the diagonal ones. However, only terms of the upper group are mixed among themselves due to the exchange interaction. This interaction shifts some terms of this group to the lower one. The numbers of terms belonging to the upper and lower groups before and after the diagonalization of $\|g_1\|$ matrix are indicated in Table 10.1.

The separation of energy levels by the Coulomb exchange interaction into two groups is illustrated in Fig. 10.1 for the energy level spectra of $4d^9 4f^N$ config-

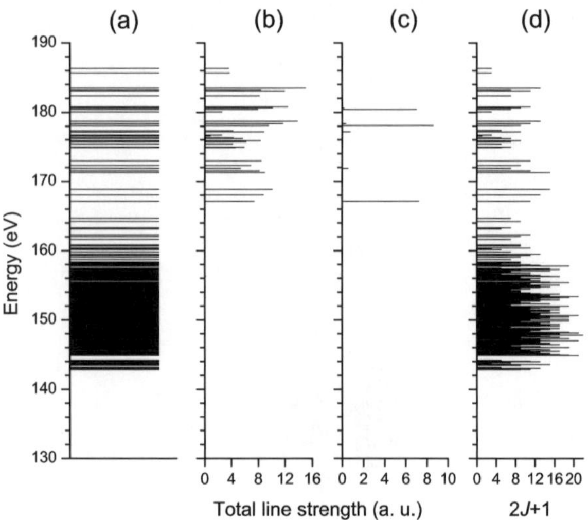

Fig. 10.2 Energy level spectrum of the Pr^{3+} $4d^9 4f^3$ (**a**) and the total line strength of transitions from each of these levels: to all levels of the $4d^{10} 4f^2$ configuration (**b**); to the ground level of this configuration (**c**); proportional to $2J + 1$ (equal from all states) (**d**). Results of calculations in the single-configuration HFR approximation

urations ($N = 1$–5). The levels of both groups are separated energetically for the configurations with a small number of electrons in the second shell. On increasing N these groups approach and overlap each other. At $N = 1$, there is only one level $^1P_1$ of the upper group, situated distantly from the other 19 levels. At $N = 2$, the upper group is formed by the six levels of three terms $(^3_2H)^2F$, $(^3_2F)^2D$ and $(^3_2H)^2G$, the total number of levels is 109. At $N = 3$, the upper group (35 levels) remains separated yet by a small energy gap from the numerous lower group (351 levels). For the larger number of $4f$ electrons the levels of both groups are distinguished only by their deexcitation rates.

Taking into account entirely the dipole part of matrix element, only levels of the upper group are involved in the electric dipole transitions. Additionally, a higher position of term correlates with a larger transition rate to/from the levels of the ground configuration. The concentration of transition rate from the levels of $4d^9 4f^3$ configuration to all levels and only to the ground level of $4d^{10} 4f^2$ configuration in the intermediate coupling is shown in Fig. 10.2. For comparison the equally probable deexcitation of all states (the deexcitation rate of level proportional to its statistical weight $2J + 1$) is shown too. Thus, the formation of NGIL is typical of the whole array of transitions between both configurations as well as of the excitations from the ground level. The first case corresponds to the emission spectrum at the statistical populations of excited levels. The second case corresponds to the photoexcitation spectrum, which is usually excited only from the ground level.

Consequently, the transitions between the excited and ground configurations tend to concentrate on a shorter wavelength side of the spectrum. Of course, the calculation in the intermediate coupling gives some probabilities of lines to/from the levels of the lower group as well, but usually their values are essentially smaller (Fig. 10.3).

The calculated distribution can be closely approximated using the new wavefunction basis, obtained by the diagonalization of $\|g_1\|$ matrix [3, 9]. In an algebraic way such a basis can be constructed using the special operator, which creates a vacancy

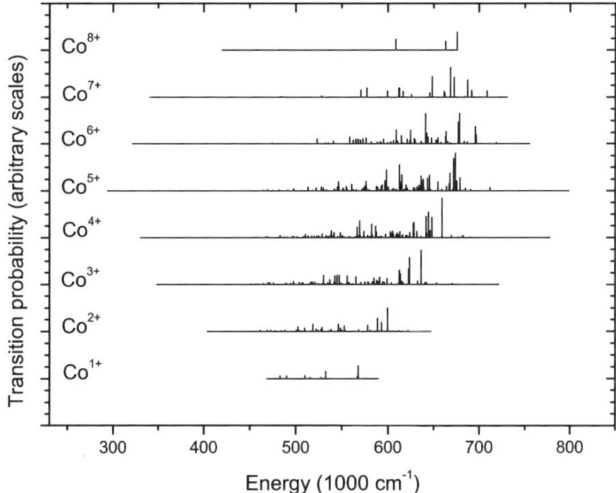

Fig. 10.3 Probabilities of radiative transitions $3p^53d^{N+1}$–$3p^63d^N$ in the Co^{q+} isonuclear sequence [10]. The *horizontal sections of graphs* indicate the intervals of possible transition energies

in one shell and an electron in the other shell, thus, the wavefunction of the excited configuration is obtained [10, 36]. Such a vacancy-electron pair is coupled to the ^1P term. The lowest configuration with the same number of electrons has no vacancies in the first shell. When acting with this operator repeatedly, the various states of the sequence of configurations with a different number of vacancy-electron pairs p can be obtained. Thus, a new quantum number p is introduced for the classification of states by the parentage in such a configuration sequence. It was shown in [10] that the additional selection rule appears in this hole-particle basis: the electric dipole transitions are possible only between the states, differing in this quantum number by unity $\Delta p = 1$.

In order to characterize the different role of energy levels of considered configuration in the transitions to the other configuration the concept of emissive zone was introduced [19]. It is defined using the statistical moments of the energy level spectrum, calculated taking into account the participation of energy levels in these transitions. In order to obtain their algebraic expressions two assumptions are made: (i) the excited levels are populated statistically and (ii) the line intensities are proportional to the line strengths. The average energy of the emissive zone for configuration K corresponding to the transitions into configuration K' was defined in [19]:

$$\bar{E}_{K'}(K) = \frac{\sum_{\gamma\gamma'}\langle K\gamma|H|K\gamma\rangle S(K\gamma, K'\gamma')}{S(K, K')}, \tag{10.7}$$

where $S(K\gamma, K'\gamma')$ is the line strength of transition between the levels γ and γ'. The quantity in the denominator is the total line strength. It is useful to determine the position of zone with respect to the average energy of configuration $\bar{E}(K)$. For the pair of the excited and ground configurations this shift of average energy has the following explicit expression [19]:

$$\delta \bar{E}_{K_0 nl_1^{N_1+1} nl_2^{N_2-1}}\left(K_0 nl_1^{N_1} nl_2^{N_2}\right)$$

$$= -\frac{N_1(4l_2+2-N_2)}{(4l_1+1)(4l_2+1)}\left\{\left[\frac{2}{3}\delta(k,1) - \frac{1}{2(2l_1+1)(2l_2+1)}\right]\right.$$

$$\times (-1)^k \langle l_1 \| C^{(k)} \| l_2 \rangle^2 G^k(nl_1, nl_2)$$

$$\left. + \sum_{k>0} \begin{Bmatrix} l_1 & l_1 & k \\ l_2 & l_2 & 1 \end{Bmatrix} \langle l_1 \| C^{(k)} \| l_1 \rangle \langle l_2 \| C^{(k)} \| l_2 \rangle F^k(nl_1, nl_2) \right\}, \quad (10.8)$$

where K_0 means the passive closed shells.

Analysis of Eq. (10.8) confirms that the dipole exchange term mainly determines the concentration of line strength in the transitions from the upper levels. The shift of emissive zone is always positive because the first term in the brackets exceeds the second negative term. This shift obtains the largest value at $N_2 = 1$ and decreases by increasing N_2.

Also the second moment of emissive zone—the variance can be introduced:

$$\sigma_{K'}^2(K) = \frac{\sum_{\gamma\gamma'}[\langle K\gamma|H|K\gamma\rangle - \bar{E}_{K'}(K)]^2 S(K\gamma, K'\gamma')}{S(K, K')}. \quad (10.9)$$

It has more complex expression which is given in [37]. The knowing of variance enables one to determine the width of zone as the full width at the half maximum of the normal distribution:

$$\Delta E_{K'}(K) = 2\sqrt{2\ln 2 \sigma_{K'}^2(K)}. \quad (10.10)$$

The examples of emissive zones, calculated using such a model, were given in [19, 37]. Certainly, the zones for the same configuration, but corresponding to the transitions to the different final configurations, do not coincide and can essentially differ. The interval of zone, determined by Eqs. (10.7), (10.9) and (10.10), corresponds rather approximately to the position of the upper group of levels; the emissive zone can be obtained even beyond the interval of energy level spectrum of configuration [19].

Similarly, the receptive zone, corresponding to the final configuration of transition can be introduced.

It is necessary to note that the same term *emissive zone* is also used in other sense: the interval of emission lines with the transition probabilities greater then a certain given value [21].

10.3 Formation of a Narrow Group of Intense Lines by the Configuration Interaction

In this section we will consider the conditions and regularities of NGIL formation for transitions (10.2) in the isoelectronic and isonuclear sequences. Spectra of ions can be recorded at various conditions: in laser-produced plasma, EBIT, vacuum spark, Large Helical Device and other devices. It is necessary to distinguish the manifestation of this effect in two types of spectra:

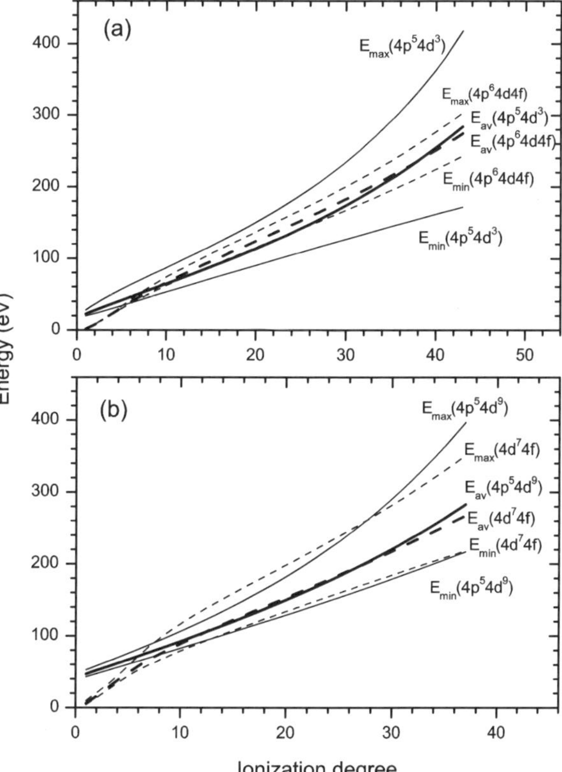

Fig. 10.4 Overlap of the energy level spectra of the $4p^5 4d^3$ and $4p^6 4d 4f$ configurations in the Sr isoelectronic sequence (**a**) and of the $4p^5 4d^9$ and $4p^6 4d^7 4f$ configurations in the Ru isoelectronic sequence (**b**) [31]. Average energy (E_{av}), upper (E_{max}) and lower (E_{min}) limits of the spectrum are given with respect to the ground level of the corresponding ion. Results of calculations in the single-configuration HFR approximation

(i) In a plasma of high density at local thermodynamic equilibrium the intensities of emission lines are proportional to the transition rates. The wavelengths of considered transitions for the ions correspond to VUV or X-ray regions, thus, the transition energy exceeds the interval of energy level spectrum and the transition rate can be approximated by the line strength.

(ii) In a plasma of low density, typical of EBIT or tokamak, only the ground level of ions is mainly populated. Thus, the important processes are the dipole excitation from the ground level of $nl^{4l+2} n(l+1)^N$ configuration as well as the subsequent emission after such an excitation.

The main interest presents the formation of NGIL in the emission and photoexcitation spectra of ions, corresponding to the transitions:

$$\left(3s 3p^{N+1} + 3s^2 3p^{N-1} 3d\right) - 3s^2 3p^N, \tag{10.11}$$

$$\left(4p^5 4d^{N+1} + 4p^6 4d^{N-1} 4f\right) - 4p^6 4d^N. \tag{10.12}$$

These cases are considered in our review.

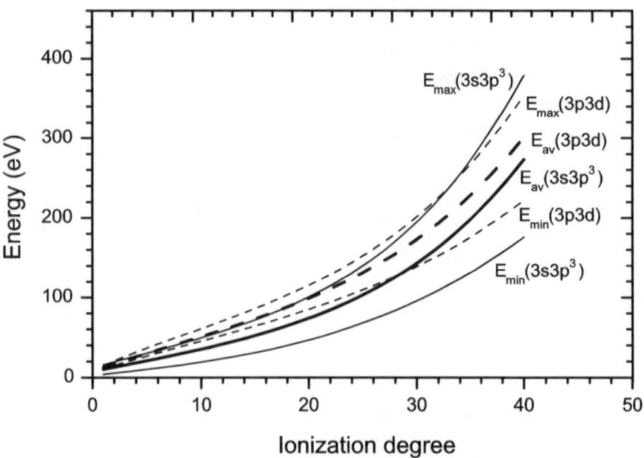

Fig. 10.5 Overlap of the energy level spectra of the $3s3p^3$ and $3s^23p3d$ configurations in the Si isoelectronic sequence [32]. Average energy (E_{av}), upper (E_{max}) and lower (E_{min}) limits of the spectrum are given with respect to the ground level of $3s^23p^2$ configuration. Results of calculations in the HFR single-configuration approximation

10.3.1 Overlap of the Energy Level Spectra of the $nl^{4l+1}n(l+1)^{N+1}$ and $nl^{4l+2}n(l+1)^{N-1}n(l+2)$ Configurations. Validity of the Two-Configuration Model

Both excited configurations overlap considerably at various numbers of electrons N and ionization degrees q (Figs. 10.4 and 10.5). At small numbers of N, the larger width has the energy level spectrum of $nl^{4l+1}n(l+1)^{N+1}$ configuration with more filled outer shell. At large numbers of N, the spectrum of $nl^{4l+2}n(l+1)^{N-1}n(l+2)$ configuration tends to expand more than the spectrum of $nl^{4l+1}n(l+1)^{N+1}$ configuration.

In neutral atoms, the average energy of $nl^{4l+1}n(l+1)^{N+1}$ configuration exceeds that of $nl^{4l+2}n(l+1)^{N-1}n(l+2)$ configuration. However, in the isoelectronic sequence on increasing ionization degree the collapse of $P_{n(l+2)}(r)$ radial orbital and its further contraction take place (Fig. 10.6). The interaction of this electron with the $n(l+1)^{N-1}$ shell becomes stronger, which upraises the energy level spectrum of $nl^{4l+2}n(l+1)^{N-1}n(l+2)$ configuration; thus, the average energies of both configurations interchange. At large numbers of q, the average energies of two configurations become closer again ($n = 3$) and even intersect the second time ($n = 4$); the reason is a different number of 4d electrons, interacting with a vacancy and an excited electron [31].

The fast contraction of the $P_{n(l+2)}(r)$ radial orbital also strongly influences the variation of the line strength of $n(l+1) - n(l+2)$ transitions in the isoelectronic sequence (Fig. 10.7). The total line strength of these transitions resonantly increases at the beginning of the sequence, obtains its maximal value and fast decreases further on. The sharper resonance is obtained for the smaller orbital quantum number of $n(l+2)$ electron and for the more filled $n(l+1)^{N-1}$ shell. It is necessary to note that the configuration mixing does not change the total line strength of both transition arrays, calculated in the single-configuration approximation.

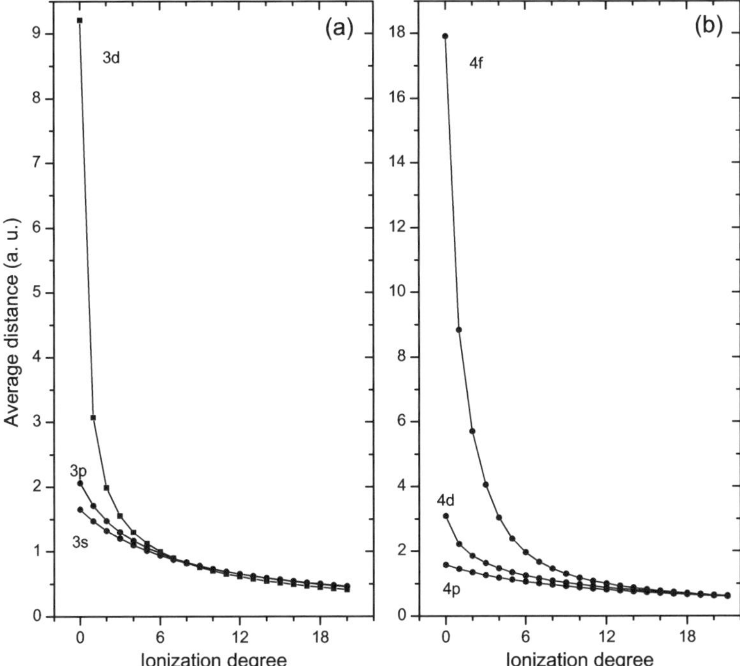

Fig. 10.6 Dependence of the average distance of electrons from the nucleus on the ionization degree: (**a**) in the S $3s^23p^33d$ isoelectronic sequence [32]; (**b**) in the Tc $4p^64d^64f$ isoelectronic sequence [31]. Results of calculations in the single-configuration HFR approximation

In neutral atoms and first ions the considered excited configurations are also mixed with other configurations, but from the ionization degree $q \approx 5$ the formation of NGIL is properly determined by SEOS in the two-configuration approximation. It was demonstrated in [24, 31, 32] by the comparison of various spectra, calculated in the single-configuration and two-configuration approximations, with the results of more exact calculations using a wider basis of wavefunctions or semi-empirical values of the radial integrals. Some examples are presented in Figs. 10.8 and 10.9. The spectrum of two transition arrays, calculated in the single-configuration approximation, does not resemble a real spectrum at all.

Though the energy level spectra of both excited configurations strongly overlap, their main lines usually are situated distantly. Taking into account the strong CI SEOS the most lines with the longer wavelengths are canceled and the line strength is concentrated in a few intense lines on the shorter wavelength side of the spectrum. Thus, a fairly good agreement with the results of more exact calculation of the spectrum is obtained.

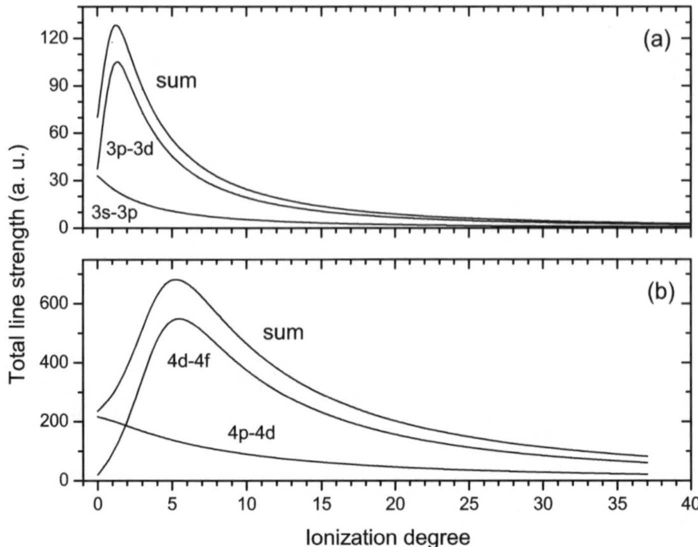

Fig. 10.7 Dependence of the total line strength of transitions (10.2) on the ionization degree: (**a**) $3s3p^5$–$3s^23p^4$, $3s^23p^33d$–$3s^23p^4$ and $(3s3p^5 + 3s^23p^33d)$–$3s^23p^4$ transitions in the S isoelectronic sequence [32]; (**b**), and $4p^54d^8$–$4p^64d^7$, $4p^64d^64f$–$4p^64d^7$ and $(4p^54d^8 + 4p^64d^64f)$–$4p^64d^7$ transitions in the Tc isoelectronic sequence [31]. Results of calculations in the single-configuration and two-configuration HFR approximations

10.3.2 Formation of NGIL in the Spectra with the Intensities of Lines Proportional to the Line Strength

The approximate explicit expression for intensity of line enables one to obtain the closed formula of the average energy and other mean characteristics of transition array. Their analysis reveals some general regularities of NGIL formation in corresponding spectra.

The average energy of such a transition array equals the difference of average energies of emissive and receptive zones for the initial and final configurations:

$$\bar{E}(K - K') = \bar{E}_{K'}(K) - \bar{E}_K(K').\tag{10.13}$$

The influence of CI SEOS on the line strength distribution can be estimated by the difference of its average energy in the two-configuration and single-configuration approximations. For transitions (10.2) it obtains the following expressions [18]:

$$\Delta E_{\text{CI}}\left(\left(K_0 ns n p^{N+1} + K_0 ns^2 np^{N-1} nd\right) - K_0 ns^2 np^N\right)$$

$$= \frac{4}{15} \frac{N(6 - N)\langle ns|r|np\rangle\langle np|r|nd\rangle}{2N\langle np|r|nd\rangle^2 + (6 - N)\langle ns|r|np\rangle^2} R^1(nsnd, npnp);\tag{10.14}$$

$$\Delta E_{\text{CI}}\left(\left(K_0 np^5 nd^{N+1} + K_0 np^6 nd^{N-1} nf\right) - K_0 np^6 nd^N\right)$$

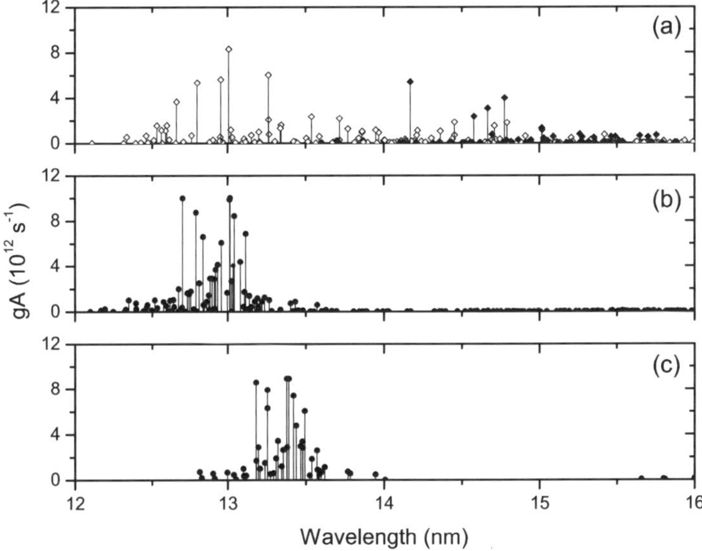

Fig. 10.8 Distribution of the probabilities of transitions $(4p^54d^3 + 4p^64d4f)$–$4p^64d^2$ for Sn^{12+}, calculated in the single-configuration RDFS approximation (**a**), two-configuration RDFS approximation (**b**) [39] and semi-empirically (**c**) [38]. In (**a**) the lines of transitions from the $4p^54d^3$ configuration levels are indicated by *solid diamonds* and from $4p^64d4f$ configuration by *open diamonds*

$$= \frac{4}{15} \frac{N(10-N)\langle np|r|nd\rangle\langle nd|r|nf\rangle}{3N\langle nd|r|nf\rangle^2 + 2(10-N)\langle np|r|nd\rangle^2}$$

$$\times \left(\frac{7}{3} R^1(npnf, ndnd) - \frac{2}{7} R^3(npnf, ndnd) \right). \tag{10.15}$$

Here, $\langle nl|r|nl'\rangle$ is the radial integral of dipole transition and R^k is the interconfigurational radial integral of the Coulomb interaction. The shift ΔE_{CI} as well as the shift (10.8) is also mainly determined by the dipole term with $k = 1$. The radial orbitals $\mathrm{P}_{nl}(r)$ and $\mathrm{P}_{n(l+1)}(r)$ have a different sign in the region of their main maxima, thus, the radial transition integral $\langle nl|r|n(l+1)\rangle$ obtains a negative value and the interconfigurational radial integral R^k—a positive value. Consequently, the shift of average energy due to CI SEOS is always positive. As was demonstrated in Figs. 10.4 and 10.5, the energy spectra of both excited configurations strongly overlap at various ionization degrees, but the intense lines of both arrays do not necessarily overlap too. If they are situated separately, SEOS configuration mixing enhances the shorter wavelength transitions from the higher configuration and quenches the longer wavelength transitions from the lower configuration; such an example is given in Fig. 10.8. Depending on the atomic number and ionization degree the intense lines of the first or the second transition arrays have longer wavelengths, thus, NGIL can be formed in a region of one or the other array. Such a role of two arrays in CI SEOS can be changed not only in the isoelectronic, but also

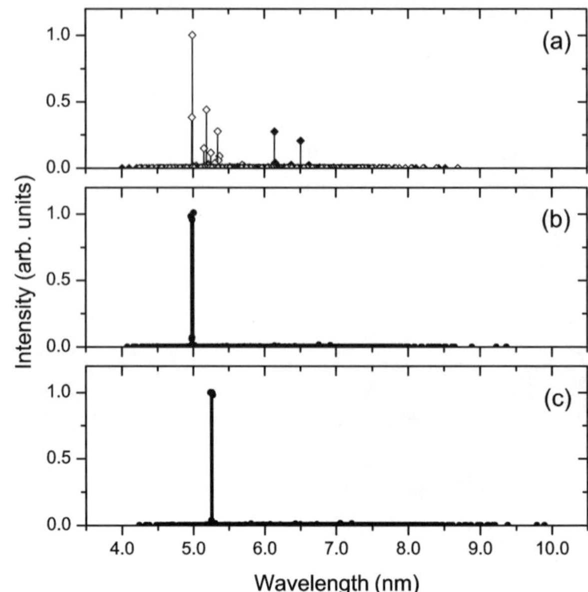

Fig. 10.9 Emission spectrum of W^{31+} around 5 nm after the photoexcitation from the ground level of $4p^64d^7$ configuration [31]. (**a**) Transitions $4p^54d^8$–$4p^64d^7$ (*solid diamonds*), $4p^64d^64f$–$4p^64d^7$ (*open diamonds*) in the single-configuration approximation. (**b**) The mixed array of $(4p^54d^8 + 4p^64d^64f)$–$4p^64d^7$ transitions in the two-configuration approximation. (**c**) Spectrum obtained taking into account eight additional configurations, namely, $4p^54d^64f^2$, $4p^54d^64f5p$, $4p^64d^44f^3$, $4p^64d^54f(5d + 5g)$, $4p^44d^84f$, $4p^44d^64f^3$ for the initial excited state, and $4p^64d^54f^2$ for the ground state. Results of the RDFS calculations

in the isonuclear sequence, for example, in the Pr^{q+} [18] or Gd^{q+} and Tb^{q+} [26] sequences on decreasing a number of electrons in the $4d^N$ shell.

Another regularity of CI SEOS influence on transition array, manifesting itself in the various isonuclear sequences, is a very slow variation of the mean wavelength on increasing the ionization degree, while the mean wavelength of one and the other array changes reasonably [18, 40, 41].

It is necessary to note that the formation of NGIL does not manifest itself at all if some levels of the lower configuration with the large deexcitation rates are not related by the interconfiguration matrix elements to the levels of the upper configuration and thus some intense lines on a long-wavelength side of the spectrum remain unaffected by CI SEOS. For example, it takes place at the beginning of isoelectronic sequences when the intense lines, originating from the $nl^{4l+1}n(l + 1)^{4l+5}$ configuration with a few terms, have the shorter wavelengths than the intense lines, originating from the more complex $nl^{4l+2}n(l + 1)^{4l+3}n(l + 2)$ configuration (Fig. 10.10).

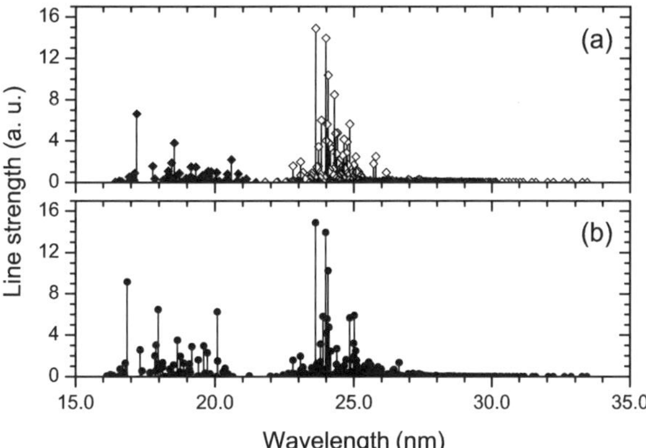

Fig. 10.10 Line strength distribution of transitions for Cd^{4+}: (**a**) $4p^5 4d^9 - 4p^6 4d^8$ (*solid diamonds*) and $4p^6 4d^7 4f - 4p^6 4d^8$ (*open diamonds*); (**b**) $(4p^5 4d^9 + 4p^6 4d^7 4f) - 4p^6 4d^8$. Calculations are made in the single-configuration and two-configuration HFR approximations

10.3.3 SEOS CI Influence on the Spectra, Corresponding to the Excitation from the Ground Level of Ion

The intensity of lines for the photoexcitation spectrum is proportional to the oscillator strength. The intensity of line for the emission spectrum is expressed [42]:

$$I(K\gamma - K_0\gamma') = \frac{\sigma_{ex}(K_0\gamma_0 - K\gamma)A(K\gamma - K_0\gamma')}{\sum_{\gamma''} A(K\gamma - K_0\gamma'')}, \tag{10.16}$$

where σ_{ex} is the photoexcitation cross section, A is the radiative transition probability, K_0 is the ground configuration, γ', γ'' are its levels and γ_0 is the lowest level, K means one of the two excited configurations in the single-configuration approximation or their superposition in the two-configuration approximation, γ is its level. The sum in the denominator corresponds to the total transition probability from γ level; for the considered cases only radiative transitions to the various levels of the ground configuration give contribution to this quantity.

In Fig. 10.11 such photoexcitation and emission spectra for the W^{q+} ions with a different number of electrons in the $4d^N$ shell are presented. Two features of them attract the attention: the similarity of both kinds of spectra and formation of NGIL only for some ions.

The similarity of the photoexcitation and emission spectra is not a self-explanatory result and means that the transitions from the excited configurations proceed mainly to the same ground level of the initial configuration. It is not typical of the spectra, calculated in the single-configuration approximation, thus, the transitions to the other levels are effectively suppressed by the two-configuration interaction.

The spectra of various W^{q+} ions demonstrate that CI SEOS does not always cause the narrowing of spectrum. This effect manifests itself very distinctly at large number of $4d$ electrons $N \geq 4$: the width of the group of intense lines does not exceed several percent of the whole energy interval of transition array (Table 10.2) and only very weak lines appear in the remaining part of spectrum. However, at

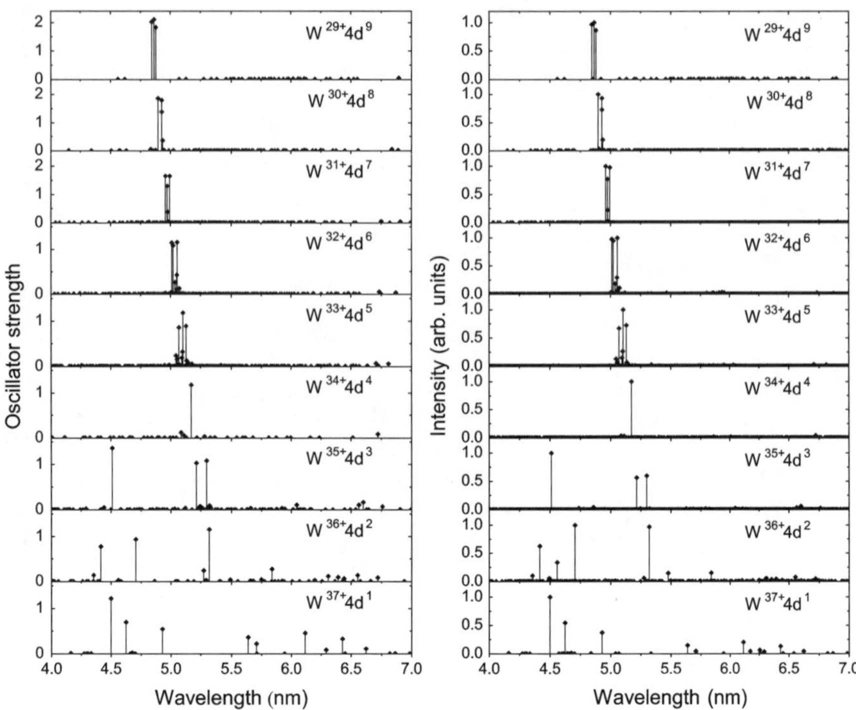

Fig. 10.11 Photoexcitation and emission spectra of W^{q+} ions, corresponding to $(4p^54d^{N+1} + 4p^64d^{N-1}4f)-4p^64d^N$ transitions, after the photoexcitation from the ground level of the corresponding ion [43]. Calculations are made in the two-configuration HFR approximation

Table 10.2 Intervals of wavelengths of the whole emission spectrum, corresponding to transitions $(4p^54d^{N+1} + 4p^64d^{N-1}4f)-4p^54d^N$, $(\Delta\lambda_s)$ and those of the main group of lines which intensities exceed 5 % of the most intense line $(\Delta\lambda_{il})$ [43]

N	$\Delta\lambda_s$ (nm)	$\Delta\lambda_{il}$ (nm)	$\Delta\lambda_{il}/\Delta\lambda_s$ (%)
9	3.03	0.03	0.99
8	4.28	0.04	0.93
7	5.50	0.04	0.73
6	6.89	0.06	0.87
5	8.82	0.09	1.01
4	9.55	0.00	0
3	9.80	2.09	21.3
2	8.02	6.12	76.3
1	5.25	1.81	34.5

small number of electrons $N \leq 3$, the formation of NGIL does not take place at all—the intense lines remain distributed within a rather wide interval of wavelengths.

The reason of such a different influence of CI SEOS on the spectra of ions of the same element with the same open shell follows from the comparison of their

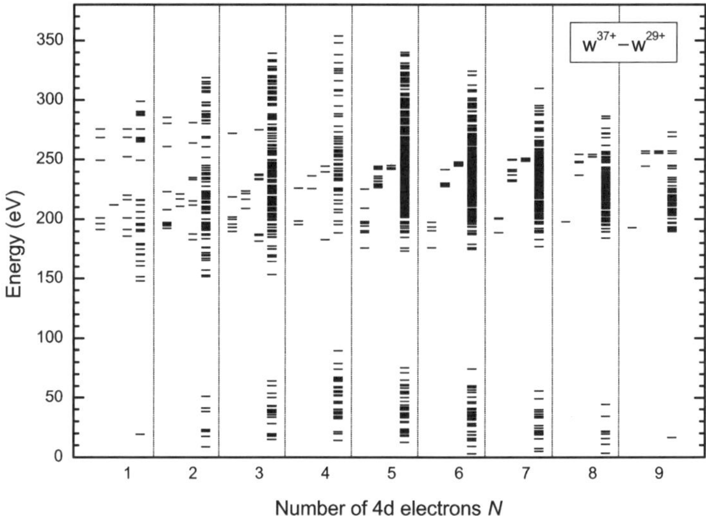

Fig. 10.12 Energy level diagram for W^{q+} ($q = 29$–37) ions [31]. The *columns of bars* at the *bottom part of diagram* represent all energy levels of the ground configuration $4p^64d^N$. At the *upper part of diagram*, the energy levels which can be excited from the lowest level of the ground configuration are plotted. For every ion, the *right column* represents all possible excited levels of $4p^54d^{N+1}$ and $4p^64d^{N-1}4f$ configurations, the other columns represent only the energy levels with excitation cross sections exceeding 5 % of this value for the most excited level: the *first (second) column* gives such levels of $4p^54d^{N+1}$ ($4p^64d^{N-1}4f$) configuration; the *third column* gives the most excited levels in the two-configuration approximation. Calculations are made in the HFR approximation

energy level spectra (Fig. 10.12). These results indicate that the formation of NGIL depends on the mutual positions of levels of the upper group, which are mainly populated by the excitation from the ground level. The necessary condition of this effect is that such levels of $4p^54d^{N+1}$ configuration should lie below such levels of $4p^64d^{N-1}4f$ configuration. This condition is fulfilled at $N \geq 4$, when the $4d_{3/2}$ subshell is closed and the $4p_{1/2}$–$4d_{3/2}$ as well as $4p_{3/2}$–$4d_{3/2}$ transitions are forbidden. On decreasing a number of electrons N from 4 up to 3, such transitions become allowed, therefore the higher levels of $4p^54d^{N+1}$ configuration are populated and transitions to them are not canceled by CI SEOS.

This regularity is confirmed by calculations for the other elements and explains the variation of spectra in the isoelectronic sequences. Due to the similarity of the photoexcitation and emission spectra we will consider only one of them.

In the first ions, the $4f$ orbital is not yet essentially contracted and the excited levels of $4p^64d^{N-1}4f$ configuration tend to lie below such levels of $4p^54d^{N+1}$ configuration, then the width of spectrum is rather large (Fig. 10.13). On increasing the ionization degree the levels of $4p^64d^{N-1}4f$ configuration are shifted down and then spectra are dominated by NGIL. At large ionization degrees, three types of the variation of spectra can be distinguished. At $N \leq 3$, the emission spectrum tends to become wider with q (Fig. 10.13a). At $4 \leq N \leq 9$, the strong concentration of lines

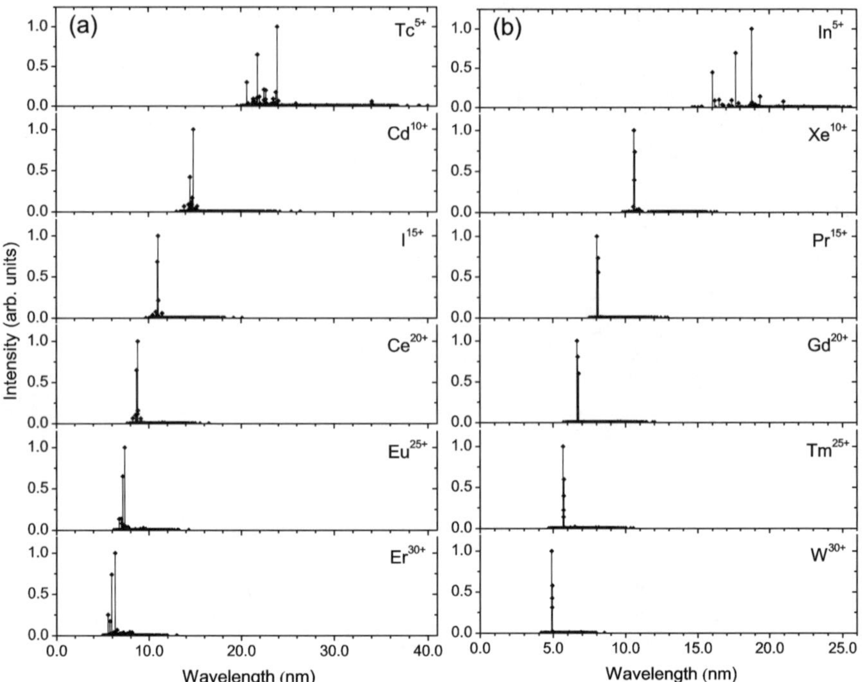

Fig. 10.13 Emission spectra, corresponding to $(4p^5 4d^3 + 4p^6 4d 4f)$–$4p^6 4d^2$ transitions, for the Sr isoelectronic sequence (**a**) and to $(4p^5 4d^9 + 4p^6 4d^7 4f)$–$4p^6 4d^8$ transitions for the Ru isoelectronic sequence (**b**) [31]. Spectra were photoexcited from the ground level. Results of calculation in the two-configuration HFR approximation

remains characteristic of highly charged ions due to the above indicated relativistic effect—the filling of the $4d_{3/2}$ subshell (Fig. 10.13b). The particularly narrow spectrum at various ionization degrees is obtained for $N = 5$ when the excitation takes place from the ground level with its total quantum number equal to zero (this essentially reduces a number of possible transitions).

These main regularities hold also for the spectra, corresponding to transitions (10.11). The necessary condition of the formation of NGIL is that all levels of $3s 3p^{N+1}$ configuration, mainly excited from the ground level, must lie below the corresponding levels of $3s^2 3p^{N-1} 3d$ configuration. However, the open shells with a smaller orbital quantum number contain considerably smaller number of levels. Thus, some levels of $3s 3p^{N+1}$ configuration at $N = 1$ and 2 are not related by the Coulomb interaction to the levels of $3s^2 3p^{N-1} 3d$ configuration; consequently, the corresponding lines are not canceled by CI (Fig. 10.14). The formation of NGIL is more expressed at a half-filled and an almost-filled $3p^N$ shell. In the isoelectronic sequences this effect begins at smaller ionization degrees than for transitions (10.12) and takes place up to about $q = 30$ (Fig. 10.15). Namely, in this interval of the isoelectronic sequence the intense lines of $3p$–$3d$ transitions are situated on the shorter wavelength side than the intense lines of $3s$–$3p$ transitions; at larger values of q,

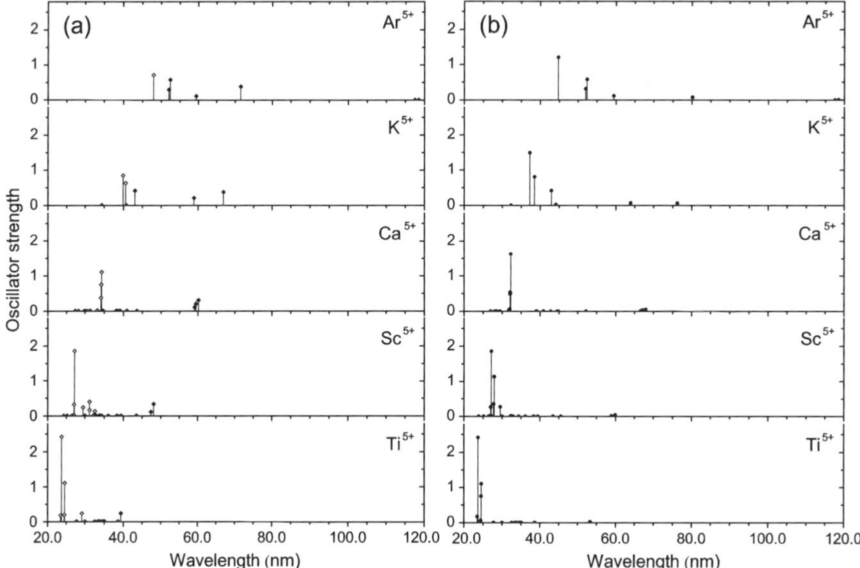

Fig. 10.14 Photoexcitation spectra corresponding to the transitions from the ground level of the $3s^23p^N$ configuration to the levels of the $3s3p^{N+1}$ and $3s^23p^{N-1}3d$ configurations at $N = 1$–5 and $q = 5$: (**a**) transitions to the levels of the $3s^23p^{N-1}3d$ (*open diamonds*) and of $3s3p^{N+1}$ (*solid diamonds*) configurations; (**b**) transitions to the same configurations in the two-configuration HFR approximation [32]

they begin to overlap and the spectrum becomes wider. The spectra of photoexcitation from the ground level and the inverse emission spectra are also rather similar due to the suppression by CI SEOS of transitions to the levels of upper terms of the ground configuration.

10.4 Conclusions

The formation of NGIL is typical of the electric dipole transitions between the configuration with a vacancy $nl^{4l+1}n(l+1)^{N+1}$ and the ground configuration $nl^{4l+2}n(l+1)^N$. Two different cases can be distinguished:

(i) The ground configuration has an open $3d^N$ or $4f^N$ shell. The strong Coulomb exchange interaction between the shells with the same principal quantum number separates the energy level spectrum of the excited configuration $nl^{4l+1}n(l+1)^{N+1}$ into two groups of levels with very different rates of transitions to/from $nl^{4l+2}n(l+1)^N$ configuration. This is the reason of the formation of NGIL on a short-wavelength side of the spectrum and manifests itself by giant resonances in the photoabsorption spectra of lanthanides and some iron group elements.

(ii) The ground configuration has an open $3p^N$ or $4d^N$ shell. Then also the configuration $nl^{4l+2}n(l+1)^{N-1}n(l+2)$ of the same complex exists and its strong

Fig. 10.15 Emission spectra of S-like ions at $q = 5$, 15, 25 and 35, corresponding to transitions $(3s3p^5 + 3s^2 3p^3 3d) - 3s^2 3p^4$, after the photoexcitation from the ground level of $3s^2 3p^4$ configuration [32]. Calculations are made in the two-configuration HFR approximation

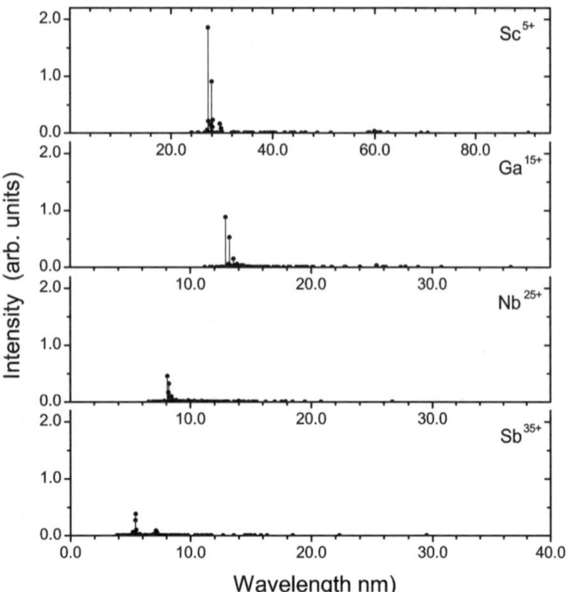

interaction with the $nl^{4l+1}n(l+1)^{N+1}$ configuration has an essential influence on both emission and photoexcitation spectra, corresponding to the transitions:

$$nl^{4l+2}n(l+1)^N - \left(nl^{4l+1}n(l+1)^{N+1} + nl^{4l+2}n(l+1)^{N-1}n(l+2)\right).$$

The main effect is also the formation of NGIL. When the whole array of transition probabilities is considered, CI SEOS enhances the shorter wavelength transitions from the higher configuration and quenches the longer wavelength transitions from the lower configuration. In the case of spectra, excited from the ground level of $nl^{4l+2}n(l+1)^N$ configuration, the necessary condition for the formation of NGIL is the following: all levels of $nl^{4l+1}n(l+1)^{N+1}$ configuration, mainly excited from the ground level, must lie below the corresponding levels of $nl^{4l+2}n(l+1)^{N-1}n(l+2)$ configuration. The exceptions are possible for simple configurations, which some terms are not related by the interconfiguration matrix elements.

There exists a similarity between the two mechanisms of the intensity concentration in the single-configuration and two-configuration approximations. In both cases the intensity is enhanced in the transitions involving the upper group of levels and is canceled in the transitions involving the lower group of levels. Both mechanisms are related to the specific structure of matrix elements of the Coulomb interaction operator: the possibility to express them in terms of transition amplitudes.

Cancellation of many lines means the existence of additional selection rule. For the single-configuration model such a rule for a number of vacancy-electron pairs was established using the hole-particle basis. The introduction of a new wavefunction basis and strict theoretical formulation of cancellation mechanism in the photoexcitation and emission spectra due to CI SEOS is an actual problem.

Acknowledgements   This work was partially supported by the Research Council of Lithuania, contract No. MIP-61/2010 and by the European Communities under the contract of association between EURATOM and Lithuanian Energy Institute.

References

1. D.R. Beck, C.A. Nicolaides, Phys. Rev. A **26**, 857 (1982)
2. T.M. Zimkina, V.A. Fomichev, S.A. Gribovskii, I.I. Zhukova, Sov. Phys., Solid State **9**, 1128 (1967)
3. J. Sugar, Phys. Rev. B **5**, 1785 (1972)
4. I. Glembockis et al., Litov. Fiz. Sb. **12**, 35 (1972) (in Russian)
5. J.P. Connerade, J.M. Esteva, R.C. Karnatak (eds.), *Giant Resonances in Atoms, Molecules and Solids* (Plenum Press, New York, 1986)
6. M. Richter et al., Phys. Rev. A **40**, 7007 (1989)
7. J.P. Connerade, Contemp. Phys. **19**, 415 (1978)
8. S. Kučas, R. Karazija, Phys. Scr. **58**, 220 (1998)
9. S. Kučas, R. Karazija, J. Phys. B **24**, 2925 (1991)
10. A. Bernotas, R. Karazija, J. Phys. B **34**, L741 (2001)
11. M. Meyer et al., Z. Phys. D **2**, 347 (1986)
12. M. Martins, K. Godehusen, T. Richter, Ph. Wernet, P. Zimmermann, J. Phys. B **39**, R79 (2006)
13. C. Froese Fischer, J. Quant. Spectrosc. Radiat. Transf. **8**, 755 (1968)
14. R.D. Cowan, J. Phys. (Paris) **31**, C4–191 (1970)
15. M. Aymar, Nucl. Instrum. Methods **110**, 211 (1973)
16. G. O'Sullivan, P.K. Carroll, J. Opt. Soc. Am. **71**, 227 (1981)
17. P.K. Carroll, G. O'Sullivan, Phys. Rev. A **25**, 275 (1982)
18. J. Bauche, C. Bauche-Arnoult, M. Klapisch, P. Mandelbaum, J.L. Schwob, J. Phys. B **20**, 1443 (1987)
19. J. Bauche et al., Phys. Rev. A **28**, 829 (1983)
20. G. O'Sullivan et al., in *EUV Sources for Lithography*, ed. by V. Bakshi (SPIE, Bellingham, 2006)
21. S.S. Churilov, R.R. Kildyarova, A.N. Ryabtsev, S.V. Sadovsky, Phys. Scr. **80**, 045303 (2009)
22. J. White et al., J. Appl. Phys. **98**, 113301 (2005)
23. S.S. Churilov, A.N. Ryabtsev, Phys. Scr. **73**, 614 (2006)
24. R. Karazija, S. Kučas, A. Momkauskaitė, J. Phys. D **39**, 2973 (2006)
25. R. D'Arcy et al., Phys. Rev. A **79**, 042509 (2009)
26. D. Kilbane, G. O'Sullivan, J. Appl. Phys. **108**, 104905 (2010)
27. 2011 International Workshop on EUV Lithography, Maui, 2011. www.euvlitho.com
28. R. Radtke et al., Phys. Rev. A **64**, 012720 (2001)
29. C. Biedermann, R. Radtke, R. Seidel, T. Pütterich, Phys. Scr. T **134**, 014026 (2009)
30. T. Pütterich et al., Plasma Phys. Control. Fusion **50**, 085016 (2008)
31. S. Kučas, R. Karazija, V. Jonauskas, A. Momkauskaitė, J. Phys. B **42**, 205001 (2009)
32. R. Karazija, A. Momkauskaitė, L. Remeikaitė-Bakšienė, J. Phys. B **44**, 035002 (2011)
33. R.D. Cowan, *The Theory of Atomic Structure and Spectra* (University of California Press, Berkeley, 1981)
34. M.F. Gu, Can. J. Phys. **86**, 675 (2008)
35. I.P. Grant, B.J. McKenzie, P.H. Norrington, D.F. Mayers, N.C. Pyper, Comput. Phys. Commun. **21**, 207 (1980)
36. A. Bernotas, Doctoral Thesis, Vilnius, 1990
37. S. Kučas, V. Jonauskas, R. Karazija, Phys. Scr. **51**, 566 (1995)
38. S.S. Churilov, A.N. Ryabtsev, Opt. Spectrosc. **101**, 169 (2006)
39. S. Kučas, V. Jonauskas, R. Karazija, A. Momkauskaitė, Lith. J. Phys. **47**, 249 (2007)

40. R. Radtke, C. Biedermann, J.L. Schwob, P. Mandelbaum, R. Doron, Phys. Rev. A **64**, 012720 (2001)
41. D. Kilbane, G. O'Sullivan, Phys. Rev. A **82**, 062504 (2010)
42. R. Karazija, *Introduction to the Theory of X-Ray and Electronic Spectra* (Plenum Press, New York, 1996)
43. V. Jonauskas, S. Kučas, R. Karazija, J. Phys. B **40**, 2179 (2007)

Chapter 11
Fluorescence in Astrophysical Plasmas

Henrik Hartman

Abstract Following the initial detection by Bowen in 1934 of the strong O III lines being due to accidental resonance with strong He II radiation, many strong spectral emission lines are explained as produced by fluorescence. Many of these are Fe II lines pumped by H Lyα, as a consequence of strong radiation from hydrogen and a favorable energy level structure for Fe II. The lines are observed in many types of objects with low density plasma components. The Weigelt condensations in the vicinity of the massive star Eta Carinae is one location where these lines are observed and can be studied in detail, as well as been used for diagnostics.

These gas condensations do not only show a spectrum indicating a non-equilibrium excitation but also non-equilibrium ionization, where the strong hydrogen radiation plays a key role. Early studies identified certain strong lines being the result of Resonance Enhanced Two-Photon Ionization (RETPI). Further investigations suggest that RETPI can be the responsible mechanism for the ionization structure of gas condensation.

We will review the resonance processes, with emphasis on the Eta Carinae spectrum. Large spectral, spatial and temporal coverage is available for this fascinating object, allowing for detailed analysis.

11.1 Fluorescence in Astrophysical Plasma

Many emission lines observed in spectra of nebulae and other low-density plasmas are thought to be formed either by collisions with electrons or by recombination. However, some levels can also be photoexcited, either by absorption of continuous radiation from a nearby star or by monochromatic light from strong emission lines present in the environment. The atom is then excited from a ground state or low excitation state. The subsequent decay occurs in one or more fluorescence channels producing emission lines.

H. Hartman (✉)
Faculty of Technology, Group of Material Science and Applied Math, Malmö University and Lund Observatory, 20506 Malmo, Sweden
e-mail: henrik.hartman@astro.lu.se

M. Mohan (ed.), *New Trends in Atomic and Molecular Physics*,
Springer Series on Atomic, Optical, and Plasma Physics 76,
DOI 10.1007/978-3-642-38167-6_11, © Springer-Verlag Berlin Heidelberg 2013

Thus, in the presence of strong radiation fields a non-LTE level population can be obtained in a low-density plasma due to selective photoexcitation. In this chapter we will consider two cases: Photoexcitation by continuum radiation (PCR) and photoexcitation by an accidental resonance (PAR). There will be a special emphasis on a PAR process where Fe II is selectively photoexcited by H Lyα.

Fluorescence appears in many fields and in many different scientific areas. Here we limit the discussion to fluorescence processes observed in low-density astrophysical plasmas.

Parity forbidden lines are, by definition, lines from levels with same parity, and the upper level long-lived metastable levels. Due to the low transition rates, metastable levels are not pumped from lower levels. The forbidden lines are thus widely used as probes of plasma density and temperature. We also discuss the possibility of populating levels through selective two-photon ionization from a lower ion, so-called Resonance Enhanced Two-Photon Ionization (RETPI).

11.2 Photoexcitation by Continuous Radiation, PCR

Some stellar spectra contain numerous permitted emission lines, where the corresponding upper level is pumped by continuum radiation from a nearby star. This is often referred to as photoexcitation by continuous radiation (PCR). Such lines are observed e.g. in satellite spectra of the symbiotic star KQ Puppis [28, 32], where high-lying excited levels in Fe II are populated by absorption of UV continuum radiation from one of the stars in the symbiotic system. A large number of levels are observed having an enhanced population and all of them are photoexcited from a low level, which is verified by observed absorption lines. And, inversely, all levels having strong channels down to the lowest levels in this wavelength region show an enhanced population. The population enhancement is determined by the excitation energy of the absorbing level and the oscillator strength of the absorption line. Some radiative energy is in this way transferred from the UV continuum to optically thin fluorescent lines at longer wavelength through the line absorption.

In analyses of the abundance pattern in Active Galactic Nuclei (AGN) it has been shown that it is important to consider PCR [31]. In low resolution spectra of AGN the iron abundance is measured as integrated flux in large UV bands of the Fe II resonance region. However, this is the region where also the PCR pumped fluorescence lines appear. It is thus important to include fluorescence in the modeling of AGN since it affects the total flux in the UV region.

11.3 Photoexcitation due to an Accidental Resonance, PAR

In spectra of some astrophysical plasmas one can observe some of the individual transitions in an LS multiplet as strong lines, whereas other components of the same multiplet are absent in spite of a similar excitation energy. The observations

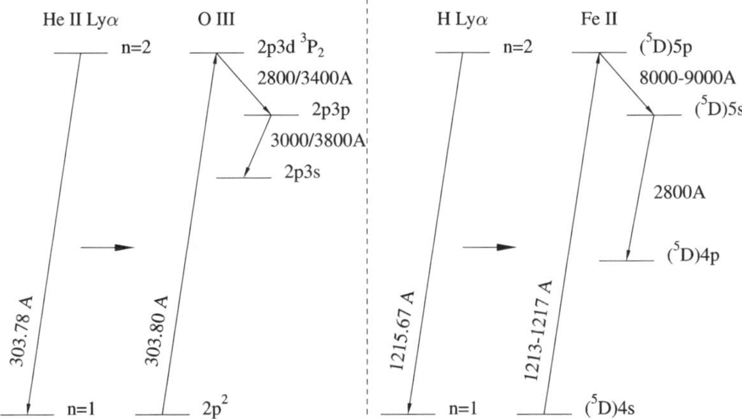

Fig. 11.1 Principle for the Bowen mechanism in O III and a similar fluorescence case in Fe II

clearly imply a large non-LTE distribution of the population of the energy levels. The excitation mechanism behind this observation has to be selective with sharp resonances in the excitation cross sections. The most probable mechanism is a photoexcitation due to an accidental resonance (PAR) [26], i.e. the wavelength of an intense line coincides with the wavelength of a transition in the pumped ion. Photons from the pumping line of an abundant element in the plasma (e.g. H, He, C, etc.) excite the pumped ion in the same plasma from a very low state, in general, to a higher state. When the photoexcited level in the pumped ion decays, the radiation is distributed in a number of decay channels (fluorescence lines) according to their transition probabilities. The fluorescence lines are often optically thin and can therefore easily escape from the plasma and act as a cooling agent for the gas.

The PAR process described above is often referred to as the Bowen mechanism since Bowen was not only first to identify forbidden [O III] lines in nebulae but also first to explain strong nebular lines as fluorescent O III lines [2]. All these identifications were based on Bowen's own laboratory analysis of O III. The strength of the [O III] lines in nebulae had been a puzzle and seemed to be correlated with the strength of optical He II lines [1]. Lines from high excitation levels in O III were also observed in nebular spectra, but only from certain energy levels. Bowen noted the coincidence in wavelength between the He II line at 303.780 Å and the O III transition from the ground state $2p^2$ ^{3}P$_2$ to the excited state 2p3d ^{3}P$_2$ at 303.799 Å. Radiative energy can thus be transferred from He II to O III, which is excited to the 2p3d ^{3}P$_2$ state. The most probable decay branch for this level is back to the $2p^2$ ground state with a new 303 Å photon as a result. But, there is a probability of a few percent that O III decays to 2p3p and subsequently to 2p3s resulting in emission of fluorescence lines in the 3000–4000 Å region [2, 3]. A simplified level diagram is given in Fig. 11.1. Since the fluorescence lines are high excitation lines, and their lower states are efficiently depleted by a prompt decay down to the ground states, they are optically thin. The radiative energy can thus be transferred from the optically thick He II line to the O III lines and thereby more easily escape from the nebula.

For the fluorescence process to work efficiently a number of requirements must be fulfilled. The pumping line must be strong, i.e. have an intense radiation field. The absorption line with the coincident wavelength in the pumped ion must have a large transition probability, and the lower state a sufficiently large population. The latter is normally achieved by collisional excitation for low states in abundant ions. A complex spectrum with many energy levels is also very line rich, and there is a high probability for a wavelength coincidence to occur.

An example is the prominent UV 34 multiplet in Fe III, where one level is pumped and yields fluorescence in the 1914 Å line [22]. The level z^7P_3 is pumped by H Lyα from the ground state a^5D_4, but the other levels, z^7P_2 and z^7P_4, do not show any sign of enhanced population. The pumping channels for the z^7P_2 are too far from H Lyα, the closest being at 1226 Å. The z^7P_4 has a transition at 1213 Å which is often within the line profile of H Lyα. However, the transition probability is too low for efficient pumping, a few orders of magnitude less than for the transitions to z^7P_3. The reason for this difference is a level mixing with the close z^5P term, which does not have a $J=4$ level but can transfer some quintet character to the $J=3$ level. Thus, the z^7P_4 level has a pure septet character resulting in a low transition probability for the intercombination transitions to a^5D.

Since the first explanation by Bowen of the nature of fluorescence lines, a number of similar processes with other elements involved have been detected. A compilation of known PAR processes, including pumping lines and pumped levels are given in Table 11.1.

Fe II line emission from a number of levels at 11 eV is observed strong in a number of stars [6, 13, 18, 19, 23]. Other levels, with similar or lower energy, do not show any emission at all. The populated levels, with configuration $(^5D)5p$, do all have strong channels down to the ground configuration $(^5D)4s$ with wavelengths all within a few Ångströms from H Lyα at 1215 Å and are thought to be pumped by the PAR process. They are discussed in more detail in the section below in connection with the identification of fluorescence lines in the symbiotic nova RR Tel.

The spectrum of Cr II is also observed to show fluorescence lines pumped by H Lyα. The atomic structure of Cr II is similar to Fe II; the lowest configurations are $3d^5$ or $3d^4(^ML)nl$ in Cr II compared to $3d^7$ or $3d^6(^ML)nl$ in Fe II. The configurations $3d^6$ and $3d^4$ give rise to the same terms which means that $3d^6(^ML)nl$ and $3d^4(^ML)nl$ in Fe II and Cr II, respectively, have the same parent terms ^ML. The energy separation in Cr II is similar to Fe II, and H Lyα matches therefore also the energy separation 4s–5p in Cr II. Infrared fluorescence lines of Cr II are seen from these levels in Eta Carinae [12, 34].

In the next chapter we will discuss the Fe II fluorescence in more detail, with the spectrum of RR Tel as a base.

Table 11.1 A number of fluorescence mechanisms have been identified since the first explanation by Bowen [1]. These are present in different plasmas, such as planetary nebulae, symbiotic stars and atmospheres of cool stars, showing emission lines. This table presents a compilation of known fluorescence processes

Ion	Pumped level	Energy (cm^{-1})	Pumping line	Ref.
O I	$2p^3(^4S)3d\,^3D$	97488	H Lyβ	B47
S I (UV9)	$3p^34s\,^3P_1$	77150	O I 1304	BJ80
S I (UV9)	$3p^34s\,^3P_2$	77181	O I 1304	BJ80
Ti I	v^1P_1	42927	Mg II 2802	T37
V I	$u^4D_{7/2}$	35379	Mg I 2852	T37
Cr I	s^5F_4	50210	Fe II 2373	T37
Cr II	$4s4p\,x^6D_{1/2}$	93968	H Lyα	ZHJ01
Cr II	$5p\,^4P_{5/2}$	93974	H Lyα	ZHJ01
Cr II	$5p\,^6P_{3/2}$	94002	H Lyα	ZHJ01
Cr II	$4s4p\,x^6D_{3/2}$	94098	H Lyα	ZHJ01
Cr II	$5p\,^6P_{5/2}$	94144	H Lyα	ZHJ01
Cr II	$5p\,^4F_{3/2}$	94256	H Lyα	ZHJ01
Cr II	$4s4p\,x^6D_{5/2}$	94265	H Lyα	ZHJ01
Cr II	$5p\,^6P_{7/2}$	94363	H Lyα	ZHJ01
Cr II	$5p\,^4F_{5/2}$	94365	H Lyα	ZHJ01
Cr II	$4s4p\,x^6D_{7/2}$	94452	H Lyα	ZHJ01
Cr II	$5p\,^4F_{7/2}$	94522	H Lyα	ZHJ01
Cr II	$4s4p\,x^6D_{9/2}$	94656	H Lyα	ZHJ01
Cr II	$5p\,^4F_{9/2}$	94749	H Lyα	JC88
Mn I	$y^6P_{7/2}$	35769	Mg II 2795	C70
Mn II	$3d^5(^6S)4f\,^5F_1$	98461	Si II 1197.57	JWG95
Mn II	$3d^5(^6S)4f\,^5F_2$	98462	Si II 1196.72	JWG95
Mn II	$3d^5(^6S)4f\,^5F_3$	98463	Si II 1195.00	JWG95
Mn II	$3d^5(^6S)4f\,^5F_4$	98464	Si II 1192.31	JWG95
Mn II	$3d^5(^6S)4f\,^5F_5$	98465	Si II 1190.41	JWG95
Fe I	z^3G_4	35767	Mg II 2795	T37
Fe I	z^3G_3	36079	Mg II 2802	T37
Fe I	y^3F_3	37162	Ca II 3968	C70
Fe I	y^3D_2	38678	Hξ	C70
Fe I	y^3D_1	38995	Fe II 2611	WC90
Fe I	x^5F_1	41130	Fe II 2484	T37
Fe I	x^5P_3	42532	Fe II 2373	T37
Fe I	x^5P_2	42859	Fe II 2382	T37
Fe I	y^5G_3	43137	Mg II 2795	T37
Fe I	y^5G_2	43210	Mg I 2852	T37
Fe I	w^5D_4	43499	Mg II 2795	T37

Table 11.1 (Continued)

Ion	Pumped level	Energy (cm^{-1})	Pumping line	Ref.
Fe I	v^{5}D$_3$	44166	Fe II 2714	WC90
Fe I	w^{5}D$_2$	44183	Fe II 2761	T37
Fe I	x^{5}G$_5$	45726	Fe II 2607	T37
Fe I	w^{3}P$_1$	50043	Hξ, H16	C70
Fe II	(^{3}H)4p z^{4}G$_{9/2}$	60807	[Si III] 1892	HJ00
Fe II	(^{3}H)4p z^{4}G$_{5/2}$	61041	[Fe IV] 2835	HJ00
Fe II	z^{4}I$_{13/2}$	61527	Fe II UV33, 207, 63	CP88
Fe II	(^{3}H)4p z^{2}G$_{9/2}$	62083	[O III] 1661	HJ00
Fe II	y^{4}G$_{5/2}$	64087	Fe II 2628/2359	CP88
Fe II	(^{3}G)4p x^{4}G$_{7/2}$	65931	[Ne V] 1575	HJ00
Fe II	x^{4}G$_{5/2}$	66078	Fe II 2599	CP88
Fe II	(^{3}G)4p y^{4}H$_{11/2}$	66463	C IV 1548	J83
Fe II	(^{3}G)4p x^{4}F$_{5/2}$	66522	[Si III] 1892	HJ00
Fe II	(a^{1}G)4p x^{2}H$_{9/2}$	72130	⟨Fe II⟩ 1776	HJ00
Fe II	(a^{1}D)4p w^{2}D$_{3/2}$	78487	C IV 1548	HJ00
Fe II	(b^{3}F)4p ^{4}G$_{9/2}$	90042	H Lyα	JJ84
Fe II	(^{5}D)5p ^{6}F$_{9/2}$	90067	H Lyα	JJ84
Fe II	(^{5}D)5p ^{6}F$_{7/2}$	90300	H Lyα	MJC99
Fe II	(^{5}D)5p ^{4}F$_{9/2}$	90386	H Lyα	MJC99
Fe II	(^{5}D)5p ^{4}D$_{7/2}$	90397	H Lyα	MJC99
Fe II	(b^{3}P)4p ^{4}S$_{3/2}$	90629	H Lyα	JJ84
Fe II	(^{5}D)5p ^{4}D$_{5/2}$	90638	H Lyα	JJ84
Fe II	(^{5}D)5p ^{4}F$_{7/2}$	90780	H Lyα	JJ84
Fe II	(b^{3}P)4p ^{4}P$_{1/2}$	90839	H Lyα	JJ84
Fe II	(b^{3}P)4p ^{4}P$_{3/2}$	90898	H Lyα	JJ84
Fe II	(^{5}D)5p ^{4}P$_{5/2}$	90901	H Lyα	JJ84
Fe II	(^{5}D)5p ^{4}D$_{3/2}$	91048	H Lyα	JJ84
Fe II	(^{5}D)5p ^{4}F$_{5/2}$	91070	H Lyα	JJ84
Fe II	(^{5}D)5p ^{4}D$_{1/2}$	91199	H Lyα	JJ84
Fe II	(^{5}D)5p ^{4}F$_{3/2}$	91208	H Lyα	JJ84
Fe II	(^{4}G)4s4p(^{3}P) x^{4}H$_{11/2}$	92166	He II 1084	JH98
Fe II	(b^{3}F)4p u^{2}G$_{9/2}$	92171	He II 1084	HJ00
Fe II	(b^{3}F)4p u^{4}F$_{3/2}$	93328	[N IV] 1487	HJ00
Fe II	(^{4}P)4s4p ^{4}D$_{3/2}$	95858	H Lyα	CP88
Fe II	(^{1}G)4p ^{2}H$_{11/2}$	98278	H Lyα	MJC99
Fe II	(^{4}P)4s4p(^{x}P) ^{2}S$_{1/2}$	103967	H Lyα	MJC99
Fe II	(^{2}D)4s4p(^{3}P) ^{4}F$_{7/2}$	104907	H Lyα/O V 1218	JC88
Fe II	(^{2}F)4s4p(^{x}P) ^{4}G$_{9/2}$	107674	H Lyα	MJC99
Fe II	(^{2}F)4s4p(^{3}P) ^{4}G$_{11/2}$	107720	H Lyα	JJ84

Table 11.1 (Continued)

Ion	Pumped level	Energy (cm^{-1})	Pumping line	Ref.
Fe II	$(^4\text{F})4\text{s}4\text{p}(^x\text{P})\,^6\text{D}_{5/2}$	108130	H Lyα	MJC99
Fe II	$(^2\text{F})4\text{s}4\text{p}(^x\text{P})\,^4\text{F}_{9/2}$	108217	H Lyα	MJC99
Fe II	$(^4\text{F})4\text{s}4\text{p}(^3\text{P})\,^6\text{D}_{7/2}$	108239	H Lyα	MJC99
Fe II	$(^4\text{G})4\text{s}4\text{p}(^1\text{P})\,^4\text{H}_{9/2}$	108868	Ne V 1146/Si III 1207	JH98
Fe II	$(\text{a}^3\text{F})5\text{p}\,^4\text{D}_{5/2}$	110568	O VI 1032	J88
Fe III	$3\text{d}^5(^6\text{S})4\text{p}\,\text{z}^7\text{P}_3$	82333	H Lyα	JZH00
Zr I	x^1G_4		Mg II 2795	B47
In I	$\text{d}\,^2\text{S}_{1/2}$		Hγ	T37

B47 = [4], BFJ81 = [7], BJ80 = [5], C70 = [9], CP88 = [8], HJ00 = [13], HLW77 = [11], J83 = [15], J88 = [16], JC88 = [17], JH98 = [25], JJ84 = [19], JWG95 = [21], JZH00 = [22], MJC99 = [27], T37 = [30], WC90 = [33], ZHJ01 = [34]

11.4 Fluorescent Fe II Lines in RR Tel

Following the more general introduction, we will discuss the fluorescent Fe II lines in more detail using the beautiful spectrum of the symbiotic star RR Tel as an example.

The continuum radiation is often weak in the ultraviolet region of symbiotic stars since the Planck radiation from the hot and the cool stars in the symbiotic system dominates at shorter and longer wavelengths, respectively. Superimposed on the weak continuum is a rich emission line spectrum, dominated by Fe II lines from high-excitation levels in addition to lines of highly-ionized light elements (see spectrum in Fig. 11.2). Many of the Fe II lines are observed from specific odd levels with an energy of about 11 eV.

All levels which show an enhanced population and fluorescence lines have a strong transition connecting to a low state. As described above, many of these coincide in wavelength with H Lyα at 1215 Å. The flux is thus transferred from H Lyα, absorbed by Fe II and reemitted in the fluorescence lines. The levels observed to be pumped by H Lyα belong to the configurations $(^5\text{D})5\text{p}$, $(\text{b}^3\text{P})4\text{p}$ and $(\text{b}^3\text{F})4\text{p}$. The small energy difference between them enhances the mixing between the levels and they share some properties. This mixing is responsible for enhancing the strength of the pumping channels for the latter two configurations. Strong lines are also seen from even parity levels. These cannot be pumped directly from any of the low even parity levels but are rather populated through cascades from the H Lyα pumped levels.

11.5 Identified Fluorescence Mechanisms

The natural way for the H Lyα pumped $(^5\text{D})5\text{p}$ levels to decay is within the $(^5\text{D})nl$ system either back to 4s or via the 3-photon route $5\text{p} \rightarrow 5\text{s} \rightarrow 4\text{p} \rightarrow 4\text{s}$ where

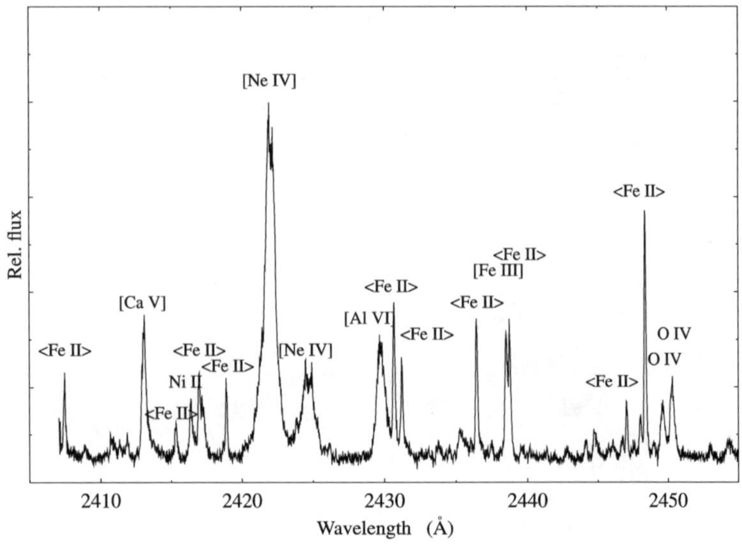

Fig. 11.2 Part of the ultraviolet spectrum of RR Tel showing numerous fluorescent ⟨Fe II⟩ and high ionization lines (HST/GHRS)

the wavelength regions for the different steps are 9000 Å, 2800 Å and 2500 Å, respectively. The secondary decay is observed from almost all of the $(^5D)5s$ levels, both from e^4D and e^6D. When decaying, the flux is spread over many channels, and the flux per transition is thus also less from levels further down in the system. But some 5s levels are fed by cascades from many upper levels which might compensate for that. The $(b^3P)4p$ and $(b^3F)4p$ levels have their primary decay in the ultraviolet (see Fig. 11.4).

Fluorescence lines pumped by H Lyα are observed in a number of astrophysical objects. In RR Tel, strong lines from additional levels in Fe II are observed as well. In general they come from low excitation states compared to the H Lyα pumped levels. As reported in [13, 15, 16] other lines are responsible for this pumping, see Table 11.1.

The pumping channels for these levels differ in wavelength, since the different $(^ML)4p$ states have different energies depending on parent term, from 60 kK for $(a^3P)4p$ up to almost 100 kK for $(b^1G)4p$ (see level diagram on Fe II in Fig. 11.3). The decay is, on the contrary, localized to the region 2500–2800 Å, since the difference between the $(^ML)4p$ and the $(^ML)4s$ states, which determines the wavelength of the fluorescence, is independent of parent term. The decay to the $3d^7$ levels falls in a wider wavelength region. This is an effect of the energy structure of the Fe II spectrum.

The strongest Fe II lines in the ultraviolet spectrum of RR Tel are the lines at 1776, 1881 and 1884 Å, all from the level $(a^3F)5p^4 D_{5/2}$. This level is pumped due to a coincidence with the O VI line at 1032 Å [16]. The discovery of the fluorescence lines was the first indication of the presence of O VI in the RR Tel system.

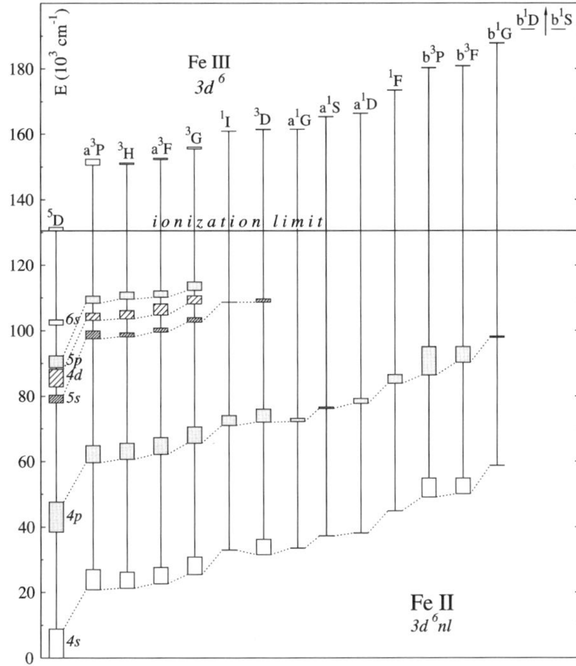

Fig. 11.3 Energy level diagram on Fe II (by courtesy of Sveneric Johansson)

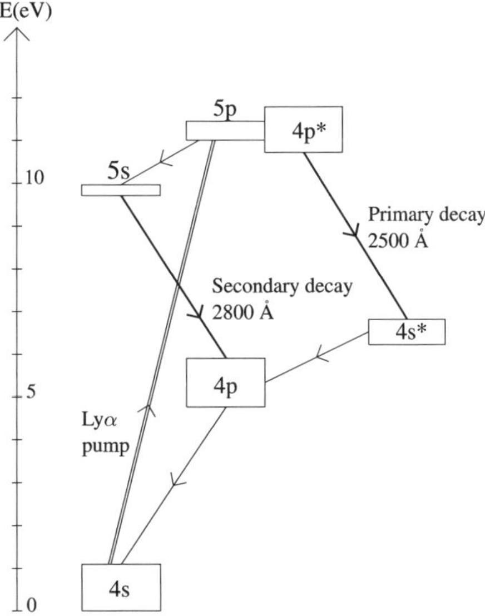

Fig. 11.4 Primary and secondary UV fluorescence channels (*bold lines*) from Fe II levels pumped by H Lyα (Notations: $nl = (^5D)nl$; $nl^* = (b^3F)nl, (b^3P)nl$)

Later Schmid [29] resolved the long-standing puzzle of two strong lines unidentified in spectra of many symbiotic stars. He suggested that the O VI lines at 1032 and 1038 Å were Raman scattered by hydrogen atoms. The energy difference between the oxygen lines and H I yβ was the same as the difference between the scattered lines and Hα. The presence of O VI lines has later been confirmed by direct far-UV observations of the λ 1032, 1038 lines. The fluorescence lines at 1776 Å might even work as a pumping line itself. This line coincides in wavelength with another Fe II line which connects a low state with a high-excitation state where the latter is seen to have an enhanced population [13].

The strong fluorescence lines pumped by C IV and O VI are not seen in Eta Carinae, although the H Lyα fluorescence is strong in this object [35] and apparently has favorable conditions for strong fluorescence. The reason is the lack of lines or ions responsible for the pumping. The mechanism producing the fluorescent Fe III in Eta Carinae mentioned in the previous section is on the contrary not working in RR Tel, although both H Lyα radiation and Fe^{2+} ions are present. One explanation can be the lack of spatial overlap for the radiation and atoms necessary for an efficient absorption.

11.6 Selective Ionization Through Resonance Enhanced Two-Photon Ionization (RETPI)

In low-density plasmas with strong radiation fields, the ionization structure is not determined by the LTE conditions and the SAHA equation, but rather by photo-ionization and recombination. A review of spectroscopy of photo-ionized plasmas is given by [10]. In analogy with the fluorescence discussed earlier, the radiation for the ionization emerge from either continuum or line emission. In general, broad continuum makes this more efficient. However, for a plasma where the emission lines such as the Lyman series of hydrogen gets very intense these lines can be as important for the ionization structure of certain elements. The intensity of the Lyman continuum radiation is converted to line radiation through photoionization of hydrogen and subsequent recombination producing the line radiation. The line radiation in turn photoionizes other elements and contribute to the ionization balance.

The RETPI process in low-density plasma was considered by [20], examining the possibilities for the process to work in C, N, O and Ar, Ne. An astrophysical plasma where this process may be important is the gas condensation close to the massive star Eta Carinae, the so-called Weigelt blobs (WBs). These are spatially resolved from the central stars using HST/STIS and the spectra indicated that the optical depth of hydrogen is sufficient to produce strong Ly line emission with the WBs. Eta Carinae exhibit a variation in the stellar radiation, where the radiation illuminating the WBs is cut of during several months each 5.5 years. These variations act as laboratory experiment with varying conditions, and allows for separating excitations processes with different time-scales. Excitation through recombination or collisions with electrons depend on the electron density, which changes on timescales of months. The radiation, on the contrary, drops on the time scale of days. Thus, data recorded during the months of less incident radiation allow for a discrimination between excitation mechanisms of the lines in the spectrum of the WBs [14]. From the recombination coefficients, the electron density can also be derive from

Fig. 11.5 RETPI pumping
scheme for the production of
strong Si III emission at
1892 Å. For energies
involved, see [24]

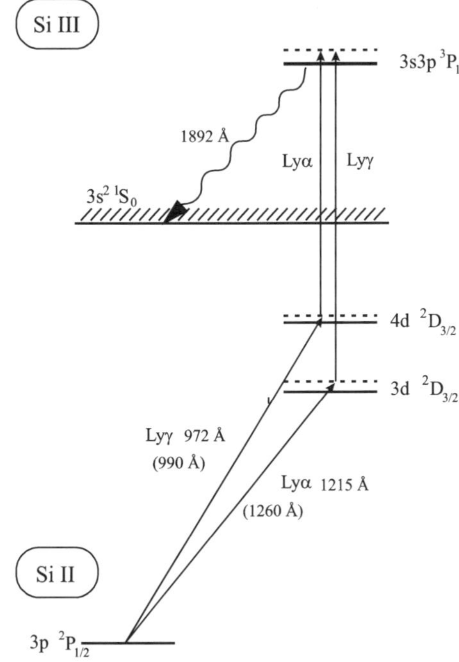

the decay time of the recombination lines, allowing for additional probes of physical conditions in addition to the standard line ratios of forbidden lines.

One of the most striking examples of RETPI is the case of the [Si III] line at 1892 Å, being one of the strongest lines in the spectrum of Eta Carinae [35]. This line also shows the most prominent variations with time, as it disappears during the minimum state of Eta Carinaes 5.5 year cycle. The excitation routes are shown in Fig. 11.5.

Acknowledgements I am grateful to the organizing committee for the opportunity to present this work at the CDAMOP conference in New Delhi, 2011. This research is supported through the Swedish research council (VR) through contract 621-2011-4206. I am grateful to Prof. Sveneric Johansson for at an early stage introducing me to the field of excitation processes in astrophysical plasmas and sharing his vast knowledge on atomic spectroscopy using laboratories and astronomical objects.

References

1. I.S. Bowen, Astrophys. J. **67**, 1 (1928)
2. I.S. Bowen, Publ. Astron. Soc. Pac. **46**, 146 (1934)
3. I.S. Bowen, Astrophys. J. **81**, 1 (1935)
4. I.S. Bowen, Publ. Astron. Soc. Pac. **59**, 196 (1947)
5. A. Brown, C. Jordan, Mon. Not. R. Astron. Soc. **191**, 37P (1980)
6. A. Brown, C. Jordan, R. Wilson, in *The First Year of IUE*, ed. by A. Willis (1979), p. 232

7. A. Brown, M. Ferraz, C. Jordan, in *The Universe at Ultraviolet Wavelengths: The First Two Years of International Ultraviolet Explorer* (1981), pp. 297–302
8. K.G. Carpenter, J.E. Pesce, R.E. Stencel et al., Astrophys. J. Suppl. Ser. **68**, 345 (1988)
9. C. Cowley, *The Theory of Stellar Spectra* (Gordon & Breach, New York, 1970)
10. G.J. Ferland, Annu. Rev. Astron. Astrophys. **41**, 517 (2003)
11. B.M. Haisch, J.L. Linsky, A. Weinstein, R.A. Shine, Astrophys. J. **214**, 785 (1977)
12. F. Hamann, D.L. Depoy, S. Johansson, J. Elias, Astrophys. J. **422**, 626 (1994)
13. H. Hartman, S. Johansson, Astron. Astrophys. **359**, 627 (2000)
14. H. Hartman, A. Damineli, S. Johansson, V.S. Letokhov, Astron. Astrophys. **436**, 945 (2005)
15. S. Johansson, Mon. Not. R. Astron. Soc. **205**, 71 (1983)
16. S. Johansson, Astrophys. J. Lett. **327**, L85 (1988)
17. S. Johansson, K.G. Carpenter, in *A Decade of UV Astronomy with the IUE Satellite*, vol. 1 (1988), pp. 361–363
18. S. Johansson, F. Hamann, Phys. Scr. T **47**, 157 (1993)
19. S. Johansson, C. Jordan, Mon. Not. R. Astron. Soc. **210**, 239 (1984)
20. S. Johansson, V. Letokhov, Science **291**, 625 (2001)
21. S. Johansson, G. Wallerstein, K.K. Gilroy, A. Joueizadeh, Astron. Astrophys. **300**, 521 (1995)
22. S. Johansson, T. Zethson, H. Hartman et al., Astron. Astrophys. **361**, 977 (2000)
23. S. Johansson, T. Zethson, H. Hartman, V. Letokhov, in *Eta Carinae and Other Mysterious Stars: The Hidden Opportunities of Emission Line Spectroscopy*, ed. by T.R. Gull, S. Johansson, K. Davidson. ASP Conf. Ser., vol. 242 (Astronomical Society of the Pacific, San Francisco, 2001), p. 297
24. S. Johansson, H. Hartman, V.S. Letokhov, Astron. Astrophys. **452**, 253 (2006)
25. C. Jordan, G.M. Harper, in *Cool Stars, Stellar Systems and the Sun: Tenth Cambridge Workshop*. ASP Conf. Ser., vol. 154 (1998), p. 1277
26. S. Kastner, A. Bhatia, Comments At. Mol. Phys. **18**, 39 (1986)
27. A.D. McMurry, C. Jordan, K.G. Carpenter, Mon. Not. R. Astron. Soc. **302**, 48 (1999)
28. A. Redfors, S.G. Johansson, Astron. Astrophys. **364**, 646 (2000)
29. H.M. Schmid, Astron. Astrophys. **211**, L31 (1989)
30. A.D. Thackeray, Astrophys. J. **86**, 499 (1937)
31. E. Verner, F. Bruhweiler, D. Verner, S. Johansson, T. Gull, Astrophys. J. Lett. **592**, L59 (2003)
32. R. Viotti, L. Rossi, A. Cassatella, A. Altamore, G.B. Baratta, Astrophys. J. Suppl. Ser. **71**, 983 (1989)
33. G.M. Wahlgren, K.G. Carpenter, in *Cool Stars, Stellar Systems, and the Sun*, ed. by G. Wallerstein. ASP Conf. Ser., vol. 9 (Astronomical Society of the Pacific, San Francisco, 1990), p. 67
34. T. Zethson, H. Hartman, S. Johansson et al., in *Eta Carinae and Other Mysterious Stars: The Hidden Opportunities of Emission Spectroscopy*, ed. by T.R. Gull, S. Johansson, K. Davidson. ASP Conf. Ser., vol. 242 (Astronomical Society of the Pacific, San Francisco, 2001), p. 97
35. T. Zethson, S. Johansson, H. Hartman, T.R. Gull, Astron. Astrophys. **540**, A133 (2012)

Chapter 12
Topological Insulators with Ultracold Atoms

Indubala I. Satija and Erhai Zhao

Abstract Ultracold atom research presents many avenues to study problems at the forefront of physics. Due to their unprecedented controllability, these systems are ideally suited to explore new exotic states of matter, which is one of the key driving elements of the condensed matter research. One such topic of considerable importance is topological insulators, materials that are insulating in the interior but conduct along the edges. Quantum Hall and its close cousin Quantum Spin Hall states belong to the family of these exotic states and are the subject of this chapter.

12.1 Topological Aspects: Chern Numbers and Edged States

Quantum engineering, i.e., preparation, control and detection of quantum systems, at micro- to nano-Kelvin temperatures has turned ultracold atoms into a versatile tool for discovering new phenomena and exploring new horizons in diverse branches of physics. Following the seminal experimental observation of Bose Einstein condensation, simulating many-body quantum phenomena using cold atoms, which in a limited sense realizes Feynman's dream of quantum simulation, has emerged as an active frontier at the cross section of AMO and condensed matter physics. There is hope and excitement that these highly tailored and well controlled systems may solve some of the mysteries regarding strongly correlated electrons, and pave ways for the development of quantum computers.

The subject of this chapter is motivated by recent breakthroughs in creating "artificial gauge fields" that couple to neural atoms in an analogous way that electromagnetic fields couple to charged particles [11]. This opens up the possibility of studying Quantum Hall (QH) states and its close cousin Quantum Spin Hall (QSH) states, which belong to the family of exotic states of matter known as *Topological Insulators*.

I.I. Satija (✉) · E. Zhao
School of Physics, Astronomy and Computational Sciences, George Mason University, Fairfax, VA 22030, USA
e-mail: isatija@gmail.com

M. Mohan (ed.), *New Trends in Atomic and Molecular Physics*,
Springer Series on Atomic, Optical, and Plasma Physics 76,
DOI 10.1007/978-3-642-38167-6_12, © Springer-Verlag Berlin Heidelberg 2013

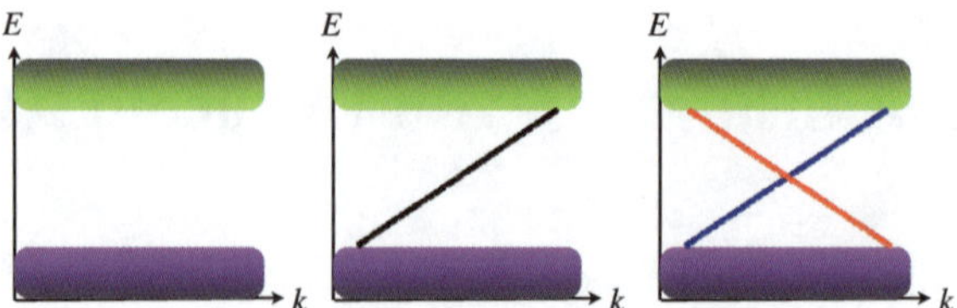

Fig. 12.1 Figure illustrates distinction between ordinary band insulator (*left*), QH (*middle*) and QSH (*right*) insulators in the energy spectrum. Connecting the conduction and valence bands are the edge states

12.1.1 Topological Aspects of QH and QSH States

Topological insulators (TI) are unconventional states of matter [6] that are insulating in the bulk but conduct along the boundaries. A more general working definition of TI, applicable to the case of neutral atoms, is that they are band insulators with a bulk gap but robust gapless boundary (edge or surface) modes (Fig. 12.1). These boundary modes are protected against small deformations in the system parameters, as long as generic symmetries such as time-reversal or charge-conjugation are preserved and the bulk gap is not closed. Such states arise even in systems of non-interacting fermions and are considerably simpler than topological states of strongly interacting electrons, such as the fractional QH effect. Despite such deceiving simplicity, as we will show in this chapter, the subject is very rich and rather intricate, with plenty of open questions.

We will consider two prime examples of TIs in two dimension, the QH and QSH states. A topological insulator is not characterized by any local order parameter, but by a topological number that captures the global structures of the wave functions. For example, associated with each integer quantum Hall state is an integer number, the Chern number. The Chern number also coincides with the quantized Hall conductance (in unit of $e^2/\hbar$) of the integer QH state. The quantization of the Hall conductance is extremely robust, and has been measured to one part in 10^9 [10]. Such precision is a manifestation of the topological nature of quantum Hall states. A quantum Hall state is accompanied by chiral edge modes, that propagate along the edge along one direction and their number equals the Chern number. Semi-classically, these edge states can be visualized as resulting from the skipping orbits of electrons bouncing off the edge while undergoing cyclotron motion in external magnetic field. The edge states are insensitive to disorder because there are no states available for backscattering, a fact that underlies the perfectly quantized electronic transport in the QHE. This well known example demonstrates the "bulk-boundary correspondence", a common theme for all topological insulators.

The QSH state is time-reversal invariant [8]. It is characterized by a binary Z_2 topological invariant $\nu = 1$. The edge modes in this case come in the form of Kramers pairs, with one set of modes being the time-reversal of the other set. In the simplest example, spin up electron and spin down electron travel in opposite direction along the edge. A topologically non-trivial state with $\nu = 1$ has odd number of Kramers pairs at its edge. Topologically trivial states have $\nu = 0$, and even number of Kramers pairs, the edge states in this case are not topological stable. This is another manifestation of the "bulk-boundary correspondence". Recent theoretical research has achieved a fairly complete classification, or "periodic table", for all possible topological band insulators and topological superconductors/superfluids in any spatial dimensions [9, 15, 16]. Several authoritative reviews of this rapidly evolving field now exist [6, 13, 14], which also cover the fascinating subject of how to describe TIs using low energy effective field theory with topological (e.g., Chern-Simons or θ-) terms.

12.1.2 Experimental Realization

QH and QSH effect usually requires magnetic field and spin-orbit coupling, respectively. In contrast to electrons in solids, cold atoms are charge neutral and interact with external electromagnetic field primarily via atom-light coupling. Artificial orbital magnetic field can be achieved by rotating the gas [2]. In the rotating frame, the dynamics of atoms becomes equivalent to charged particles in a magnetic field which is proportional to the rotation frequency. To enter the quantum Hall regime using this approach requires exceedingly fast rotation, a task up to now remains a technical challenge. In this regard, another approach to engineer artificial magnetic field offers advantages. This involves engineering the (Berry's) geometric phase of atoms by properly arranging inhomogeneous laser fields that modify the eigenstates (the dressed states) via atom-light coupling [3]. Such geometric phase can mimic for example the Aharonov-Bohm phase of orbital magnetic field. Because the dressed states can be manipulated on scales of the wavelength of light, the artificial gauge fields can be made very strong. For example, the phase acquired by going around the lattice unit cell can approach the order of π. This gives hope to enter the quantum Hall regime. The strength of this approach also lies in its versatility. Artificial gauge fields in a multitude of forms, either Abelian (including both electric and magnetic field) or non-Abelian (including effective spin-orbit coupling), can be introduced for continuum gases or atoms in optical lattices [3]. Recent NIST experiments have successfully demonstrated artificial magnetic field [11] as well as spin-orbit coupling [12] for bosonic Rb atoms and experiments are underway to reach QH and QSH regimes.

12.2 The Hofstadter Model and Chern Number

For a moderately deep square optical lattice, the dynamics of (spinless) fermionic atoms in the presence of an artificial magnetic field is described by the tight binding Hamiltonian known as the Hofstadter model [7],

$$H = -\sum_{nm} t_x c^{\dagger}_{n+1,m} e^{i\theta_{n+1,m:n,m}} c_{n,m} + t_y c^{\dagger}_{n,m+1} e^{i\theta_{n,m+1:n,m}} c_{n,m} + h.c. \qquad (12.1)$$

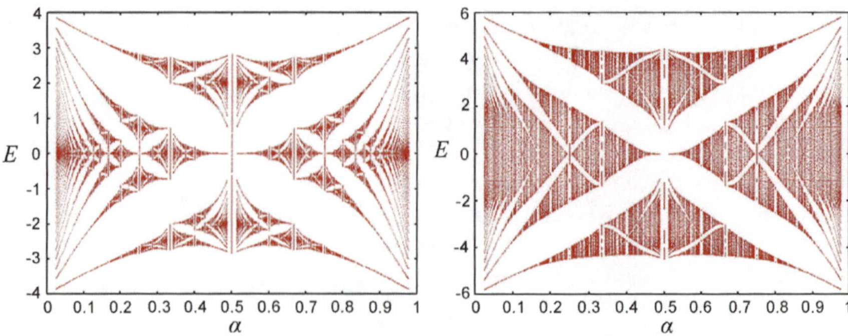

Fig. 12.2 The energy spectrum as function of the magnetic flux α per plaquette, commonly referred as the butterfly spectrum, for the Hofstadter model with $\lambda = t_y/t_x = 1, 2$ respectively

where $c_{n,m}$ is the fermion annihilation operator at site (n, m), with n/m is the site index along the x/y direction, t_x and t_y are the nearest-neighbor hopping along the x and the y-direction respectively. The anisotropy in hopping is described by dimensionless parameter $\lambda = t_y/t_x$. The phase factor $\theta_{n,m:n',m'}$ is given the line integral of the vector potential $\mathbf{A}$ along the link from site (n', m') to (n, m), $(2\pi e/ch) \int \mathbf{A} \cdot d\mathbf{l}$. The magnetic flux per plaquete measured in flux quantum is labeled α, and represents the strength of magnetic field perpendicular to the 2D plane. The energy spectra for $\lambda = 1, 2$ are shown in Fig. 12.2. At commensurate flux, i.e., α is a rational number $\alpha = p/q$ with p and q being integers, the original energy band splits into q sub-bands by the presence of magnetic field, and the spectrum has $q - 1$ gaps. For incommensurate flux, namely when α is an irrational number, the fractal spectrum contains a hierarchy of sub-bands with infinite fine structures. As shown in Fig. 12.2, varying anisotropy changes the size of the gaps without closing any gap.

In the Landau gauge, the vector potential $A_x = 0$ and $A_y = -\alpha x$ and the unit cell is of size $q \times 1$ and the magnetic Brillouin zone is $-\pi/q \leq k_x \leq \pi/q$ and $-\pi \leq k_y \leq \pi$ (the lattice spacing is set to be 1). The eigenstates of the system can be written as $\Psi_{n,m} = e^{ik_x n + ik_y m} \psi_n$ where ψ_n satisfies the Harper equation [7],

$$e^{ik_x} \psi_{n+1}^r + e^{-ik_x} \psi_{n-1}^r + 2\lambda \cos(2\pi n\alpha + k_y)\psi_n^r = -E\psi_n^r. \tag{12.2}$$

Here $\psi_{n+q}^r = \psi_n^r$, $r = 1, 2, \ldots, q$, labels linearly independent solutions. The properties of the two-dimensional system can be studied by studying one-dimensional equation (12.2). For chemical potentials inside each energy gap, the system is in an integer quantum Hall state characterized by its Chern number c_r and the transverse conductivity $\sigma_{xy} = c_r e^2/h$, where the Chern number of r-filled bands is given by,

$$c_r = \text{Im} \sum_{l=1}^{r} \int dk_x dk_y \sum_{n=1}^{q} \partial_{k_x}\left(\psi_n^l\right)^* \partial_{k_y} \psi_n^l. \tag{12.3}$$

Here the integration over k_x, k_y is over the magnetic Brillouin zone.

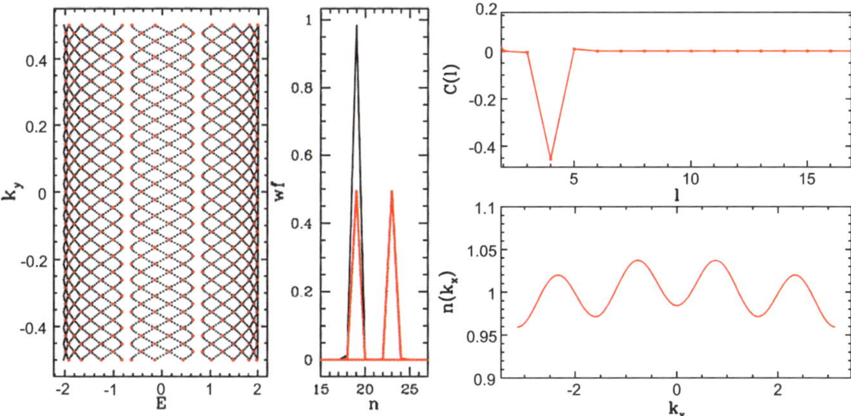

Fig. 12.3 *Left*: energy spectrum for $\lambda = 10$, $\alpha = 13/21$. The *red dots* indicate dimerized states at the band edges. *Middle*: the wave function of a state localized at a single site, with energy away from the band edge (*black*), and the dimerized wave function that localizes at two distinct sites at a distance of four-lattice spacing apart for state at band edge near a gap with Chern number 4 (*red*). Figure on the *right* shows the two point correlation function $C(l)$ that vanishes unless $l = 4$ resulting in almost sinusoidal momentum distribution with period 4

12.3 Chern-Dimer Mapping in the Limit of Large U

As illustrated in Fig. 12.2, the hopping anisotropy changes the size of the gaps, without ever closing the gap. Hence the topological aspects of the underlying states are preserved, as λ is continuously tuned. In the limit $\lambda \gg 1$, the problems becomes tractable using degenerate perturbation theory [4]. For $\lambda \to \infty$, the q bands have dispersion $E_r(k_y) = -2t_y \cos(2\pi\alpha r + k_y)$, $r \in \{1, \ldots, q\}$. These eigenstates are extended along y but localized to single sites as a function of x, with wave function $\psi_n^r = \delta_{n,r}$. These cosine bands shifted from each other intersect at numerous degeneracy points k_y^*. For finite λ, the hopping term proportional to t_x hybridizes the two states at each crossing, resulting an avoided crossing as shown in Fig. 12.3. Near each degeneracy point the system is well described by a two-level effective Hamiltonian [4],

$$H(\mathbf{k}) = \Delta_\ell \big[\cos(k_x\ell)\sigma_x + \sin(k_x\ell)\sigma_y\big] + v_\ell\big(k_y - k_y^*\big)\sigma_z. \tag{12.4}$$

Here $\sigma_{x,y,z}$ are the Pauli matrices in the space of localized states $|r\rangle$ and $|r'\rangle$, and $\ell = r' - r$ is the spatial separation between these two localized states. The expression for Δ_ℓ and v_ℓ are given in Ref. [18]. The eigenstates are equal-weight superpositions of $|r\rangle$ and $|r'\rangle$,

$$\psi_n^\pm = \frac{1}{\sqrt{2}}\big(\delta_{n,r} + e^{i\beta_\pm}\delta_{n,r'}\big), \tag{12.5}$$

where $\beta_+ = -\ell(k_x + \pi) + \pi$ and $\beta_- = -\ell(k_x + \pi)$ are relative phases for the upper and lower band edges, respectively. We call such dimerized state "Chern-dimers". Starting from Eq. (12.5), it is straightforward to show [4] that the corresponding

Chern number is related to the phase factor $e^{ik_x\ell}$, and simply given by $c_r = \ell$. Thus, for sufficiently anisotropic hopping, dimerized states form at the band edges with its spatial extent along x equal to the Chern number of the corresponding gap. Such one-to-one correspondence between the Chern number and the size of the real space dimer (in the Landau gauge) will be referred to as "Chern-dimer mapping". While calculation of Chern number in this limit is not new, this mapping has not been elucidated and emphasized until our work [18].

The formation of Chern-dimers of size $\ell = c_r$ at the edges of the r-th gap implies a "hidden spatial correlation". The correlation function

$$C(l) = \sum_{n,m} \left\langle c_{n,m}^{\dagger} c_{n+l,m} \right\rangle \tag{12.6}$$

will peak at $l = \ell = c_r$. For chemical potential in the r-th gap, as $\lambda \to \infty$ (Fig. 12.3), $C(l)$ asymptotes to a delta function $C(l) \to \delta(l - \ell)$, due to the contribution of Chern-dimers at the lower edge of the r-th gap. Note that the net contribution to $C(l)$ from all other dimerized states associated with gaps fully below the Fermi energy is zero, because $\beta_+ - \beta_- = \pi$ for each of these crossings. Remarkably, $C(l)$, or more precisely its Fourier transform $n(k_x)$, is directly accessible in the time-of-flight (TOF) image, and routine observable in cold atom experiments.

In the time-of-flight experiments, the artificial gauge field is first turned off abruptly. This introduces an impulse of effective electric field that converts the canonical momentum of atoms (when gauge field was on) into the mechanical momentum (these subtleties are discussed in [18]). Then the optical lattice and trap potential are turned off, the density of the atomic cloud $n(\mathbf{x})$ is imaged after a period of free expansion t_{TOF}. The image is a direct measurement of the (mechanical) momentum distribution function with $\mathbf{k} = m\mathbf{x}/\hbar t_{\text{TOF}}$,

$$n(\mathbf{k}) = \sum_{\mathbf{r},\mathbf{r}'} \left\langle c_{\mathbf{r}}^{\dagger} c_{\mathbf{r}'} \right\rangle e^{i\mathbf{k}\cdot(\mathbf{r}-\mathbf{r}')}, \tag{12.7}$$

where the site label $\mathbf{r} = m\hat{x} + n\hat{y}$, and for simplicity we neglect the overall prefactor and the Wannier function envelope. Thus, from TOF images, the 1D momentum distribution can be constructed by integrating $n(\mathbf{k})$ over k_y, $n(k_x) = \int_{-\pi}^{\pi} dk_y n(\mathbf{k})$, which is nothing but the Fourier transform of $C(l)$. Thus, in the limit of large λ, $C(l)$ becomes a delta function, and the 1D momentum distribution function becomes

$$n(k_x) = r \pm \frac{1}{2}\cos(c_r k_x) + \cdots. \tag{12.8}$$

It assumes a sinusoidal form, with oscillation period given by the Chern number. Note the sign depends on which gap the chemical potential lies in [18], and the offset r comes from the contribution from the filled bands below the Fermi energy. In summary, we have arrived at the conclusion that Chern numbers can be simply read off from $n(k_x)$ as measured in TOF, in the large anisotropy limit.

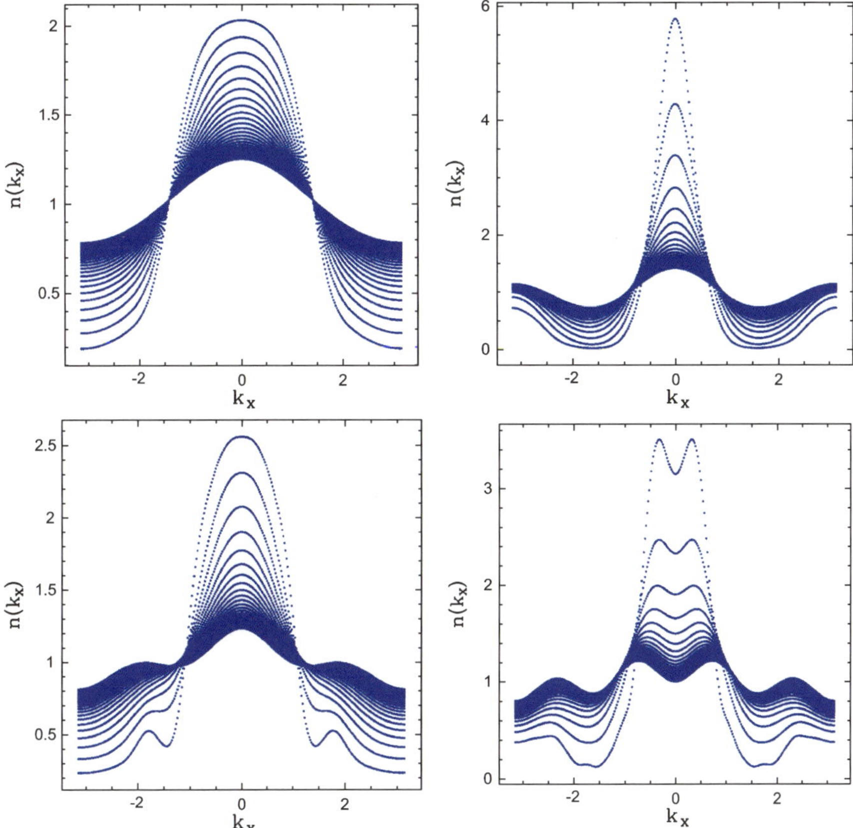

Fig. 12.4 Momentum distribution as anisotropy parameter varies from 1–10, for $\alpha = 0.45$ for Chern numbers 1 (*top-left*), 2 (*top-right*), 3 (*bottom-left*), and 4 (*bottom-right*)

12.4 Traces of Chern Number in Time-of-Flight Images

This begs the question whether it is still possible to extract c_r from TOF when away from the large λ limit. Note that, while the Chern number does not change inside the same gap as λ is tuned, $n(k_x)$ is not topologically invariant. There is a prior no reason to expect Chern number continues to leave its fingerprints in $n(k_x)$. Motivated by these considerations, we have carried out detailed numerical studies of momentum distribution in the entire $\alpha - \lambda$ parameter plane, going well beyond the "Chern-dimer regime". Figure 12.4 highlights our results with various λ but fixed α and Chern number. As λ is reduced, $n(k_x)$ gradually deviates from the sinusoidal distribution. Yet, the local maxima (peaks) persist. In the limit of isotropic lattice, $n(k_x)$ shows pronounced local peaks. For a wide region of parameters, we can summarize our finding into an empirical rule: *the number of local peaks in $n(k_x)$ within the first Brillouin zone is equal to the Chern number*. Note that this rule is not always precise, as seen in Fig. 12.4 for the case of Chern number 3 for intermediate

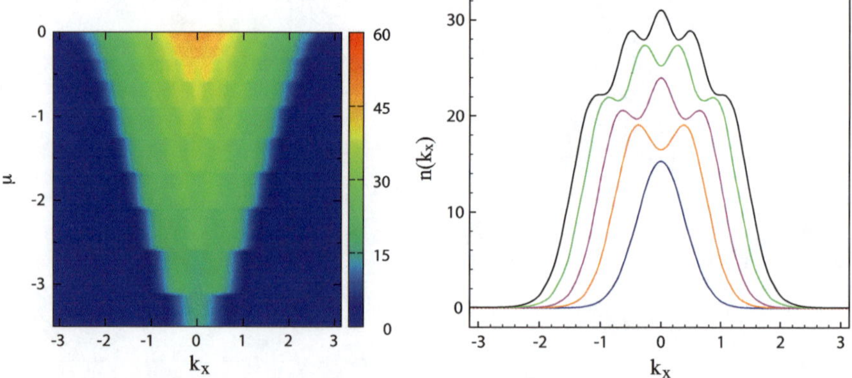

Fig. 12.5 *Left*: $n(k_x, \mu)$ in false color for a 50×50 lattice with isotropic hopping and flux $\alpha = 0.05$. *Right*: $n(k_x)$ for chemical potential $\mu/t = -3.4, -2.8, -2.4, -1.8, -1.5$ respectively (*bottom* to *top*). The corresponding Chern number is 1, 2, 3, 4, and 5

values of λ where local peaks become a broad shoulder-like structure. Despite this, it is valid in extended regions including both large λ and $\lambda = 1$. In fact, given the reasoning above, our empirical rule seems to work surprisingly well. But this is an encouraging result for cold atoms experiments. From these distinctive features in $n(k_x)$, the Chern number can be determined rather easily by counting the peaks, especially for the major gaps with low Chern number.

The Chern-dimer mapping provides a simple, intuitive understanding of $n(k_x)$ in the limit of large λ. For the isotropic case, there is no dimerization, why are there pronounced peaks in $n(k_x)$, and why is the peak number precisely equal to the Chern number? To obtain further insight, we considered the low flux limit, $\alpha \to 0$, where the size of the cyclotron orbit is large compared to the lattice spacing, and the system approaches the continuum limit. The spectrum then can be approximated by discrete Landau levels. For r filled Landau levels, $n(k_x)$ can be obtained rather easily,

$$n(k_x) = \sum_{v=0}^{r-1} \left(2^v v!\right)^{-1} H_v^2(k_x) e^{-k_x^2}, \tag{12.9}$$

where H_v is the Hermite polynomial, and k_x is measured in the inverse magnetic length $\sqrt{eB/\hbar c}$. $n(k_x)$ is given by the weighted sum of consecutive positive-definite function, $H_v^2(k_x)$. The node structure of H_v is such that $n(k_x)$ has *exactly* r local maxima (peaks) for r filled Landau levels. This nicely explains the numerical results for the Hofstadter model at low (but finite) flux $\alpha = 1/q$, where the Chern number $c_r = r$ in the r-th gap, and $n(k_x)$ has precisely r peaks (see Fig. 12.5). Note that this calculation does not directly explain the feature shown in Fig. 12.4 for $\alpha = 0.45$ and $\lambda = 1$. As α increases, for μ residing within the same energy gap, the distinctive features of $n(k_x)$ are generally retained, so the counting rule is still valid. Compared to the low flux regime, however, the local peaks (except the one at $k_x = 0$) move to higher momenta.

12.5 Chern Tree at Incommensurate Flux

While it is numerically straightforward to investigate the $n(k_x)$ or σ_{xy} for QHE at incommensurate flux α, analytical analysis of the Chern number for this case represents a conceptual as well as a technical challenge. In fact, for irrational α, the system no longer has translational symmetry, and the notion of magnetic Brillouin zone or Bloch wave function lose their meaning. The incommensurability acts like *correlated disorder*. The TKNN formula for a periodic system is not directly applicable. In addition, the energy spectrum becomes infinitely fragmented (a Cantor set). One expects then a dense distribution of Chern numbers even within a small window of energy. Then two questions arise. First, will these Chern numbers largely random or are there deterministic patterns governing their sequence as μ is continuously varied? Second, what is the experimental outcome of $n(k_x)$ or σ_{xy} when measured with a finite energy resolution?

We address some of these questions here for the isotropic lattice ($\lambda = 1$) using the golden-mean flux, $\alpha = (\sqrt{5} - 1)/2$, as an example. We will employ a well known trick, and approach the irrational limit by a series of rational approximation $\alpha_n = F_n/F_{n+1}$, where F_n is the n-th Fibonacci numbers. Explicitly, the series $3/5, 5/8, 8/13, 13/21, 21/34, 34/55, 55/89, \ldots$ approaches the golden-mean with increasing accuracy. At each stage of the rational approximation, the energy spectrum and the corresponding sequence of Chern numbers can be determined. Higher Chern numbers are typically associated to smaller gaps in the asymptotically fractal spectrum. At first sight, the Chern number associated with the r-th gap seems an irregular function of r as shown in Fig. 12.6 for $\alpha = 34/55$. This is in sharp contrast with the case of $\alpha = 1/q$, where $c_r = r$ for the r-th gap.

Now we show analytically that there actually exists a deterministic pattern underlying the sequence of Chern numbers. This pattern follows from the solutions of the Diophantine equation (DE) [17], $r = c_r p + s_r q$, where r is the gap index, c_r is the Chern number, at each level of the rational approximation, $p = F_n, q = F_{n+1}$. We first classify all the Chern numbers into two classes, the Fibonacci Chern numbers and the non-Fibonacci Chern numbers. All the solutions to the DE, which determine the Chern number, can be exhausted by following two simple rules, labeled as I, II and explicitly stated below.

(I) Fibonacci Chern numbers occur at gaps labeled by a Fibonacci index $r = F_m$: for $\alpha = F_{n-1}/F_n$, $c_{r=F_m} = -(-1)^{n-m} F_{n-m}$. Here $r = 1$ corresponds to the first gap near the band edge with Chern number $F_{n-2} = (-1)^{n_f}(q - p)$, while the last Fibonacci gap near the band center has a Chern number $(-1)^{n_f}$ where n_f counts the number of Fibonacci numbers less than $|q/2|$. The Fibonacci Chern numbers increases monotonically in magnitude as one moves from the center towards the edge of the band (see Fig. 12.6).

(II) The rest are the non-Fibonacci Chern numbers. The Chern number for the r-th gap and its neighboring gaps ($r \pm 1$) obey the recursion relation,

$$c_r = (c_{r+1} + c_{r-1})/2,$$
$$c_{r+1} = c_r + (-1)^{n_f}(q - p),$$
$$c_{r-1} = c_r - (-1)^{n_f}(q - p).$$

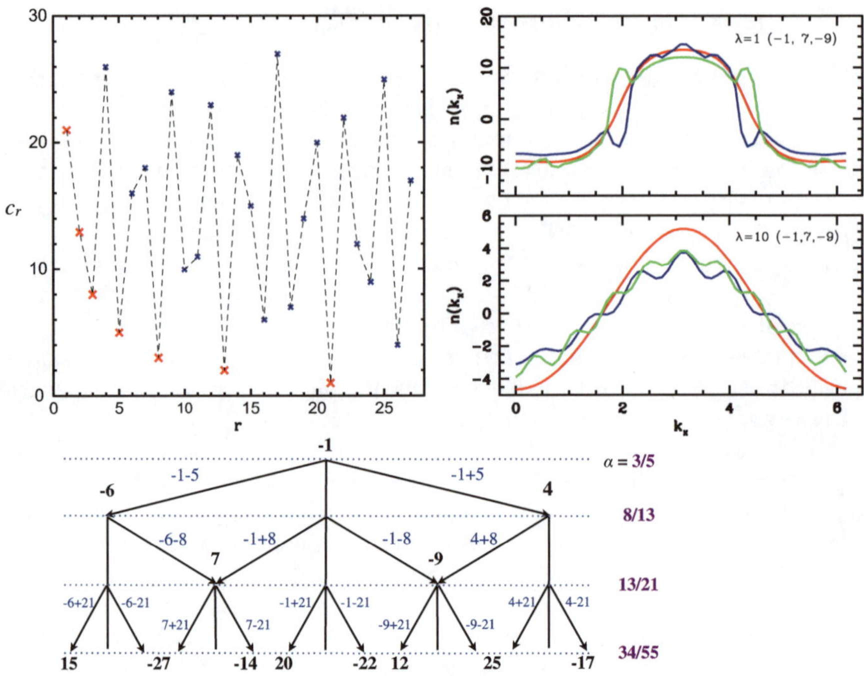

Fig. 12.6 The Chern tree at golden-mean flux showing four consecutive rational approximations: $\alpha = 3/5, 8/13, 13/21, 34/55$ starting with Chern number -1. The *horizontal axis* is not to scale and only shows the correct order of Chern numbers. *Top-left* figure shows Fibonacci (*red*) and non-Fibonacci (*blue*) Chern numbers vs the gap index r for $\alpha = 34/55$. *Top-right* shows momentum distribution for three different topological states characterized by Chern numbers 1 (*red*), 7 (*blue*) and -9 (*green*). Figure illustrates self-averaging aspect of the momentum distribution: states 7 and -9 that reside in close proximity to the state with Chern number unity have ripples that tend to cancel each other

Following these rules, we can construct an exhaustive list of all Chern numbers at each level of rational approximation. We have numerically checked that the Chern numbers computed from the TKNN formula are consistent with the predictions from the DE above. This procedure yields a hierarchy of Chern numbers, which can be neatly organized into a tree structure, thanks to the recursion relation above. A subset of this tree hierarchy is shown in Fig. 12.6. Most notably, we find the average of the left and right neighboring Chern number is exactly the Chern number in the middle.

This property helps to understand the measurement outcomes of real experiments, which inevitably involves average of states with energy very close to each other. In the context of cold atoms, due to the overall trap potential, the local chemical potential slowly varies in space. These local contributions sum up in TOF images according to the local density approximation. The self-averaging property to some degree ensures that such average yields definite result, and the infinitely fine details in the energy spectrum are largely irrelevant. This aspect is illustrated in Fig. 12.6

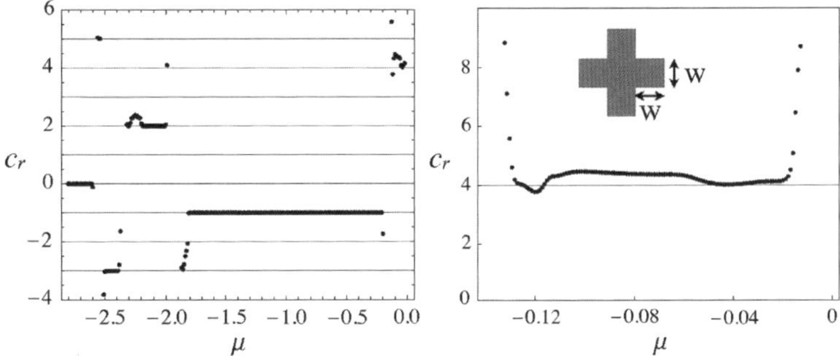

Fig. 12.7 *Left*: the quantized conductance $c_r = \sigma_{xy}/(e^2/h)$ versus the chemical potential μ for $\lambda = 1$ and $\alpha = (\sqrt{5} - 1)/2$. *Right*: the blow up for μ close to zero showing an extended plateau at $c_r = 4$. c_r is computed for a four-terminal quantum Hall bar of width $W = 20a$ (*inset*). The device is connected to four semi-infinite leads of the same material with inelastic scattering rate $0.005t$

for the Chern tree shown in Fig. 12.6. The momentum distribution $n(k_x)$ in the average sense for $\alpha = 3/5, 8/13, 13/21, 34/55$ will be essentially the same.

Similar consideration applies to σ_{xy}. One may see a robust plateau, which may not be perfectly quantized, in a region of chemical potential due to the self-averaging. We have numerically computed σ_{xy} using the Landauer-Buttiker formalism [1] for a mesoscopic quantum Hall bar (inset of Fig. 12.7) at golden-mean flux and connected to four semi-infinite leads. The transmission coefficients T_{ij} are obtained by solving for the Green's function of the sample and the level width functions of each leads. The transverse conductivity at golden-mean flux is shown in Fig. 12.7. One clearly sees a conductance plateau with $c_r = 4$, in addition to $c_r = 1, 2, 3$.

12.6 Quantum Spin Hall Effect for Ultracold Atoms

In solid state, the QSH effect is closely related to spin-orbit coupling. In a recent study [5], it was shown that the effect of spin-orbit coupling can be achieved for cold atoms by creating $SU(2)$ artificial gauge fields.

The theoretical model that captures the essential properties of this system is given by a Hamiltonian describing fermions hopping on a square lattice in the presence of an $SU(2)$ gauge field

$$\mathcal{H} = -t \sum_{mn} \mathbf{c}^{\dagger}_{m+1,n} e^{i\theta_x} \mathbf{c}_{m,n} + \mathbf{c}^{\dagger}_{m,n+1} e^{i\theta_y} \mathbf{c}_{m,n} + h.c.$$

$$+ \lambda_{stag}(-1)^m \mathbf{c}^{\dagger}_{m,n} \mathbf{c}_{m,n} \tag{12.10}$$

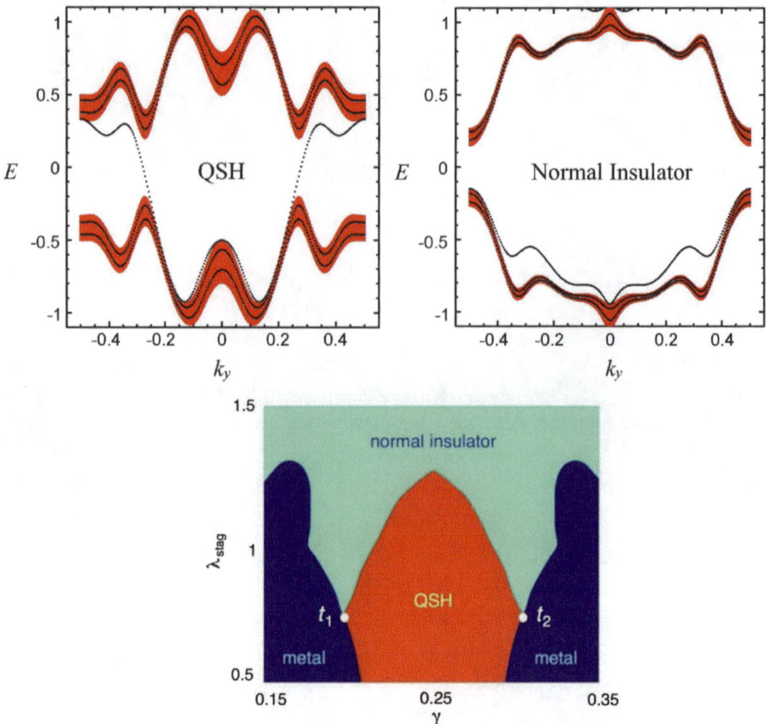

Fig. 12.8 Shows the bulk (*red*) and surface spectrum (*black*), near half-filling, for $\alpha = 1/6$, $\gamma = 0.25$, $\lambda = 1$, and $\lambda_{\text{stag}} = 0.5$ (*left*) and $\lambda_{\text{stag}} = 1.5$ (*right*). The presence of gapless edge modes for $\lambda_{\text{stag}} = 0.5$ and their absence for $\lambda_{\text{stag}} = 1.5$ shows that these two respectively correspond to topologically non-trivial and trivial phases. The *bottom caption* shows the corresponding Phase diagram in $\gamma - \lambda_{\text{stag}}$ plane

where $\mathbf{c}_{m,n}$ is the 2-component field operator defined on a lattice site (m, n), and λ_{stag} is the onsite staggered potential along the x-axis. The phase factors $\theta_{m,n}$ corresponding to $SU(2)$ gauge fields are chosen to be,

$$\theta_x = 2\pi\gamma\boldsymbol{\sigma}_{\mathbf{x}},$$

$$\theta_y = 2\pi x\alpha\boldsymbol{\sigma}_{\mathbf{z}}, \tag{12.11}$$

where $\boldsymbol{\sigma}_{\mathbf{x},\mathbf{z}}$ are Pauli matrices. The control parameter γ is a crossover parameter between QH and QSH states as $\gamma = 0$ describes uncoupled double QH systems while $\gamma = \pi/2$ corresponds to the two maximally coupled QH states. It should be noted that the Hamiltonian (12.10) is invariant under time-reversal. The two components of the field operators correspond in general to a pseudo-spin $1/2$, but could also refer to the spin components of atoms, such as ^{6}Li, in an $F = 1/2$ ground state hyperfine manifold. For $\gamma = 0$ case, spin is a good quantum number and the system decouples into two subsystems which are respectively subjected to the magnetic fluxes α and $-\alpha$.

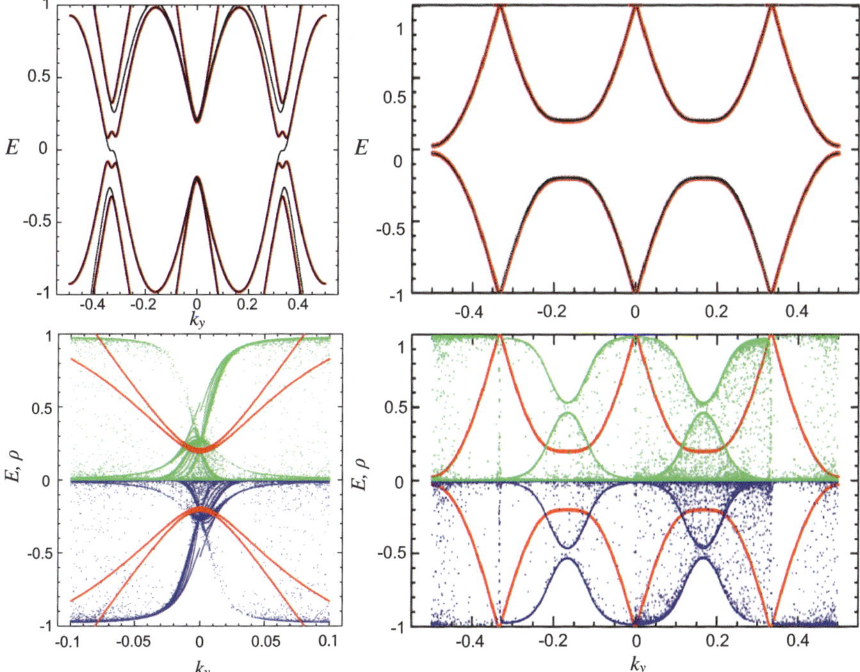

Fig. 12.9 Figure illustrates quantum phase transition from topologically non-trivial state (indicated by gapless edge modes) to normal insulating state: for $\gamma = 0.25$, $\lambda = 5$, and $\lambda_{\text{stag}} = 5$ (*top-left*) and $\lambda_{\text{stag}} = 10$ (*right*). In the *bottom caption, green* and *blue* show the local density of the two-component state at all sites. For normalized spinor wave function, a density of 0.25 shows the existence of a dimerized state for *each component* of the spinor

For isotropic lattices, this Multiband system has been found to exhibit [5] a series of quantum phase transitions between QSH, normal insulator and metallic phases in two-dimensional parameter space for various values of chemical potential. Figure 12.8 shows the landscape of the phases near half-filling for $\alpha = 1/6$. The transitions from topologically non-trivial quantum spin Hall (QSH) to topologically trivial normal band insulator is signaled by the disappearance of gapless Kramer-pair of chiral edge modes.

Motivated by the fact that anisotropic lattices, reveal a novel type of encoding of non-trivial topological states in the form of "Chern-Dimers", for $U(1)$ gauge systems, we present our preliminary results of QSH states for anisotropic lattices. As shown in Fig. 12.8, anisotropic lattices subjected to $SU(2)$ gauge fields exhibit quantum phase transitions from QSH to normal insulating state by tuning the staggered potential. In contrast to the QH system, the avoided crossings in QSH system need not be isolated as seen near $k_y = 1/3$. However, there are states, such as those near $k_y = 0$ where one sees an isolated avoided crossing.

Our preliminary investigation as summarized in Fig. 12.9 shows that for anisotropic lattices, in QSH phase, both components of the spinor wave function are dimerized near an isolated avoided crossings. It is the existence of this double or spinor-dimer that distinguishes QSH state from a normal insulator. In other words, each component of the wave function of topologically non-trivial insulating phase localizes at two distinct sites, which is in contrast to the topologically trivial phase, where the components of the wave functions localize at a single site. For normalized states, the existence of a double-dimer can be spotted by density approaching 0.25 as illustrated in Fig. 12.9. The distinction between the QSH and normal insulator, characterized by spinor-dimer and its absence suggests that such topological phase transitions may be identified by time of flight images.

12.7 Open Questions and Outlook

Identifying topological states in time of flight images would be a dream measurement for cold atom experimentalists. Therefore, our result that the momentum distribution carries fingerprints of non-trivial topology is important.

Our studies raise several open questions regarding topological states and its manifestations. Firstly, our detailed investigation suggests that the features of topological invariant (Chern number) in momentum distribution are quite robust. Is there a rigorous theoretical framework that can establish this connection even for parameters beyond Chern-dimer regime? Secondly, the sign of the Chern number can not be directly determined from the momentum distribution. It remains an open question whether some other observable can detect its sign.

Another natural question is: how generic is the existence of dimerized states in topological states of matter? Somewhat analogous to edge states, the size of the dimerized states in our example is a robust (topologically protected) quantity. Furthermore, are edge states and dimerized states related to each other? Naively, one expects a dimer of size N in bulk will result in N-possible types of defects at open boundaries. These "defects" quite possibly will manifest as edge states inside the gap. It is conceivable that the topological aspect of the dimers of size N is intimately related to N-edge states in QH systems. Further work is required to test these speculations, even in special cases.

To investigate the generality of the dimer framework in topological states, we have studied QSH states for anisotropic lattices. Our preliminary studies suggest that the topological phase transition from QSH to normal insulating state is accompanied by the absence of dimerized states and hence may be traceable in time of flight images. Detailed characterization of Z_2 invariance within the framework of spatially extended objects, however remains appealing but open problem.

The subject of topological states of matter is an active frontier in physics. Lessons learned from their realizations in ultracold atoms will help to advance the fundamental understanding, and inspire new material design or technological applications. Ultracold topological superfluids, Mott insulators with non-trivial topology, and non-equilibrium topological states are most likely to bear fruit in reasonably near future.

Acknowledgements We thank Noah Bray-Ali, Nathan Goldman, Ian Spielman, and Carl Williams for collaboration and their contributions during the course of this research. We acknowledge the funding support from ONR and NIST.

References

1. M. Buttiker, Phys. Rev. B **38**, 9375 (1988)
2. N. Cooper, Adv. Phys. **57**, 539 (2008)
3. J. Dalibard, G. Juzeliuna, P. Öhberg, arXiv:1008.5378 (2010)
4. E. Fradkin, *Field Theories of Condensed Matter Systems* (Addison-Wesley, Reading, 1991), pp. 287–292
5. N. Goldman, I. Satija, P. Nikolic, A. Bermudez, M.A. Martin-Delgado, M. Lewenstein, I.B. Spielman, Phys. Rev. Lett. **105**, 255302 (2010)
6. M.Z. Hasan, C.L. Kane, Rev. Mod. Phys. **82**, 3045 (2010)
7. D.R. Hofstadter, Phys. Rev. B **14**, 2239 (1976)
8. C.L. Kane, E.J. Mele, Phys. Rev. Lett. **95**, 146802 (2005)
9. A. Kitaev, Am. Inst. Phys. Conf. Ser. **1134**, 22 (2009)
10. K.v. Klitzing, G. Dorda, M. Pepper, Phys. Rev. Lett. **45**, 494 (1980)
11. Y.-J. Lin, R.L. Compton, K.J. Garcia, J.V. Porto, I.B. Spielman, Nature **462**, 628 (2009)
12. Y.-J. Lin, K. Jimenez-Garcia, I.B. Spielman, Nature **471**, 83–86 (2011)
13. X.-L. Qi, S.-C. Zhang, Phys. Today **63**, 33 (2010)
14. X. Qi, S. Zhang, arXiv:1008.2026 (2010)
15. X.-L. Qi, T.L. Hughes, S.-C. Zhang, Phys. Rev. B **78**, 195424 (2008)
16. A.P. Schnyder, S. Ryu, A. Furusaki, A.W.W. Ludwig, Phys. Rev. B **78**, 195125 (2008)
17. D.J. Thouless, M. Kohmoto, M.P. Nightingale, M. den Nijs, Phys. Rev. Lett. **49**, 405 (1982)
18. E. Zhao, N. Bray-Ali, C. Williams, I. Spielman, I. Satija, Phys. Rev. A **84**, 063629 (2011)

Chapter 13
Modulation Transfer Through Coherence and Its Application to Atomic Frequency Offset Locking

B.N. Jagatap, Ayan Ray, Y.B. Kale, Niharika Singh, and Q.V. Lawande

Abstract We discuss the process of modulation transfer in a coherently prepared three-level atomic medium and its prospective application to atomic frequency offset locking (AFOL). The issue of modulation transfer through coherence is treated in the framework of temporal evolution of dressed atomic system with externally superimposed deterministic flow. This dynamical description of the atom-field system offers distinctive advantage of using a single modulation source to dither passively the coherent phenomenon as probed by an independent laser system under pump-probe configuration. Modulation transfer is demonstrated experimentally using frequency modulation spectroscopy on a subnatural linewidth electromagnetically induced transparency (EIT) and a sub-Doppler linewidth Autler-Townes (AT) resonance in Doppler broadened alkali vapor medium, and AFOL is realized by stabilizing the probe laser on the first/third derivative signals. The stability of AFOL is discussed in terms of the frequency noise power spectral density and Allan variance. Analysis of AFOL schemes is carried out at the backdrop of closed loop active frequency control in a conventional master-slave scheme to point out the contrasting behavior of AFOL schemes based on EIT and AT resonances. This work adds up to the discussion on the subtle link between dressed state spectroscopy and AFOL, which is relevant for developing a master-slave type laser system in the domain of coherent photon-atom interaction.

13.1 Introduction

Coherent pump-probe spectroscopy performed in an alkali vapor cell has received much attention in recent years, particularly in the context of coherent population trapping (CPT), electromagnetically induced transparency (EIT) and electromagnetically induced absorption (EIA) [1, 4, 15]. Quantum coherence and interference

B.N. Jagatap (✉) · Y.B. Kale · N. Singh · Q.V. Lawande
Atomic & Molecular Physics Division, Bhabha Atomic Research Centre, Mumbai 400 085, India
e-mail: bnj@barc.gov.in

A. Ray
Radioactive Ion Beam Group, Variable Energy Cyclotron Centre, Kolkata 700 064, India

M. Mohan (ed.), *New Trends in Atomic and Molecular Physics*,
Springer Series on Atomic, Optical, and Plasma Physics 76,
DOI 10.1007/978-3-642-38167-6_13, © Springer-Verlag Berlin Heidelberg 2013

in a coherently driven three-level atomic system are central to these phenomena. Apart from generating renewed interest in the understanding of subtle effects in coherent photon-atom interactions, these studies have created a new technological frontier based on photon engineering in a dressed atomic medium. Both CPT and EIT exhibit subnatural linewidth resonances even in an inhomogeneously broadened atomic medium, and these provide sharp frequency discriminator suitable for metrological applications. Some related examples include atomic frequency standard [12], miniaturized atomic clock [14], ultra-sensitive magnetometry [21] and atomic frequency offset locking (AFOL) [3, 9–11, 17]. In an alkali vapor medium, a three-level system can be conveniently constructed from the hyperfine manifold of $nS_{1/2} \rightarrow nP_{1/2}$ (D$_1$) and $nS_{1/2} \rightarrow nP_{3/2}$ (D$_2$) transitions. Different techniques like two independent lasers [24], two independent but phase correlated lasers [3, 17] and sidebands of a single laser [13] are employed to generate the requisite bichromatic pump-probe configuration.

Generally a three level system in Λ configuration is preferred in applications related to metrology. In such a system, the coherence in the medium is established through interplay of optical pumping, spontaneous decay and non-radiative decay between dipole forbidden levels. Note also that even if two independent lasers are used as pump and probe, they eventually get coupled due to the coherence established in the medium and as a result the modulation present in the pump laser gets transferred to the probe. The modulation transfer through this coherent linkage can be exploited beneficially to devise schemes for AFOL [10, 11]. However the technical noise present in the pump can also get reflected on the probe during the transfer process, thus complicating the closed loop structure for frequency control. In effect the study of the correlation between two lasers is important in view of estimating the frequency stability of an AFOL scheme. It may be recounted that EIT is intimately connected to the mechanism of AC Stark shift, which gives rise to the Autler-Townes (AT) doublet in the probe absorption spectrum. Experimentally it is also possible to generate ultra-narrow AT resonances by a proper choice of atom-field interaction parameters, and these can in turn be employed to build up correlation between pump and probe lasers used in AFOL. At a first glance the nature of the coherent connection introduced between two lasers in a medium exhibiting EIT and AT seems to be of similar nature. However our recent works on this subject [9–11] establish the fact that AFOL based on AT is characteristically different from that centered on EIT. Also AFOL may emerge as the prospective application area of AT, which until now has remained relatively unexplored.

In this paper we present specific case studies of the probe laser frequency stability while locked on sub-natural EIT and sub-Doppler AT resonance, the later being formed both within and outside of the Doppler broadened absorption profile. Three-level system used in this work is the Λ configuration of the hyperfine levels of D$_2$ transition of cesium ($\sim$852 nm). To investigate the process of modulation transfer, a single color sinusoidal wave is superimposed on the pump frequency and its effect on probe absorption spectrum in the domain of AT and EIT is studied. The coherent coupling between pump and probe lasers is further explained through a conceptual sketch of the control loop and its possible effect on the probe laser frequency stability is discussed. We obtain EIT signal of line width $\sim$2 MHz and sub-Doppler AT resonances of linewidth $\sim$10 MHz ($\sim$8 MHz) outside (inside) Doppler broadened absorption profile, and use them as frequency reference for probe laser stabilization. The frequency stability of the probe laser is extracted through frequency noise power ($S_{\Delta v}(f)$) and Allan variance ($\sigma(2, \tau)$), and is compared with that of the pump laser that is stabilized on a saturated absorption signal (SAS). We explicitly demonstrate here that the pump laser, which dresses the atomic medium, limits the frequency stability of the probe laser when locked on sub-Doppler dressed level absorption (AT) signal. On the other hand the probe laser frequency stability can surpass that of the pump laser under EIT locking. This contrasting nature of the control loop is explained in terms of the very mechanisms underlying AT and EIT resonances in a Doppler broadened atomic medium. This study helps us to draw the order of superior frequency stability in a dressed atom medium, i.e., EIT locking > SAS locking > AT locking.

13.2 A Generalized Overview of Λ System and Its Connection to AFOL

We focus our attention to the stability of the frequency offset lock using probe absorption in a three-level dressed atomic system (Fig. 13.1(a)). Here a signal having linewidth below or in the range of the natural linewidth serves as a useful discriminator for replicating the AFOL configuration. The key issue in such AFOL scheme is the coherent cross-talk, i.e., $S_{\Delta v,\text{transferred}}(f)$, in the atom-field interaction scenario [9–11]. In this context one may refer to the steady state solution of induced polarizability (ρ_{12}) given by [16]

$$\rho_{12}(v) = i\Omega_c\big[(\Gamma - i\Delta_p - ik \cdot v) + \Omega_p^2[\gamma_{31} - i(\Delta_p - \Delta_c)]^{-1}\big]^{-1} G(v) \quad (13.1)$$

where Δ_p (Δ_c) and Ω_p (Ω_c) are respective the detuning and Rabi frequency of probe (pump) laser, Γ is the natural linewidth, γ_{31} is the non-radiative decay rate between dipole forbidden lower levels, $G(v)$ is the distribution of atomic velocity (v) and $k = k_p (\approx k_c)$ is the wave vector of the probe laser. Each detuning is defined as the difference between laser frequency and corresponding atomic transition frequency. Eqution (13.1) describes the linkage established due to atomic coherence between pump and probe lasers, which are otherwise independent of each other. The probe absorption spectrum may be obtained by averaging Eq. (13.1) over the Maxwellian distribution of atomic velocities. The EIT appears when Raman resonance condition i.e. $\lim(\Delta_p - \Delta_c) \to 0$ is satisfied. The narrow transparency window appears at the position of minimum absorption, i.e., where the imaginary part of the susceptibility (χ) tends to zero. This is shown in Fig. 13.1(b) for a select set of experimental parameters.

The connection between EIT and AT can be understood from Fig. 13.1(a–b) under the purview of dressed atom picture. For a stationary atom at exact pump resonance the separation between two dressed states (AT doublet) is Ω_c where no light absorption takes place and that represents the transparency window. In a Doppler

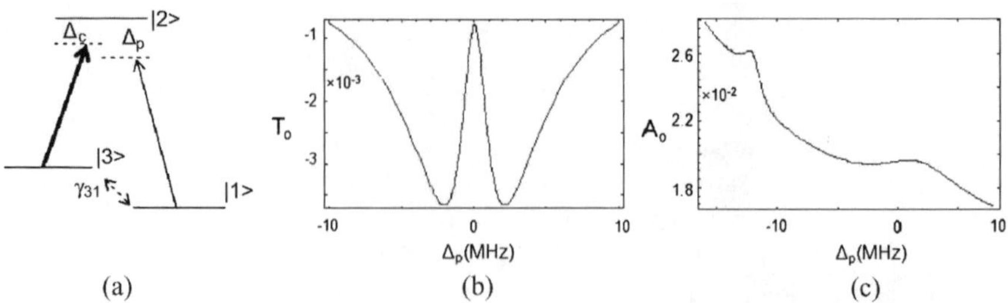

Fig. 13.1 Basic aspects of AFOL. (**a**) Model three-level Λ system, (**b**) simulated sub-natural EIT resonance plotted as transmission (T_0) versus probe detuning for $\Delta_c = 0$, $\Gamma = 5.3$ MHz, $\gamma_{13} = 20$ KHz, $\Omega_c = 4$ MHz and $\Omega_p = 0.5$ MHz and (**c**) probe absorption (A_0) corresponding to the AT doublet simulated for $\Delta_c = -10$ MHz. Other data is same as in (**b**). For (**c**) the Doppler width of the medium is assumed to be 20 MHz for the purpose of illustration

broadened medium taking into account $k \cdot v$ (cf. Eq. (13.1)) we may infer that one of the AT components is drawn closer to the central frequency while the other is pushed further away. Velocity averaging thus results in narrowing of the linewidth (Γ_{EIT}), which for low pump powers and at exact pump resonance is given by [8]

$$\Gamma_{\mathrm{EIT}} = \Omega_c (2\gamma_{31}/\Gamma)^{1/2}. \tag{13.2}$$

This expression and the related discussion clearly show that EIT is a frequency discriminator that is unaffected by moderate fluctuations of pump laser parameters [25]. This is unlike the situation in AT, which is a result of absorption, i.e., $\chi \neq 0$. At detuned Raman resonance ($\Delta_c \neq 0$), the AT signals become asymmetric and the EIT window gets partially or fully obscured due to non-cancellation of absorption. The frequency position ($\Delta_p^{\pm}$) and linewidth ($\Gamma_{\mathrm{AT}}^{\pm}$) of AT resonances are obtained as [22]

$$\Delta_p^{\pm} = \left[\Delta_c \pm \left(\Delta_c^2 + \Omega_c^2\right)^{1/2}\right]/2 \tag{13.3}$$

$$\Gamma_{\mathrm{AT}}^{\pm} = \left[1 \mp \Delta_c\left(\Delta_c^2 + \Omega_c^2\right)^{-1/2}\right](\Gamma + 2D)/4, \tag{13.4}$$

where D is the width of the velocity distribution. Thus for suitable pump detuning ($\Delta_c \gg \Omega_c$), one of the AT components has ultra-narrow linewidth, and it can be observed in a Doppler broadened medium provided the effect of Doppler background is suitably reduced. Figure 13.1(c) shows simulated AT spectrum (cf. Eq. (13.1)) for an experimental realization (cf. Sect. 13.3) where the Doppler width is reduced to $\sim$20 MHz and it gives rise to a narrow AT resonance of $\sim$8 MHz linewidth.

To scrutinize the coherent coupling between pump and probe laser systems in a simplistic manner we introduce pump frequency modulation of type $\Delta_c \rightarrow \Delta_0 + A\sin(\omega t)$ in Eq. (13.1) and study its effect on EIT and AT resonances. Here Δ_0 is the nominal pump detuning, and A and ω are respectively the amplitude and frequency of sinusoidal perturbation. We choose $\omega = 0.1$ MHz and keep $A = 0.25$. The additive inclusion of external slow periodic function is justified in view of the involvement of other physical processes with faster relaxation time constants in the atom-field interaction picture. For example, the spontaneous decay rate (Γ) is 5.3 MHz and the optical pumping rate $[\Omega_c^2 \Gamma/(\Gamma^2 + \Delta_c^2)]$ varies within $0.7 \to 5$ MHz for different values of pump laser parameters as used in our experiment (cf. Sect. 13.3). These two mechanisms are mainly responsible for building up of coherence in the Λ level scheme and they are much faster than the external perturbation. As a consequence the external perturbation is seen to act on the dressed atomic system, which has already settled into equilibrium. The relaxation term γ_{31} is mainly contributed by the transit time broadening in an alkali vapor cell media where no buffer gas is used. We estimate $\gamma_{31} \sim 20$ KHz under our experimental condition. Hence γ_{31} may be considered as the slowest physical process having a broad time envelope under which other physical processes are integrated over many cycles, i.e. Ω^2/Γ, Γ^{-1} and ω^{-1}. The results of these simulations are given in Fig. 13.2 and Fig. 13.3 for EIT and AT resonances.

The inherent assumption behind the present phenomenological representation of coherent coupling is based on considering the steady state profile $\rho_{12}(\Delta_p)$ as the starting point and then considering $A\sin(\omega t)$ as a deterministic flow externally introduced in the scenario. In analogy with the dynamical systems, it may be seen that ω^{-1} provides the necessary time scale over which the system evolves with external deterministic character superposed on it. The dynamical evolution of the system in phase space is defined by ρ_{12}, ω^{-1} and Δ_p. Here the single color sinusoidal perturbation on pump frequency only presents an outline of the phenomenon of coherence assisted modulation transfer. The effect of perturbation shows its presence on the probe absorption; see the 3D plots as well as associated contour plots in Figs. 13.2 and 13.3. It is evident from the contour plots that the modulation peaks up around the transparency window at the two photon resonance condition. In case of detuned AT, maximum influence of the perturbation is seen on the relatively narrower AT component whereas the broader AT component is perturbed the least. Intuitively we may conclude that degree of transfer of perturbation from pump to the probe laser due to coherent crosstalk, $S_{\Delta v,\text{transferred}}(f)$, happens in the order of EIT window > narrow AT resonance > broad AT resonance.

It may be noted here that the induced perturbation in a dressed level system can also be stochastic in nature. It is worthwhile to mention that the most generalized approach of introducing fluctuations in the dressed atom system involves consideration of stochastic character of the concerned density matrix elements [7]. In this regard it is instructive to refer to the result which shows that the noise transfer around EIT is proportional to $\Gamma_{\text{EIT}}^2[\Gamma_{\text{EIT}}^2 + (f - \Delta)^2]^{-1} f^2[\Gamma_{\text{EIT}}^2 + f^2]^{-1}$, where $\Delta = \Delta_p - \Delta_c$ and f is the Fourier frequency (frozen in time) of noise [18]. This has two immediate consequences. Firstly the noise transfer is efficient when noise frequency bandwidth $\delta f \sim \Delta$, i.e. probe absorption spectrum centered around two photon resonance acts as a high Q band-pass filter and secondly the phase noise correlation is minimum for $f < \Gamma_{\text{EIT}}$ within the EIT window, i.e. only high frequency phase noise get correlated in an EIT medium. This shows that the conclusion drawn from the behavior in Fig. 13.2 is valid only for a certain band of frequencies.

We now turn our attention to the AT resonance, which corresponds to a detuned system with $\chi \neq 0$. Here the pump-probe combination sees a resonant medium described by the absorption lineshape $L(v - v_0)$. Hence the phase noise of the laser

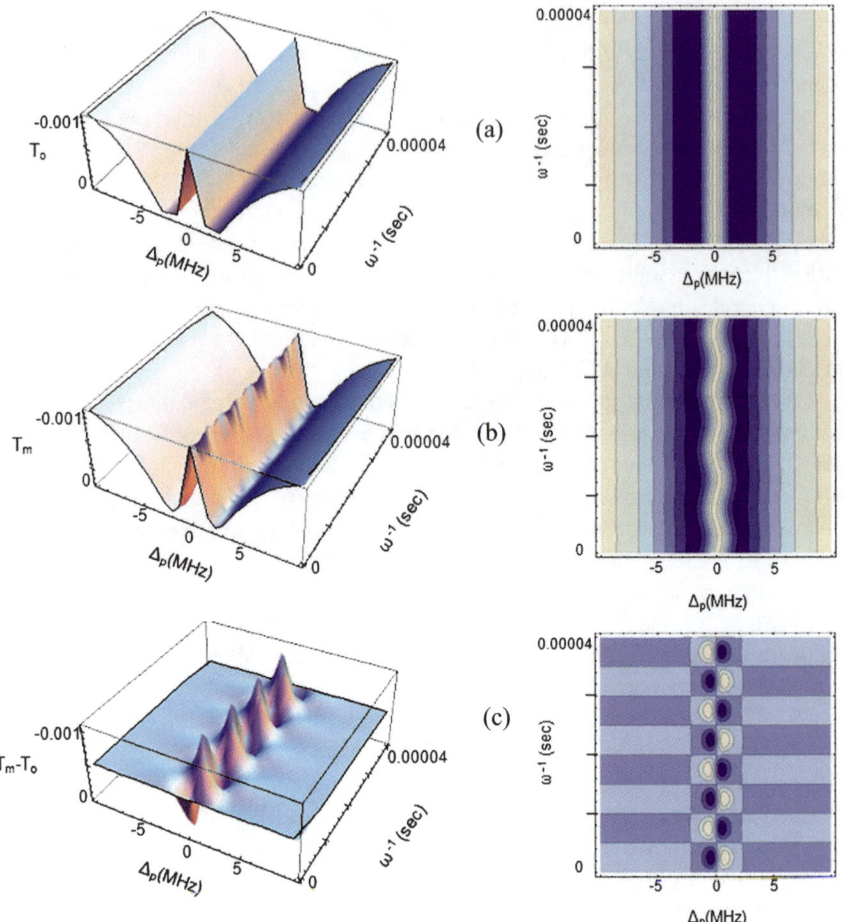

Fig. 13.2 Modulation transfer in the EIT medium. The basic data is same as in Fig. 13.1(b), and modulation frequency (ω) and amplitude (A) are taken as 0.1 MHz and 0.25 MHz respectively. Figures (**a**)–(**c**) show respectively the probe transmission in the absence of modulation (T_0), in the presence of modulation (T_m) and the difference ($T_m - T_0$). Each frame on the *right* is the contour plot corresponding to the frame on the *left*. The modulation transfer is seen to peak at the EIT position

induces changes in susceptibility of the atomic system and this further induces inevitable fluctuation in the transmitted normalized laser intensity $\delta I / \langle I \rangle$; see phase modulation (PM) to amplitude modulation (AM) conversion [5], i.e.,

$$\delta I / \langle I \rangle \approx \langle \delta \nu_{\text{laser,rms}} \rangle \left| \partial L(\nu - \nu_0) / \partial \nu \right|. \tag{13.5}$$

so that the linewidth $\sim 2.35 \nu_{\text{laser,rms}}$. Since phase noise of the laser mainly originates from its linewidth, the induced intensity noise is also a function of linewidth. Compared to high frequency phase noise correlation around two-photon resonance, the PM-AM conversion process is slow as it is involves an intermediate step, which

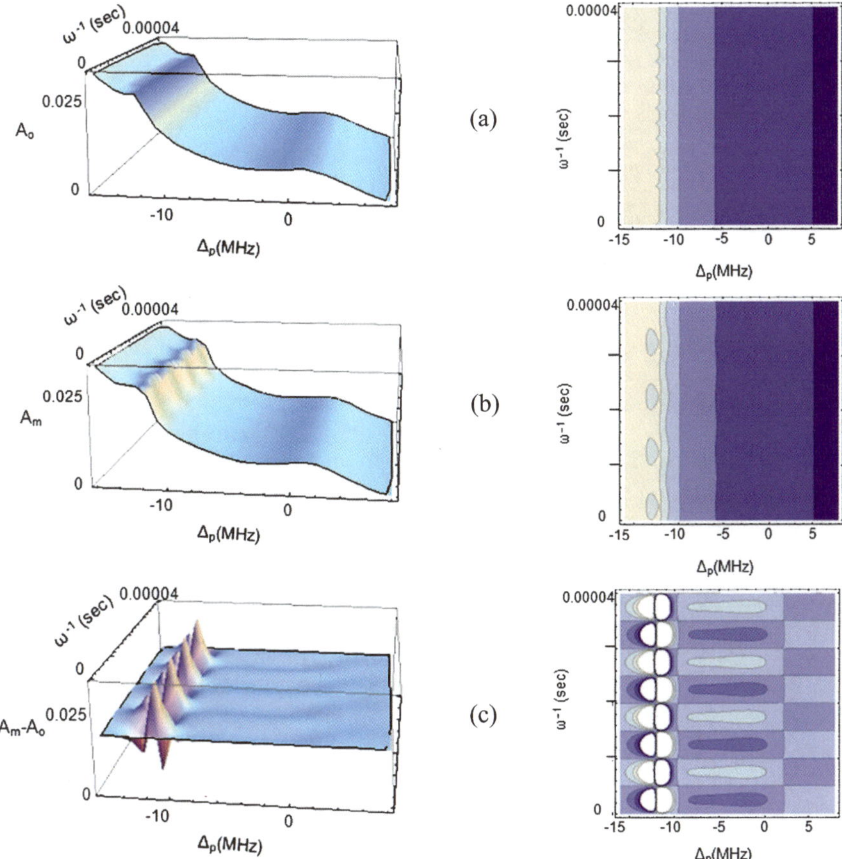

Fig. 13.3 Modulation transfer in a medium exhibiting AT splitting. The basic data is same as in Fig. 13.1(c) and modulation parameters (ω and A) are same as in Fig. 13.2. Figures (**a**)–(**c**) show respectively the probe absorption in the absence of modulation (A_0), in the presence of modulation (A_m) and the difference ($A_m - A_0$). Each frame on the *right* is the contour plot corresponding to the frame on the *left*. The modulation transfer is maximum at narrow AT resonance and negligible for the broader AT resonance

induces finite changes in χ. Aforementioned discussion provides the distinctive features of frequency stability of probe laser stabilized on EIT and AT resonances. These are experimentally verified in the sections those follow.

13.3 Experimental

We use hyperfine manifold of D_2 transition of Cs, shown schematically in Fig. 13.4, to construct various three-level Λ systems of interest. For EIT based AFOL config-

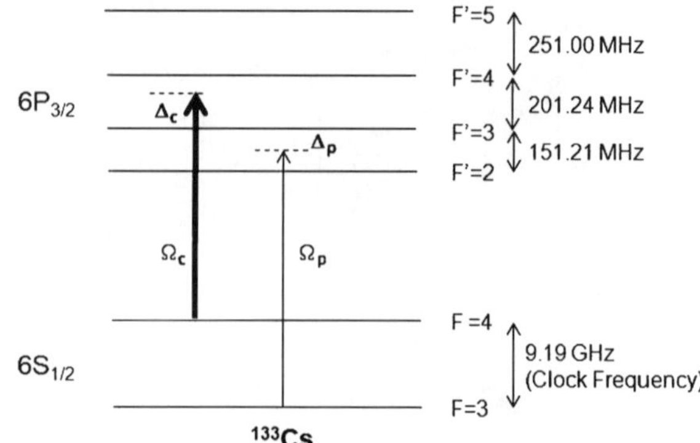

Fig. 13.4 Schematic representation of the hyperfine manifold of D$_2$ transition of ^{133}Cs. Here the pump and probe lasers act on $F = 4 \rightarrow F' = 3$ and $F = 3 \rightarrow F' = 3$ transitions respectively. Rabi frequency and detuning of the pump (probe) laser are denoted by Ω_c (Ω_p) and Δ_c (Δ_p) respectively

uration, the frequency offset corresponds exactly to the atomic clock frequency. On the other hand the AT based AFOL schemes provide tunability to the offset lock.

The experimental scheme for the AFOL techniques is shown in Fig. 13.5. A pair of external cavity diode lasers (ECDLs) is used as strong control laser (ECDLc) and weak probe laser (ECDLp). The laser beams are combined by two polarizing cubic beam splitters (PBSs) so as to make them co-propagate within a Cs vapor cell. Very elfin part ($\sim$100 µW) of both lasers is used for respective saturated absorption spectroscopy (SAS) setups. Also a moderate part of the probe laser is used (only in AT based AFOL scheme) in counter propagating geometry to partially eliminate the Doppler background and increase the S/N ratio of the AT spectrum. The Cs vapor cell (without any buffer gas) is placed within two PBSs, which are used to merge/separate the pump-probe beams. The cell is provided with μ-metal shield to minimize influence of stray magnetic field. The control laser is current modulated (modulation frequency $\omega_m = 0.03$ MHz) and stabilized at the 3rd derivative of desired transition obtained through SAS. The lock in amplifier (LIAc) and servo controller complete the control loop for the ECDLc. The acousto-optic modulator (AOM) serves the purpose of frequency shifter for control laser frequency. The probe absorption signal is detected by photodetector (PD) and the probe frequency is locked on desired resonance, either EIT or AT. With some minor modifications in the experimental arrangement of Fig. 13.5, we investigate three AFOL schemes, (1) EIT based AFOL with control and probe lasers respectively at $F = 4 \rightarrow F' = 3$ and $F = 3 \rightarrow F' = 3$ transitions, (2) AT based AFOL where the narrow AT resonance is outside the Doppler profile and (3) AT based AFOL with the narrow AT resonance inside the Doppler profile.

For EIT based AFOL scheme, the experimental arrangement of Fig. 13.4 is modified by removing the mirror M0. The pump Rabi frequency (Ω_c) is kept nearly $\sim 2\Gamma$ while the probe Rabi frequency (Ω_p) is maintained at $\sim 0.5\Gamma$. It is important to note here that the choice of Ω_c is crucial so as to make Γ_{EIT} depend linearly on Ω_c. The robustness of an AFOL scheme depends on the steepness of the slope of the frequency discriminator, the third derivative of EIT in the present case, and the associated instrumentation in feedback loop. The pump laser is locked to transition $F = 4 \rightarrow F' = 3$ using SAS set-up and the error signal obtained by taking its

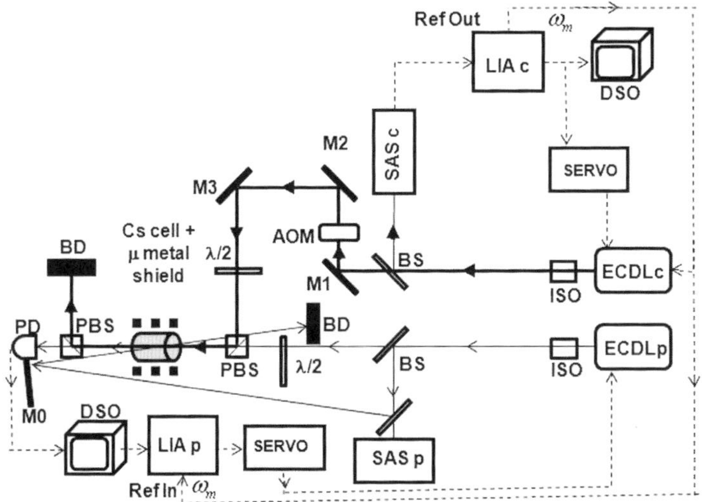

Fig. 13.5 Schematic representation of the experimental arrangement. *ECDL*: external cavity diode laser, *M*: mirror, *BS*: beam splitter, *ISO*: isolator, *BD*: beam dump, *PD*: photodetector, $\lambda/2$: wave plate, *LIA*: lock-in amplifier, *AOM*: acousto optic modulator, *DSO*: digital storage oscilloscope and *SAS*: saturation absorption setup. *AOM* is used as frequency shifter to adjust the pump laser detuning for the various AFOL schemes. The pump beam is made to co-propagate with probe beam by small adjustment of *M2, M3*. The Mirror *M0* is used for the AFOL scheme with AT inside the Doppler to reduce the probe Doppler background by providing a counter propagating part of the probe beam. Diode lasers and *SAS* used for the pump and probe lasers are indicated by '*c*' and '*p*' respectively

third derivative is fed back to the pump laser as correction signal. The probe laser is scanned in the vicinity of F $= 4 \rightarrow$ F$' = 3$ transition to obtain EIT resonance of sub-natural linewidth. Figure 13.6 shows the EIT spectrum with $\Gamma_{\text{EIT}} \sim 2$ MHz. The third derivative spectrum is simultaneously recorded by demodulating the probe signal by frequency $3\omega_m$ by a lock-in-amplifier. The probe laser is then locked to the precisely at the third derivative of EIT signal so that the frequency offset corresponds exactly to the Cs clock frequency. The error signals of both SAS locked pump and the EIT locked probe are recorded from further analysis.

In case of AT based AFOL, we first consider the situation where the narrower AT resonance is formed outside the Doppler width. To this end, the pump laser is blue detuned by 360 MHz with respect to F $= 4 \rightarrow$ F$' = 4$ transition, which makes it 560 MHz blue detuned with respect to F $= 4 \rightarrow$ F$' = 3$ transition, and the probe laser is scanned across F $= 3 \rightarrow$ F$' = 2, 3, 4$ transitions. Probe absorption spectrum as recorded in this experiment is shown in Fig. 13.7(a) and it exhibits a narrow AT signal of 8 MHz linewidth outside the Doppler profile. The other AT resonance being broader merges with the broad Doppler profile. For the purpose of AFOL, the probe laser is stabilized on the narrow AT resonance. The frequency offset in this case corresponds to 560 MHz over the clock frequency. The error signals of both pump and probe lasers are recorded for further analysis.

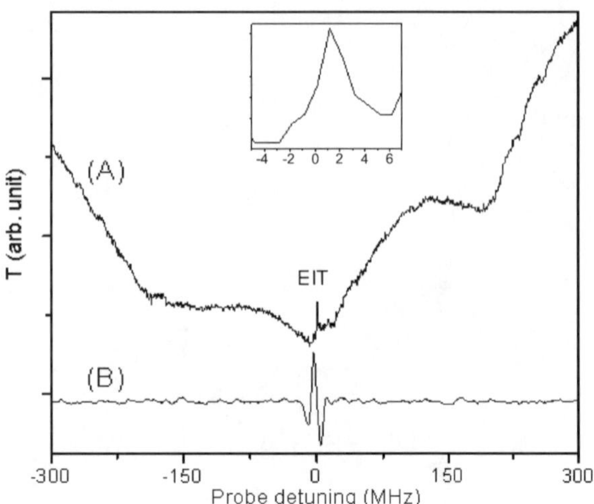

Fig. 13.6 Simultaneous recording of EIT (*curve A*) and its third derivative signal (*curve B*). Pump laser is on $F = 4 \rightarrow F' = 3$ with $\Delta_c = 0$. The *inset* shows the sub-natural EIT window (FWHM ~ 2 MHz) on which the probe is stabilized. See text for details. Probe detuning is measured from $F = 3 \rightarrow F' = 3$ resonance

For small pump detuning, the AT resonances lie inside the Doppler width and that makes it difficult to isolate them experimentally. The experimental set-up therefore is modified as follows: The larger part (60 %) of the probe beam is counter propagated using a beam splitter and the mirror M0, and used as 'saturating' beam for the probe. This in effect generates a saturated absorption set-up to probe dressed level transition in the presence of the strong pump, and that reduces the Doppler background substantially. The pump is red detuned from the $F = 4 \rightarrow F' = 3$ transition by 90 MHz and the probe beam is scanned to obtain desired resonances as shown in

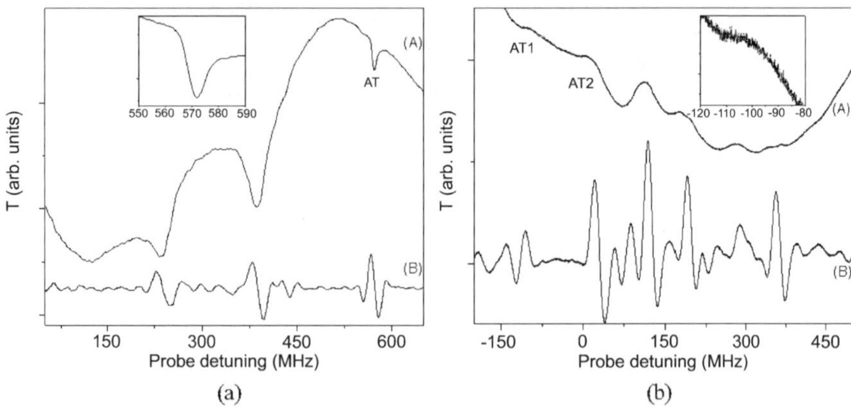

Fig. 13.7 (**a**) Sub-Doppler AT resonance (*curve A*) outside the Doppler profile and its third derivative (*curve B*). The inset shows the AT resonance of ~ 8 MHz linewidth (**b**) sub-Doppler AT resonances (*AT1*, *AT2*) inside the Doppler profile (*curve A*) and their respective third derivatives (*curve B*). The inset shows the AT resonance of ~ 10 MHz linewidth. Distortion in the shape of the AT resonance due to residual Doppler broadening is seen. See text for details. Probe detuning is measured from $F = 3 \rightarrow F' = 3$ resonance

Fig. 13.7(b). The linewidth of the narrow AT resonance is $\sim$10 MHz. It may also be noted that the resonances are not Lorentzian and that may be attributed to the residual Doppler width. For the purpose of frequency stabilization, we adopt a different strategy to establish the nature of AT based AFOL in a general way. Here the pump laser is not modulated for the purpose of simplifying the closed loop architecture by removing the additional modulation transfer process. Instead the probe is modulated and stabilized on the derivative signal of AT resonance.

13.4 Results and Discussion

The process of modulation transfer to the probe in a coherently prepared atomic medium is shown schematically in Fig. 13.8. The EIT or AT based AFOL where only the pump laser is frequency modulated by current modulation, can be visualized as coupled oscillators. Thus if we compare the AFOL with the traditional master-slave frequency offset locking scheme, the pump control loop is equivalent to the master that slaves the probe.

It may be recalled that frequency and phase noise power spectral densities, $S_{\Delta\nu}(f)$ and $S_{\phi}(f)$ respectively, are interlinked for an oscillator, i.e., $S_{\Delta\nu}(f) = (f/\nu_0)^2 S_{\phi}(f)$. Hence it is possible to gather information about the coherent coupling between pump-probe laser systems in the dressed medium by monitoring the residual error frequency fluctuation under the closed loop condition. Following the schematic illustration [6] the frequency noise spectral density $S_{\Delta\nu}(f)$ of the offset locked probe laser can be written as

$$S_{\Delta\nu,cl,\text{offsetlock}} = \left(S^2_{\Delta\nu,\text{laser}} + S^2_{\Delta\nu,\text{transferred}}\right)^{1/2} / |1 + H(f)|, \qquad (13.6)$$

where $S_{\Delta\nu,cl,\text{offsetlock}}(f)$ is the frequency noise power of the coupled laser, $H(f)$ is the transfer function of the feedback loop, $S_{\Delta\nu,\text{laser}}(f)$ is the noise power of the probe laser itself and $S_{\Delta\nu,\text{transferred}}(f)$ is the noise introduced by the pump laser to the probe laser feedback loop (cf. Fig. 13.8(a)). It may be noted here that the time evolution of laser frequency stability is based on the statistical averaging of the frequency fluctuations. Therefore, the systematic errors, e.g., shift of the pump laser lock point, and the random errors, e.g. flicker noise in the pump laser, will contribute to probe laser lock stability in addition to the probe laser's inherent noise. As a result of $S_{\Delta\nu,\text{transferred}}(f)$ the servo system starts to write the pump laser noise on the probe laser system and that points to a situation where the pump laser becomes the key factor in determining the stability of the probe laser.

Note here that Eq. (13.6) is constructed by assuming $S_{\Delta\nu,\text{laser}}(f)$ and $S_{\Delta\nu,\text{transferred}}(f)$ are in quadrature. This assumption is made for the sake of presenting simplified expression representing the noise power spectrum of the offset locked laser. Actual situation may be more complex for such coupled laser system. However in an AFOL scheme the pump laser is linked to probe laser via coherence and the situations are quite different for discriminators based on dressed level resonance and EIT resonance. This forbids us from drawing any general conclusion

Fig. 13.8 (**a**) Schematic representation of a simplified model for modulation transfer in coherently prepared atomic medium. Here $SAS =$ saturated absorption set-up, $ECDL =$ external cavity diode laser and $ACTU =$ piezo voltage controller. Indices 'p' and 'c' correspond to probe and pump lasers respectively. Relevant noise elements are defined in the text. While the scheme is shown for EIT, it is valid also for AT resonance. (**b**) Equivalent probe control loop when pump and probe are coherently coupled. See text for details

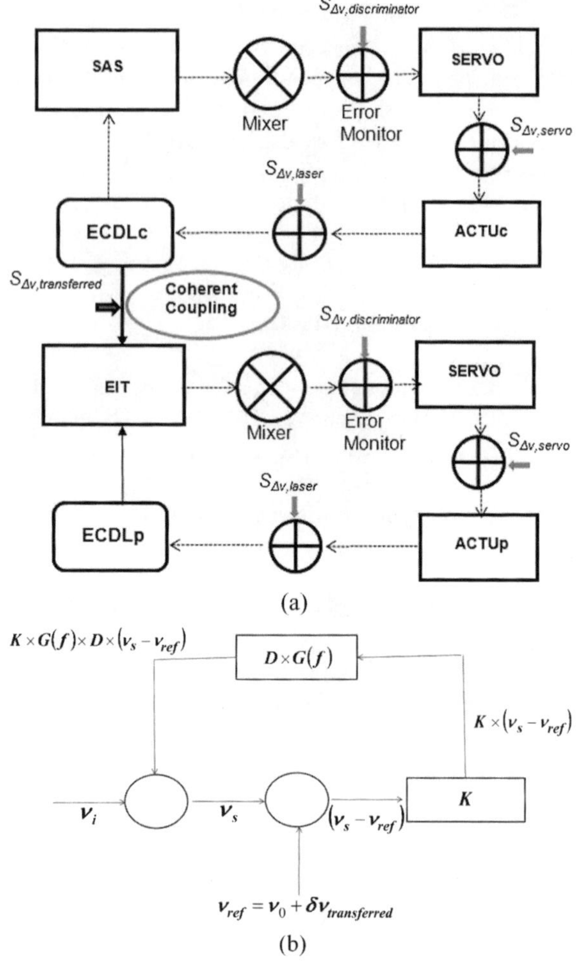

on the stability of the probe laser, which is locked to EIT and AT in the line of conventional frequency offset locking.

The closed loop for the probe laser locked on the coherently generated resonance (either EIT or AT) can be seen in a perspective as shown in Fig. 13.8(b). The initial frequency v_i of the free running probe laser is processed through the servo loop. The servo, with a frequency dependent gain $G(f)$ corrects the laser frequency to a value v_i so that the error frequency is $v_i - v_{\text{ref}}$. The frequency discriminator $(\delta V/\delta v)$ converts the error signal to an equivalent error voltage with an amplification factor K. The error voltage is processed as $k \times G(f) \times D \times (v_i - v_{\text{ref}})$ by the servo loop and the actuator to correct the laser frequency through negative feedback. Here D is the sensitivity of the actuator. The resulting negative feedback $\delta v/v = [1 + k \times G(f) \times D \times (v_i - v_{\text{ref}})]^{-1}$ tries to stabilize the probe laser frequency to the reference point. However in the presence of coherent coupling another factor

δv_{trans} is introduced to the system which is unknown to the feedback network and accounts for reflection of pump laser frequency noise on the probe. This very factor is expected to depend on the nature of the coherent resonance, whether EIT or AT.

On the milieu of the modulation transfer in a coherently prepared medium, we now discuss the results on error frequency analysis of pump and probe lasers. Figure 13.9 shows the error signals for the probe that is locked respectively on sub-natural EIT ($\sim$2 MHz) and the narrow AT resonance ($\sim$8 MHz) outside the Doppler profile. The error signal of the SAS locked pump is also shown for comparison. It is clearly seen from Fig. 13.9 that the error signal fluctuations (peak-peak) for EIT locked probe are minimum ($\sim$1.5 MHz) compared to that of AT locked probe ($\sim$8 MHz) and SAS locked pump ($\sim$4 MHz). The laser frequency noise produced in RIT based AFOL scheme is thus smallest compared to the other cases. Similar conclusions may be drawn from the Fourier frequency noise spectra shown in Fig. 13.10. The frequency noise spectra for the probe laser locked on a narrow AT resonance formed inside the Doppler profile is shown separately, since this case is somewhat different from the AT resonance outside the Doppler profile.

Figure 13.10 shows that the value of $S_{\Delta v}(f)_{\text{EIT}}$ is considerably less than that of $S_{\Delta v}(f)_{\text{SAS}}$ and $S_{\Delta v}(f)_{\text{AT}}$ for low Fourier frequencies (up to $f = 20$ Hz). In the region of 20 to 40 Hz the pump and the probe laser noises become equivalent irrespective of the lock point of the laser and for the frequencies higher than 40 Hz, $S_{\Delta v}(f)_{\text{EIT}}$ is again lower. This can be explained by considering the role of the LIA and servo controller (PI) in the negative feedback loop. Unlike the SAS or AT case, the EIT discriminator is much sharper thus the integral (I) gain in the servo loop for the EIT locked probe will lead to the instability of the probe laser. Thus to avoid that, the additional loop filter is used to roll off the gain (I) thus the contribution of the proportional (P) gain is more in overall feedback for the frequencies 20 Hz $< f <$ 40 Hz. The noise suppression beyond 40 Hz is jointly done by P control and LIA. This clearly shows that the EIT locked laser is always more stable than SAS locked pump laser and AT locked probe laser. In a similar manner, the noise power spectra for the AT based AFOL scheme, when AT doublet is inside the Doppler profile, shows that the noise power of the pump (SAS locked) is more than probe (AT locked) laser for the frequency $f >$ 100 Hz, but for the lower Fourier frequencies ($f <$ 30 Hz), the probe noise is more.

The error signal analysis describes laser stability in the Fourier frequency domain. We now confirm this behavior in the time domain. Note here that the closed loop method of frequency stability analysis represents lower limit of Allan variance because it does not consider the noise of the frequency discriminator [2, 6]. However the error signal is weighted by servo loop transfer function $H(f)$, so in principle it is possible to extract laser frequency stability from the error signal [20]. Despite of this limitation the closed loop method can effectively explore the relative frequency stability of the laser. This can be done by calculating the Allan variance,

$$\sigma^2(2, \tau) = \frac{2}{v_0^2} \int_0^\infty S_{\Delta v}(f) \frac{\sin^4(\pi f \tau)}{(\pi f \tau)^2} df, \qquad (13.7)$$

where τ is integration time and v_0 is the nominal frequency (3.518×10^{14} Hz). Figure 13.11 shows the behavior of the square root of Allen variance, $\sigma(2, \tau)$, as

Fig. 13.9 Error signal in closed loop condition recorded for (**a**) EIT locked probe, (**b**) SAS locked pump and (**c**) AT resonance (outside the Doppler) locked probe. The figures clearly vindicate that the peak to peak frequency fluctuations are considerably reduced for EIT locked probe compared to that for the other cases

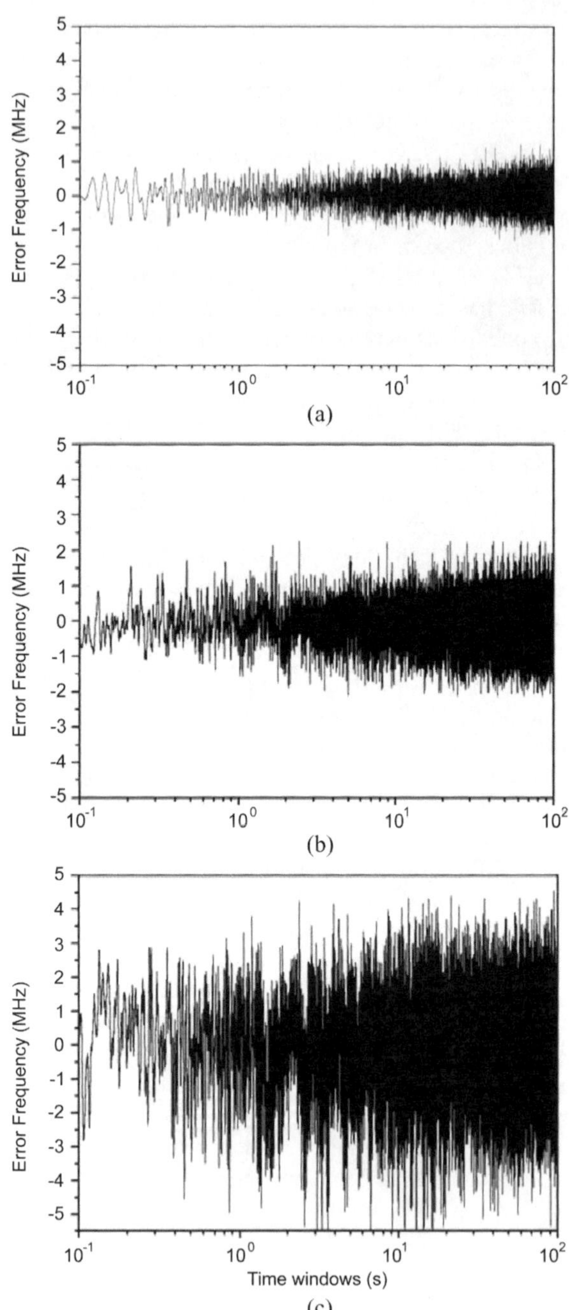

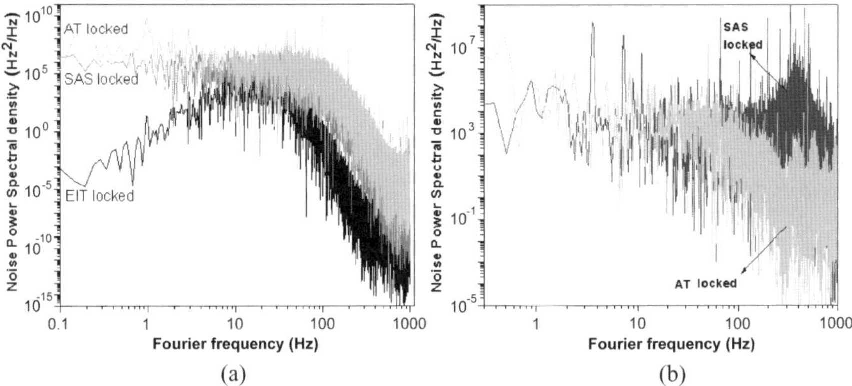

Fig. 13.10 Comparison of frequency noise power spectra for pump and probe lasers locked under different AFOL schemes. (**a**) EIT locked probe (*black*), SAS locked pump (*gray*) and AT locked probe (*light gray*). Here the AT resonance is outside the Doppler profile. (**b**) SAS locked pump (*black*) and AT locked probe (*light gray*). In this case the AT resonance is inside the Doppler profile. See text for details

a function of integration time for the AFOL schemes studied in our work. It may be seen from Fig. 13.11(a) that the EIT locked probe attains the lowest value of σ ($\sim 2 \times 10^{-13}$) after 10 sec compared to SAS locked pump ($\sim 2 \times 10^{-12}$) and AT (outside the Doppler profile) locked probe ($\sim 6 \times 10^{-12}$). Allan variance for the probe locked on AT resonance inside the Doppler profile is shown in Fig. 13.11(b). It is interesting to see that the value of σ for the AT locked probe laser is $\sim 2 \times 10^{-12}$ for $\tau < 0.1$ sec while for $\tau > 0.2$ sec it crosses that of the SAS locked pump. That means that although the short term stability the probe laser is more than that of the pump, the situation is reverse in case of the long term stability. For the $\tau > 10$ seconds, the pump laser appears to be more stable than the probe. Thus the error signal analysis in both frequency and time domain shows similar results and the performance of EIT based AFOL is always better than AT based AFOL schemes.

It is important to note here that we measure residual frequency fluctuations at the mixer output under closed loop condition; hence it is apparent that correct measure of frequency fluctuations may be larger than that is obtained experimentally. This is because the frequency noise may actually be present in the system but gets suppressed by correlated intensity fluctuations that may be introduced by feedback from a coherent media which favors PM-AM conversion. We have taken a crude account of the situation by measuring intensity fluctuations of the offset locked laser with an out-of-loop photodetector and converting it to frequency scale with the slope of the discriminator. The subsequent FFT analysis indicates that the estimated frequency noise correlated with intensity noise is prominent at the low Fourier frequency region ($f \leq 10$ Hz) where it lies above the level of the frequency noise measured at the mixer output. This fact clearly indicates that $\sigma(2, \tau)$ for AT locked laser will be influenced by PM-AM process at relatively larger values of integration time (τ). It may also be noted here that PM-AM conversion is also present within EIT resonance

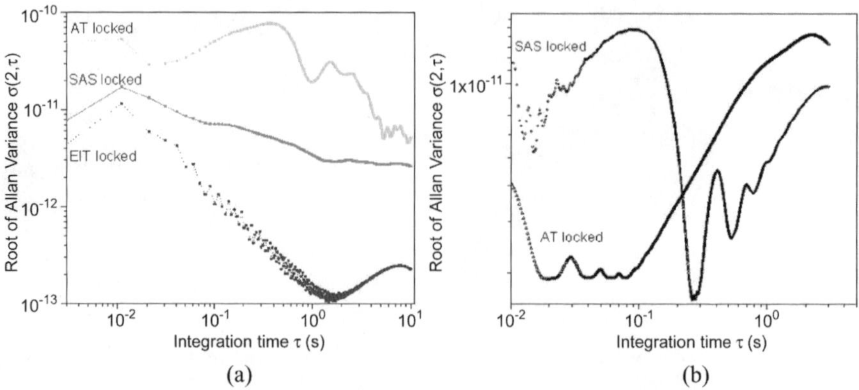

Fig. 13.11 Comparative plots for the square root of Allan variance vs integration time. Frames (**a**) and (**b**) correspond to the situations of Figs. 13.10(a) and 13.10(b) respectively

but in a much reduced scale owing to relative dominance of two photon detuning over its one photon counterpart.

In terms of the frequency stability, the EIT based AFOL is certainly better than the AT based AFOL. However, EIT based AFOL is always constrained to give 'fixed' frequency offset, clock frequency in the present work, irrespective of the choice of the experimental parameters. On the other hand, AT based AFOL can provide tunability since the AT resonances can be tuned by varying pump detuning. In a larger perspective, it is thus a tradeoff between accuracy and tunability that governs the suitability of AFOL schemes discussed in this work. It may also be pointed out that coherence assisted transfer process is beneficial since it allows the reflection of the modulation from pump to probe laser; thus augmenting the use of a single modulation source in an AFOL scheme. At the same time it can also harm the probe laser stability due to additional noise that is introduced to the respective control loop in due course. This calls for a suitable trade off among several parameters, e.g., modulation frequency and depth, laser linewidth, technical noise etc. to establish a prudent AFOL scheme.

13.5 Conclusion

In this work, we have studied EIT and AT based AFOL schemes using three level Λ schemes in the hyperfine manifold of D_2 transition of Cs. The techniques and analysis developed here can be extended to multi-level schemes, e.g., N-resonance, where more than two lasers can be offset locked with desired frequency differences. These lasers can further be phased locked using the technique of beat note. In line with earlier reports [19, 23] our aim here is to implement novel AFOL schemes rather than significantly reducing the linewidth of the EIT or AT resonances. In case of EIT signal, we have ensured that $\Gamma_{\mathrm{EIT}} \propto \Omega_c$ and that means a narrower EIT may be obtained by reducing pump Rabi frequency. We have further attempted to introduce some tunability in the AFOL scheme by stabilizing the probe on a narrow AT resonance.

While discussing the robustness of the AFOL schemes it is important to look at the frequency stability of each laser. The general conclusion that can be drawn is that the laser stability is highly dependent on the slope of the frequency discriminator of the feedback loop of the laser. The important parameter that decides the slope of the discriminator is the modulation transfer in case where only pump laser is frequency modulated. We may conclude that the modulation transfer is maximum at two-photon resonance condition ($\Delta_p = \Delta_c$) and as we move farther away it becomes weaker. The discussion of the tunability of AFOL, obviously shows that the AT based AFOL can be tuned to wide range whereas EIT based AFOL have fixed frequency offset. Thus there is visible tradeoff between robustness and tunability while constructing the present AFOL schemes. To overcome this problem we are currently working on the EIT based tunable AFOL (TAFOL) scheme that can be equally or more robust than the present schemes and can provide large tunable range without compromising the robustness. Such futures TAFOL schemes may open new opportunities of research in the field such metrology as well as ultra precision coherent spectroscopy.

References

1. E. Arimondo, Coherent population trapping in laser spectroscopy, in *Progress in Optics*, vol. XXXV (Elsevier, Amsterdam, 1996), p. 258
2. J.A. Barnes, A.R. Chi, L.S. Cutler, D.J. Healey, D.B. Leeson, T.E. McGunigal, J.A. Mullen, W.L. Smith, R.L. Sydnor, R.F.C. Vessot, G.M.R. Winkler, Characterization of frequency stability. IEEE Trans. Instrum. Meas. **IM-20**, 105 (1971)
3. S.C. Bell, D.M. Heywood, J.D. White, J.D. Close, R.E. Scholten, Laser frequency offset locking using electromagnetically induced transparency. Appl. Phys. Lett. **90**, 171120 (2007)
4. K.J. Boller, A. Imamoglu, S.E. Harris, Observation of electromagnetically induced transparency. Phys. Rev. Lett. **66**, 2593 (1991)
5. J.C. Camparo, J.G. Coffer, Conversion of laser phase noise to amplitude noise in a resonant atomic vapor: The role of laser linewidth. Phys. Rev. A **59**, 728 (1999)
6. T. Day, E.K. Gustafson, R.L. Byer, Sub-hertz relative frequency stabilization of two-diode laser-pumped Nd:YAG lasers locked to a Fabry-Perot interferometer. IEEE J. Quantum Electron. **28**, 1106 (1992)
7. M. Fleischhauer, Correlation of high-frequency phase fluctuations in electromagnetically induced transparency. Phys. Rev. Lett. **72**, 989 (1994)
8. A. Javan, O. Korcharovskaya, H. Lee, M.O. Scully, Narrowing of electromagnetically induced transparency resonance in a Doppler-broadened medium. Phys. Rev. A **66**, 013805 (2002)
9. Y.B. Kale, A. Ray, Q.V. Lawande, R. D'Souza, B.N. Jagatap, Atomic frequency offset locking in a Λ type three-level Doppler broadened Cs system. Appl. Phys. B **100**, 505 (2010)
10. Y.B. Kale, A. Ray, N. Singh, Q.V. Lawande, B.N. Jagatap, Modulation transfer in Doppler broadened Λ system and its application to frequency offset locking. Eur. Phys. J. D **61**, 221 (2011)
11. Y.B. Kale, A. Ray, Q.V. Lawande, R. D'Souza, B.N. Jagatap, Coherent spectroscopy of a Λ atomic system and its prospective application to tunable frequency offset locking. Phys. Scr. **84**, 035401 (2011)

12. J. Kitching, S. Knappe, N. Vukičević, L. Hollberg, R. Wynands, W. Weidmann, A microwave frequency reference based on VCSEL-driven dark line resonances in Cs vapor. IEEE Trans. Instrum. Meas. **49**, 1313 (2000)

13. M. Klein, M. Hohensee, Y. Xiao, R. Kalra, D.F. Phillips, R.L. Walsworth, Slow-light dynamics from electromagnetically-induced-transparency spectra. Phys. Rev. A **79**, 053833 (2009)

14. S. Knappe, V. Shah, D.P. Schwindt, L. Hollberg, J. Kitching, L. Liew, J. Moreland, A microfabricated atomic clock. Appl. Phys. Lett. **85**, 1460 (2004)

15. A. Lezama, S. Barreiro, A.M. Akulshin, Electromagnetically induced absorption. Phys. Rev. A **59**, 4732 (1999)

16. Y. Li, M. Xiao, Electromagnetically induced transparency in a three-level Λ-type system in rubidium atoms. Phys. Rev. A **51**, R2703 (1995)

17. H.S. Moon, L. Lee, K. Kim, J.B. Kim, Laser frequency stabilizations using electromagnetically induced transparency. Appl. Phys. Lett. **84**, 3001 (2004)

18. A.F. Huss, R. Lammegger, C. Neureiter, E.A. Korsunsky, L. Windholz, Phase correlation of laser waves with arbitrary frequency spacing. Phys. Rev. Lett. **93**, 223601 (2004)

19. U.D. Rapol, A. Wasan, V. Natarajan, Subnatural linewidth in room-temperature Rb vapor using a control laser. Phys. Rev. A **67**, 053802 (2003)

20. A. Ray, Study of the frequency fluctuations of a semiconductor diode laser. Can. J. Phys. **86**, 351 (2008)

21. M. Stähler, S. Knappe, C. Affolderbach, W. Kemp, R. Wynands, Chip-scale atomic magnetometer. Europhys. Lett. **54**, 323 (2001)

22. G. Vemuri, G.S. Agarwal, B.D. Nageswara Rao, Sub-Doppler linewidth with electromagnetically induced transparency in rubidium atoms. Phys. Rev. A **53**, 2842 (1996)

23. J. Wang, Y. Wang, S. Yan, T. Liu, T. Zhang, Observation of sub-Doppler absorption in the Λ-type three-level Doppler-broadened cesium system. Appl. Phys. B **78**, 217 (2004)

24. R. Wynands, A. Nagel, Precision spectroscopy with coherent dark states. Appl. Phys. B **68**, 1 (1999)

25. C.Y. Ye, A.S. Zibrov, Width of the electromagnetically induced transparency resonance in atomic vapor. Phys. Rev. A **65**, 023806 (2002)

Chapter 14
Quantum Physics Inspired Optical Effects in Tight-Binding Lattices

Clinton Thompson and Gautam Vemuri

Abstract We theoretically investigated the propagation of light inside an array of single-mode evanescently coupled waveguides that can be described by the tight-binding Hamiltonian. We show that directed photonic transport can be achieved with phase-displaced inputs. In addition, the form of a parity-symmetric waveguide-dependent coupling constant can tune the dynamics of the photon's wavepacket. Lastly, we examine the statistical aspects of the output light for different input fields when disorder is present in the waveguide array. We find that the light will undergo Anderson localization independent of the type of field and that the intensity fluctuations of the output light will increase with disorder at the initial waveguide.

14.1 Introduction

Historically, many ideas in quantum physics had their pre-cursors in classical wave optics. With the availability of integrated optical structures, it is now possible to realize quantum effects in seemingly classical systems. The wave equation in the paraxial approximation, which is normally used to describe the propagation of light in discrete structures, is nearly identical to the Schrödinger equation. Since the wave equation and the Schrödinger equation are mathematically similar, light propagation in integrated structures, such as an array of coupled waveguides, allows the possibility of studying quantum physics inspired effects over lengths scales ranging from micrometer to nanometer, hence making it easier to observe and measure these effects. Some of these effects are *Anderson localization* [1, 12], *Bloch oscillations* [2, 29, 31], *quantum random walk of photons* [27, 30], and the *Dirac Zitterbewegung* [20].

It has been shown previously that the Hamiltonian for the evolution of a wavepacket in an array of coupled waveguides is well described by the tight-binding

C. Thompson · G. Vemuri (✉)
Department of Physics, Indiana University Purdue University Indianapolis (IUPUI), Indianapolis, IN 46202-3273, USA
e-mail: gvemuri@iupui.edu

C. Thompson
e-mail: clithomp@iupui.edu

M. Mohan (ed.), *New Trends in Atomic and Molecular Physics*,
Springer Series on Atomic, Optical, and Plasma Physics 76,
DOI 10.1007/978-3-642-38167-6_14, © Springer-Verlag Berlin Heidelberg 2013

Hamiltonian that was originally used to describe the time-evolution of an electron in a metallic medium. This in fact was the reason for success in realizing, using waveguide arrays, many of the condensed matter effects which in addition to these mentioned in the first paragraph, include studies of defects in lattices [22, 23], rectification of light [19]. Further, many quantum optical effects can be studied using waveguide arrays.

This is a result of the fact that with current technology, any one-dimensional tight-binding phenomena can be modeled by changing the indices of refraction and changing the spacing between waveguides during the fabrication process. The fabrication process consists of growing wafers of the desired composition via molecular beam epitaxy (MBE) and then using lithography and etching to create the waveguide array from the wafer. Recently, two materials that have been used for the waveguides have been silica [32] and aluminum gallium arsenide (AlGaAs) [7]. To minimize absorption, one needs to operate a laser 800 nm for silica while AlGaAs has low linear and nonlinear absorption at approximately 1500 nm [8].

The shapes of the waveguides can also be changed by applying different masks during the lithography process. When the waveguides are curved, dynamical localization of the wavepacket has been experimentally observed [24], *Bloch oscillations* have been theoretically predicted to occur [14] and experimentally observed [6], and coherent population transfer has been predicted to occur [17]. Bent waveguides have been theoretically proposed to be an all-optical switch [25] and an analog of Raman chirped adiabatic passage [18].

Waveguide arrays have also been used to investigate Hamiltonians with parity ($\mathscr{P}$) and time-reversal ($\mathscr{T}$) symmetry. One method of experimentally investigating $\mathscr{P}\mathscr{T}$-symmetry is to amplify the light in one waveguide and insert impurities that will absorb light into its symmetric counterpart [34]. This is achieved through engineering the complex part of the index of refraction where positive values correspond to gain and negative values correspond to loss. The critical value of complex part for which the Hamiltonian is $\mathscr{P}\mathscr{T}$-symmetric has been shown to waveguide dependent [36].

We will investigate phase-controlled *photonic transport*, the evolutions of a wavepacket in a waveguide array with parity-symmetric tunneling, and quantum statistical aspects of *Anderson localization*.

14.2 Phase-Controlled Photonic Transport

In this section we describe the theoretical basis for *photonic transport* that is controlled by interference. We consider an array of waveguides, as shown in Fig. 14.1, in which the nearest neighboring waveguides are evanescently coupled to each other. The Hamiltonian for this system is known to be isomorphic to the Hamiltonian in the tight-binding model, and is given by

$$H = \hbar \sum_j \beta(j) a_j^\dagger a_j + \hbar C \sum_j \left(a_{j+1}^\dagger a_j + a_{j-1}^\dagger a_j \right) \tag{14.1}$$

Fig. 14.1 A large waveguide array with the input fields shown

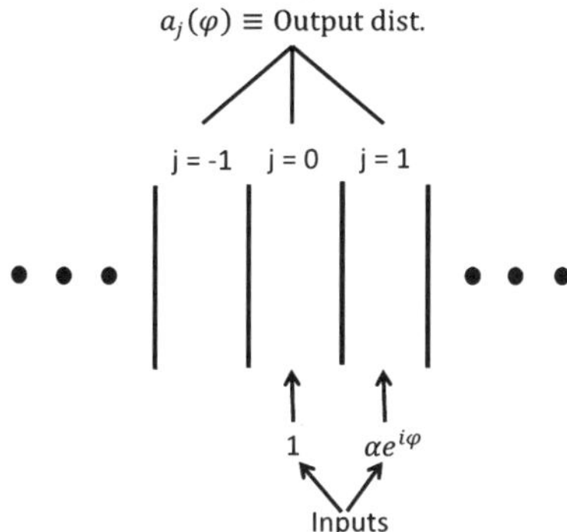

where the individual waveguides in the array are labeled by the index j and $\beta(j)$ is related to the refractive index of the jth waveguide.

In Eq. (14.1), C is the coupling between adjacent waveguides, and $a_j^\dagger(a_j)$ is the creation (annihilation) operator for the jth waveguide. These operators obey the *Heisenberg equations*, given by

$$\dot{a}_j = -i\beta(j)a_j - iC(a_{j+1} + a_{j-1})$$
(14.2)

We now turn to a classical description and so a_j's are regarded as numbers.

Equation (14.2) can now be written in the Fourier space, as

$$\dot{\tilde{a}}(k) = -2iC\tilde{a}(k)\cos k - i\beta\left(\frac{\partial}{\partial(-ik)}\right)\tilde{a}(k)$$
(14.3)

Clearly, Eq. (14.3) now has a term describing a *periodic potential*. There is also a term that is quadratic in p, i.e. a term like $\frac{\partial^2}{\partial x^2}$. We now choose

$$\beta(j) = j\beta$$
(14.4)

and then Eq. (14.2) can be written as

$$\dot{\tilde{a}}(k) = -2iC\tilde{a}(k)\cos k - ij\beta\frac{\partial}{\partial(-ik)}\tilde{a}(k)$$
(14.5)

The choice of the waveguide index, as given by Eq. (14.4), has been experimentally realized in studies on *Bloch oscillations* in waveguide arrays. It is important to emphasize here that for the waveguide structure under investigations, the *periodic potential* is in the Fourier space, k. Therefore, to study the *photonic transport* properties of this system, we study the directed motion in the site-space j.

The input conditions on the field amplitude are chosen as

$$a_j(t = 0) = \delta_{j,0} + \alpha\delta_{j,1}e^{i\varphi}$$
(14.6)

We assume that the array index runs from $j = -\infty$ to ∞ (see Fig. 14.1), where $j = 0$ is the middle waveguide. The $|a_j|^2$ play the role of density distributions. The *directed transport* that we are interested in is obtained from a calculation of $|a_j|^2$ for non-zero α and φ. The possibility of additional interference effects due to a non-zero α leads to *photonic transport*, and can be viewed as the classical analog of a *quantum ratchet*.

We write the solution to the coupled equations, Eq. (14.2), in the form

$$a_j = \sum_{j'} G_{j,j'} a_{j'}(0) \tag{14.7}$$

where $a_j(0)$ is given by Eq. (14.6) and the *Green's function*, $G_{j,j'}$ is given by Eq. (21) from [31]

$$G_{j,j'} = \exp\left[i\beta z + \frac{i(j - j')(\beta z - \pi)}{2} \right]$$
$$\times J_{j'-j}\left[\frac{4C}{\beta} \sin\left(\frac{\beta z}{2} \right) \right] \tag{14.8}$$

Using the initial condition of Eq. (14.6), the output intensity from the waveguide array can be written as

$$I_j = |G_{j,0}|^2 + |\alpha G_{j,1}|^2 + \alpha G_{j,0} G_{j,1}^* e^{-i\varphi} + \alpha G_{j,0}^* G_{j,1} e^{i\varphi}$$
$$= \left| J_{-j}\left[\frac{4C}{\beta} \sin\left(\frac{\beta z}{2} \right) \right] \right|^2 + \left| \alpha J_{1-j}\left[\frac{4C}{\beta} \sin\left(\frac{\beta z}{2} \right) \right] \right|^2$$
$$- 2\alpha J_{-j}\left[\frac{4C}{\beta} \sin\left(\frac{\beta z}{2} \right) \right] J_{1-j}\left[\frac{4C}{\beta} \sin\left(\frac{\beta z}{2} \right) \right]$$
$$\times \sin\left(\frac{\beta z}{2} - \varphi \right) \tag{14.9}$$

Using Eq. (14.9), the expectation value of the site position can be written as (after using properties of Bessel functions)

$$\sum_{j=-\infty}^{\infty} j I_j = |\alpha|^2 + \frac{4\alpha C}{\beta} \sin\left(\frac{\beta z}{2} \right) \sin\left(\frac{\beta z}{2} - \varphi \right) \tag{14.10}$$

Similarly, the expectation value for the analog of energy is given by

$$\sum_{j=-\infty}^{\infty} j^2 I_j = |\alpha|^2 + \frac{4\alpha C}{\beta} \sin\left(\frac{\beta z}{2} \right) \sin\left(\frac{\beta z}{2} - \varphi \right)$$
$$+ \frac{1 + |\alpha|^2}{2} \left(\frac{4C}{\beta} \sin\left(\frac{\beta z}{2} \right) \right)^2 \tag{14.11}$$

For small values of z, the average position, Eq. (14.10), is proportional to $-\alpha \sin(\varphi)z$, which shows that the direction of transport for the photons is dependent on the relative phase of the inputs.

We now describe the results on *photonic transport* that can arise in a waveguide array when phase-displaced inputs are utilized. For all results shown, the inputs are at the $j = 0$ and $j = 1$ waveguides. Figure 14.2(a) shows the output intensity from the array, given by Eq. (14.9), when α is equal to zero. This results in profiles which are symmetric about the center of the array, and one sees the breathing modes. However, when α is non-zero, the transport effect is immediately evident. For example,

(a)   (b)   (c)

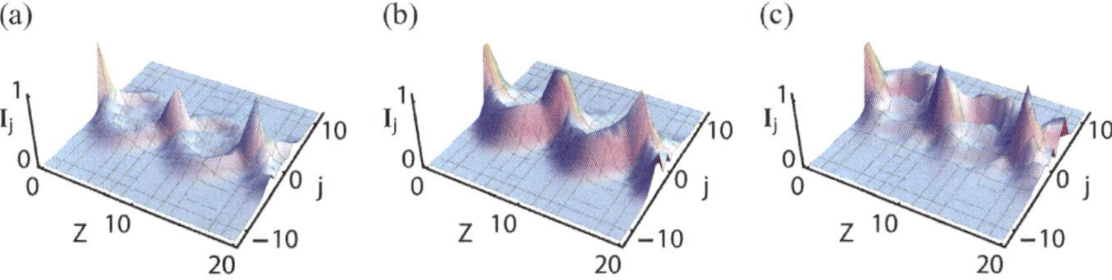

Fig. 14.2 Output intensity distribution, I_j, as a function of z for $j = -12, \ldots, 12$. The values of the parameters are $\beta/C = 073$ with (**a**) $\alpha = 0$, (**b**) $\alpha = 1$, $\varphi = 37°$, and (**c**) $\alpha = 1$, $\varphi = 217°$

in Fig. 14.2(b), we display the *intensity profiles* when $\varphi = 37°$. These profiles are asymmetric about the j-axis, indicating transport to the lower index side of the array. However, when $\varphi = 217°$, as in Fig. 14.2(c), the direction of the asymmetry is reversed, indicating that the transport is now to the high index side of the array. These results suggest that the *photonic transport* is a consequence of interference, and that one can control the direction of the transport through the relative phase of the inputs. In particular, when φ lies between $0°$ and $180°$, one gets a transport to the low index side, and for phi between $180°$ and $360°$, to the high index side.

The results described here exploit asymmetric inputs in a system with a symmetric, *periodic potential*, to obtain *photonic transport*. Thus, this work is quite different from other work wherein either a single waveguide was excited, or several waveguides were excited simultaneously. The novel feature here is the phase-displaced inputs, which gives us an additional parameter, viz. the relative phase of the inputs, to control the direction of transport. For further details, see [40] and for a full quantum mechanical description, see [33].

14.3 Waveguide Lattices with Non-uniform Parity-Symmetric Tunneling

We now consider an array of N waveguides described by the Hamiltonian, Eq. (14.1) that have $\beta(j) = 0$ and a waveguide-dependent tunneling function

$$C_\alpha(k) = C_0\big[k(N-k)\big]^{\alpha/2} = t_\alpha(N-k) \tag{14.12}$$

where $C_\alpha(k)$ is tunneling amplitude between sites k and $k+1$. The tunneling amplitude $C(k)$ is determined by the evanescent coupling between waveguides k and $k+1$, and can be tuned by varying the width of the barrier between the two waveguides [28]. A Hamiltonian eigenfunction $|\psi^n\rangle = \sum_k \psi_k^n a_k^\dagger |0\rangle$ with energy E^n satisfies the difference equation

$$C_\alpha(k-1)\psi_{k-1}^n + C_\alpha(k)\psi_{k+1}^n = -E_\alpha^n \psi_k^n \tag{14.13}$$

where we have considered a constant index of refraction n_R, which results in a constant shift in the energy eigenvalues. The *eigenvalue spectrum* for Eq. (14.13) is

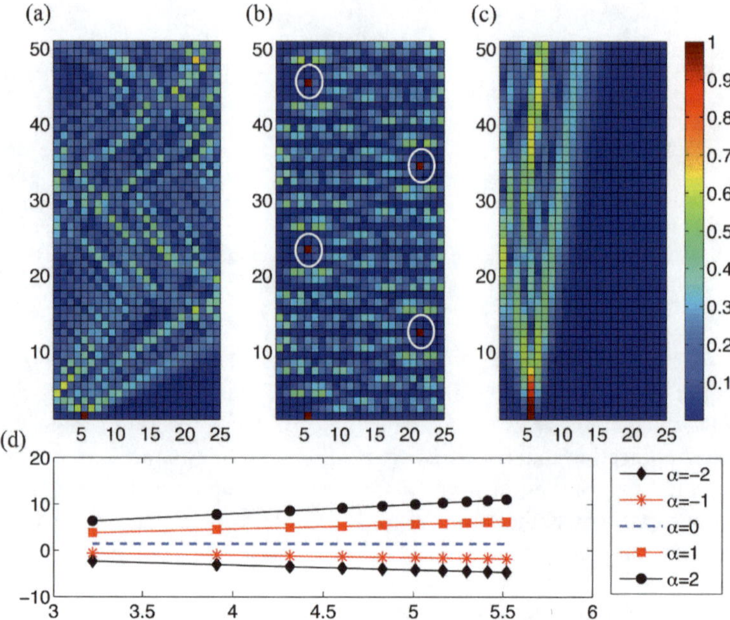

Fig. 14.3 Time-evolution in an array with a fixed length $L/(\hbar c/n_R C_0) = 50$ shown along the *vertical axis*. *Top three panels* show the probability-amplitude time-evolution plots of a single photon injected in waveguide $m_0 = 5$ in an array of $N = 25$ waveguides with (**a**) $\alpha = 0$, (**b**) $\alpha = +1$, and (**c**) $\alpha = -1$. The *vertical axis in each panel* represents the distance along the waveguide. When $\alpha = 0$ the wavepacket, initially localized at $m_0 = 5$, spreads as it travels along the waveguide. For $\alpha = 1$, because the energy levels are equidistant, the wavepacket is periodically localized at mirror-symmetric positions $(N + 1 - m_0) = 21$ and $m_0 = 5$ (*white circles*). When $\alpha = -1$, the wavepacket spread is noticeably smaller over the same length of the waveguide. Panel (**d**) shows the dimensionless bandwidth $\Delta_\alpha(N)/C_0$ (*vertical*) vs. N (*horizontal*) on a logarithmic scale for $25 \leq N \leq 250$. We see that when $\alpha \geq 1$, $\Delta_\alpha(N) \propto N^\alpha$, whereas for $\alpha \leq -1$, $\Delta_\alpha(N) \propto N^{-\alpha/2}$. Therefore, a sample with a given physical length represents "short-time" evolution when $\alpha < 0$ and "long-time" evolution when $\alpha > 0$

symmetric about zero [9]. Hence, the bandwidth of the spectrum, defined as the difference between the maximum and minimum eigenvalues, is $\Delta_\alpha = E_{\max} - E_{\min} = 2E_{\max}$. Note that when $\alpha > 0$, the tunneling function $C_\alpha(k)$ is maximum at the center of the waveguide array whereas when $\alpha < 0$, it is maximum at the ends. As a result, when $N \gg 1$ the bandwidth $\Delta_\alpha(N)$ of the Hamiltonian H_α increases monotonically with α. It is natural to use the inverse-bandwidth as the characteristic time, $C_\alpha = 2\hbar/\Delta_\alpha$, and $L_\alpha = cC_\alpha/n_R$ as the characteristic distance along the waveguide where $\hbar = h/(2\pi)$ is the scaled Planck constant and c/n_R is the speed of light in a waveguide. Note that since C_α and L_α are monotonically decreasing functions of α, a waveguide array with a fixed physical length will correspond to "short-time" scenario when $\alpha < 0$ and "long-time" scenario when $\alpha > 0$.

Figure 14.3 shows the *time-evolution* of a wavepacket that is initially localized in waveguide $m_0 = 5$ in an array of $N = 25$ waveguides. The vertical axis denotes distance along the waveguide *for a fixed physical length of the waveguide* $L/(\hbar c/nC_0) = 50$. The three vertical panels correspond to $\alpha = 0$ (left), $\alpha = 1$ (center) and $\alpha = -1$ (right). When $\alpha = 0$, the traditional model, the wavepacket broadens as it travels down the length of the waveguide array. When $\alpha = 1$, the energy levels are given by $E_n = \pm C_0(N-1), \pm C_0(N-3), \ldots$; the level spacing is constant and the bandwidth is $\Delta_{\alpha=1}(N) = 2(N-1)C_0$ [10, 21]. Therefore we obtain perfect reconstruction of the wavepacket, shown by white circles, at mirror-symmetric positions $(N+1-m_0) = 21$ and $m_0 = 5$. It should be noted that the wavepacket first reconstructs at the mirror-symmetric waveguide, i.e. 21st site, and then alternates between sites 5 and 21. For $\alpha = -1$ (right panel), *the wavepacket spread over the same distance along the waveguide is significantly smaller*, consistent with the smaller bandwidth of the Hamiltonian.

To explore the intrinsic α-dependence of the *time-evolution*, in the rest of the paper, we consider waveguides with the *same normalized length $L/L_\alpha = 100$*; physically, this will correspond to waveguides with different α-dependent lengths. Figure 14.4 shows the *time-evolution* of a wavepacket initially at $m_0 = 5$ for $\alpha = 1$ (left panel), $\alpha = 2$ (center panel) and $\alpha = -1$ (right panel) in an array of $N = 25$ coupled waveguides. The vertical axis shows distance (time) in the units of $L_\alpha(C_\alpha)$. Apart from the perfect reconstruction at mirror-symmetric points that occurs when $\alpha = 1$, we see that, in contrast to the behavior in Fig. 14.3 the spread of the wavepacket is qualitatively similar for all α over the normalized length-scales (or time-scales). The bottom panel shows, for $\alpha = -1$, the *time-evolution* of a single photon injected near the edge, $m_0 = 2$; the horizontal axis shows the normalized distance (time). In this case, the photon remains at the edge due to localized edge eigenstates that are generically present when $\alpha < 0$ [10]. For further details, see [11].

We now explore the effects of the tunneling function $t_\alpha(k)$ on the two-particle (number) correlation function defined by $\Gamma^\alpha_{mn}(t) = \langle a_m^\dagger(t)a_n^\dagger(t)a_n(t)a_m(t)\rangle$. This function encodes the Hanbury-Brown-Twiss *quantum correlations* in coincidence detections in waveguides m and n [4]. For an initial state where the two particles are localized at sites (m_0, n_0), the correlation function becomes

$$\Gamma^\alpha_{mn}(t) = \left| G_{mm_0}(t)G_{nn_0}(t) \pm G_{mn_0}(t)G_{nm_0}(t) \right|^2 \tag{14.14}$$

where $G_{pq}(t) = [\exp(-iH_\alpha t/\hbar)]_{pq}$ is the *time-evolution* operator and $\pm$ signs correspond to bosons and fermions respectively. When $\alpha = 0$, the traditional model, properties of this correlation function and its dependence on the initial state have been extensively investigated [13]. Since $\Gamma^\alpha_{mn}(t)$ is determined by the *time-evolution* operator, it follows that the bosonic and fermions correlations will be qualitatively different when $\alpha \neq 0$. In particular, when $\alpha = 1$, the constant energy level spacing implies that $\Gamma^{\alpha=1}_{mn}(t)$ is periodic in time or, equivalently, in the distance along the waveguide; since the maximum spread of a wavepacket initially confined at position $1 \leq m_0 \leq N/2$ is approximately $2m_0$, it follows that the spatial extent and shape of the correlation function in the (m, n) plane can be controlled by appropriate initial conditions.

Figure 14.5 shows Γ^α_{mn} for an array with $N = 40$ waveguides and $(m_0, n_0) = (1, 2)$. The left panels show the results for $\alpha = 1$ for bosons (top) and fermions (bottom) at time $t/T_\alpha = 25$. In contrast to the $\alpha = 0$ case [13], the correlation function is

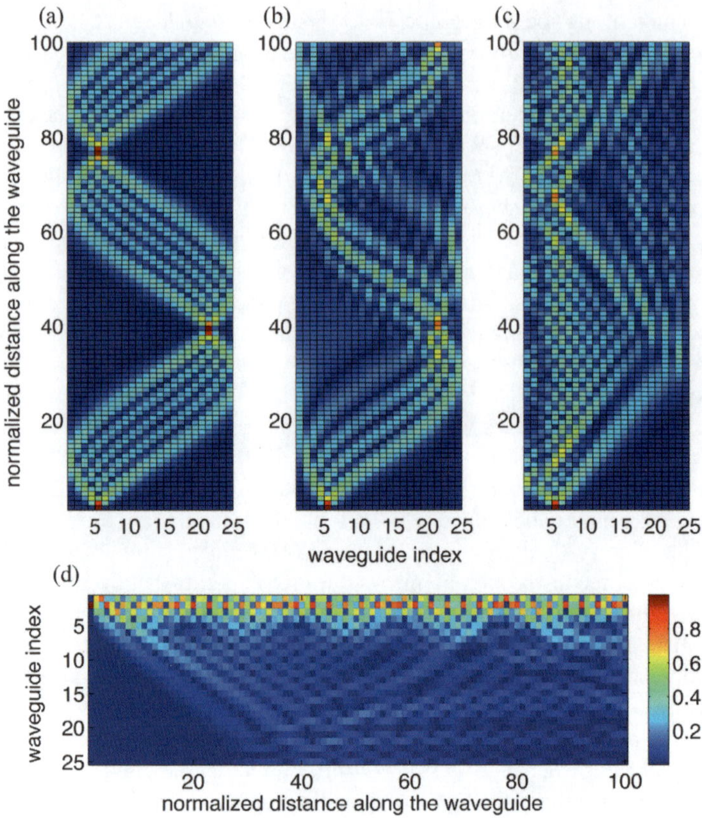

Fig. 14.4 *Top panels* show the time-evolution of a photon injected at $m_0 = 5$ in an array of $N = 25$ waveguides with (**a**) $\alpha = 1$, (**b**) $\alpha = 2$, and (**c**) $\alpha = -1$. The wavepacket spread is similar over normalized length-scales. Panel (**d**) shows that for $\alpha = -1$, the time-evolution of a photon injected near the edge, $m_0 = 2$. The strong localization of the photon near the edge is due to the presence of localized edge eigenstates that occur when $\alpha < 0$

strongly localized at all times, and has only two peaks with a single nodal line separating them. The right panels correspond to $\alpha = 2$ and $t/T_\alpha = 55$. At this time, the *bosonic* correlation function (top) is *localized near the second edge*, with a nearby parabolic nodal region. On the other hand, the *fermionic* correlation function (bottom) is sharply *localized in one direction and extended in the other*, with a broad nodal region around the diagonal. (We recall, from the central panel in Fig. 14.4, that when $\alpha = 2$, a wavepacket starting near the edge localizes substantially near the other edge when $t/T_\alpha \sim 50$.) These results show that the *quantum statistics* lead to nontrivial correlations for $\alpha = 2$ case that are dramatically different from the $\alpha = 0$ case [13] or the $\alpha = 1$ case.

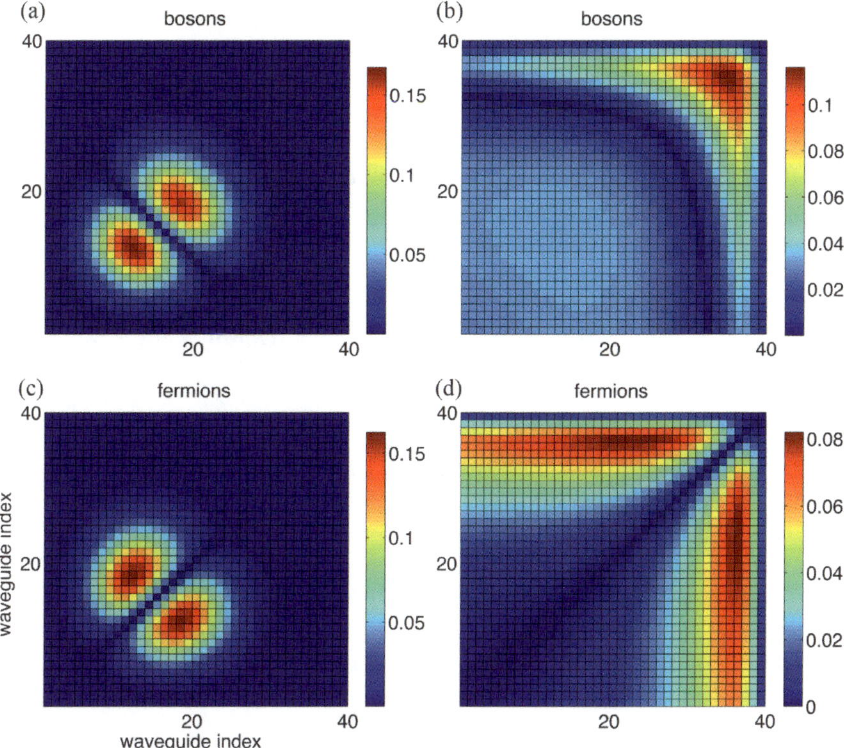

Fig. 14.5 Panels (**a**) and (**c**) show the correlation function $\Gamma^{\alpha}_{mn}(t)$ for an array of $N = 40$ waveguides at $t/T_{\alpha} = 25$ and $\alpha = 1$. The initial state of the system has two particles in the first two waveguides. The correlation function remains localized, and develops only two symmetric peaks with a single nodal line. Panels (**b**) and (**d**) correspond to $\alpha = 2$ and $t/T_{\alpha} = 55$ with the same initial conditions. We see that the bosonic correlation function (*top*) is localized near the second edge, whereas its fermionic counterpart (*bottom*) is localized in one direction and extended in the other direction

14.4 Anderson Localization with Second Quantized Fields: Quantum Statistical Aspects

Lastly, we consider an array of single-mode waveguides with neighboring waveguides that are described my Eq. (14.1) with $\beta(j) = \beta_j$. In order to study *Anderson localization* of photons, the β_j are taken to be real and random. Specifically, the deviation in β_j from its mean is assumed to be a random variable. We further assume that the random variables β_j are independent of each other. In this paper, we assume that the disorder of the medium follows a Gaussian distribution, which is given by $P(\beta) = \frac{1}{\sqrt{2\pi\,\Delta^2}} \exp(\frac{-\beta^2}{2\Delta^2})$ where Δ^2 is the variance of the distribution and is a measure of the disorder in the medium.

Note that the *Green's function* G depends on the parameters C and β_j and is random in nature due to disorder in β_j. Also, the *Green's function* in Eq. (14.6) depends on the propagation distance, z, over which the light distribution evolves. All the physical quantities at the output would require averaging of the *Greens function* and its powers. The quantized nature of the fields enters through the input Heisenberg operators.

Now we discuss what could be measurable quantities. Clearly mean intensities, I_l, at the output are obvious measurable quantities and these are given by

$$I_l = \langle a_l^\dagger a_l \rangle = \sum_p \sum_q \langle G_{l,p}^* G_{l,q} \rangle \langle a_p^\dagger(0) a_q(0) \rangle \tag{14.15}$$

where the product of *Green's functions* is to be averaged over the ensemble of distributions of β_j. In this paper we focus on the input light in a single waveguide, labeled as zero-then Eq. (14.15) simplifies to

$$I_l = \langle G_{l,0}^* G_{l,0} \rangle \langle a_0^\dagger a_0 \rangle \tag{14.16}$$

In order to examine the quantum statistical aspects of localization we can study the fluctuations in the intensity at the output. Glauber introduced the function $g^{(2)}$ defined by

$$g^{(2)} = \langle a^{\dagger^2} a^2 \rangle / \langle a^\dagger a \rangle^2 \tag{14.17}$$

Note that values of $g^{(2)}$ greater [smaller] than one correspond to bunching [anti-bunching]. Using the solution in Eq. (14.7), $g^{(2)}$ can be written in terms of the *Greens function* as

$$g^{(2)} = \frac{\langle |G_{l,0}|^4 \rangle \langle a_0^{\dagger^2} a_0^2 \rangle}{\langle G_{l,0}^* G_{l,0} \rangle^2 \langle a_0^\dagger a_0 \rangle^2} \tag{14.18}$$

Note that the quantum statistical quantity $g^{(2)}$ involves the averages of fourth powers of the *Greens function*. It is at this point that we start getting newer aspects of *Anderson localization* with quantized fields.

To further probe the effect of input light statistics on the quantum statistical aspects of Anderson localization, we calculate *site-to-site correlations* defined by

$$\langle I_l I_p \rangle = \langle |G_{10}|^2 |G_{p0}|^2 \rangle \langle a_0^{\dagger^2} a_0^2 \rangle. \tag{14.19}$$

The physical quantities introduced above do require the nature of the input fields. We will consider three types of input fields—(i) a coherent field, i.e. a field in a coherent state α_0, (ii) a *thermal field* with average photon number n_0, and (iii) a nonclassical field, such as a field in a squeezed state [37]. For all these fields, the quantities that we need in the above calculations are given as follows:

for a coherent field,

$$\langle a_0^{\dagger^2} a_0^2 \rangle = |\alpha_0|^4 \quad \text{and} \quad \langle a_0^\dagger a_0 \rangle = |\alpha_0|^2 \tag{14.20}$$

for a *thermal field*,

$$\langle a_0^{\dagger^2} a_0^2 \rangle = 2n_0^2 \quad \text{and} \quad \langle a_0^\dagger a_0 \rangle = n_0 \tag{14.21}$$

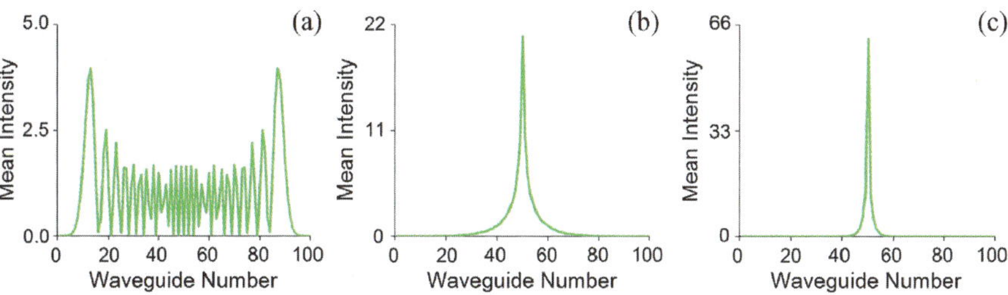

Fig. 14.6 Mean intensity vs. waveguide number for disorders of (**a**) $\Delta/C = 0$, (**b**) $\Delta/C = 1$, and (**c**) $\Delta/C = 3$. Each plot shows three indistinguishable curves for the three different input *photon statistics* (coherent, thermal and squeezed). Mean photon number for all three input fields is 100

and for a single-mode, *squeezed field*

$$\left\langle a_0^{\dagger 2} a_0^2 \right\rangle = \sinh^2 r \left(1 + 3\sinh^2 r\right) \quad \text{and} \quad \left\langle a_0^{\dagger} a_0 \right\rangle = \sinh^2 r \qquad (14.22)$$

In Eq. (14.22), r is the squeezing parameter. When we compare final results for different input fields we will assume that all fields have the same average photon number, i.e. $n_0 = |\alpha_0|^2 = \sinh^2 r$.

The *Anderson localization* for the cases described by Eq. (14.20) and Eq. (14.21) are well documented in the literature [5] whereas the results for nonclassical light as described by Eq. (14.22) are new. We first note some immediate consequences of the formula Eq. (14.18) which we can write as

$$\left\langle a^{\dagger 2} a^2 \right\rangle_{out} = g \left\langle a^{\dagger 2} a^2 \right\rangle_{in} \qquad (14.23)$$

where the subscript *out* (*in*) corresponds to the output (input) light from (into) the medium. Note that g is due to disorder of the medium, which is of classical origin, and hence is greater than one. Thus, the output field has fluctuations enhanced over the input fluctuations. This could be useful in enhancing the effects of radiation matter interaction. For example, the *two-photon absorption* probability is directly proportional to $\langle a^{\dagger 2} a^2 \rangle$ and thus can be enhanced by fields which have undergone *Anderson localization*.

For the numerical results below, we assume that the medium consists of 100 waveguides, and that the input light is coupled into the 50th waveguide. The random disorder is generated as follows: computer generated, uniformly distributed random numbers are used to generate 100 Gaussian distributed random numbers with zero mean and desired variance via the *Box-Mueller algorithm* [3]. Equation (14.2) is solved numerically, with each value of β_j corresponding to one waveguide. The output intensity is computed from the expression $I_j = \langle a_j^{\dagger} a_j \rangle$ where the angular brackets represent averaging over 1,000 realizations of the disorder and the quantum mechanical average over the input fields.

This section describes the results of our studies on the quantum statistical aspects of AL for different input *photon statistics*. For numerical calculations, we take $C = 1$, and the average value of β as zero. We begin with the injection of a classical

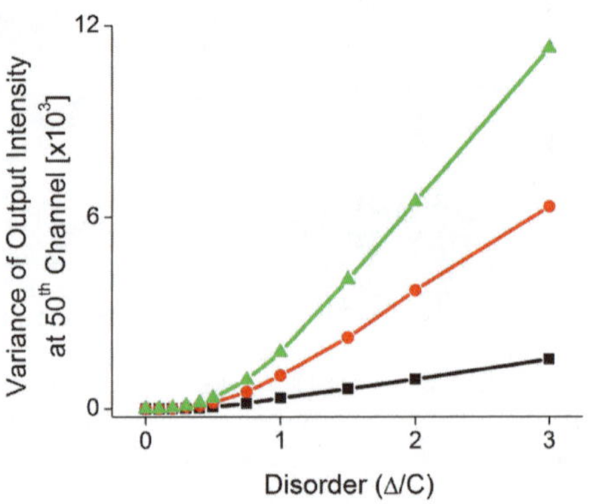

Fig. 14.7 Variance at the output of the 50th waveguide vs. disorder for a Gaussian disorder. Mean photon number for all three input fields is 100. Data shown are for input *photon statistics* of coherent field (*black*), *thermal field* (*red*) and single-mode, *squeezed field* (*green*)

coherent field, a *thermal field* and a *squeezed field* at waveguide 50. Figure 14.6(a)–14.6(c) show the mean intensity as a function of the waveguide position for all three input photon statistics. Note that each figure has three light distributions, but that the distributions for the three different field statistics are indistinguishable. We assume that the mean number of photons in the input field is 100, and that the disorder of the medium has a Gaussian distribution. Clearly, as the disorder is increased, the output field is increasingly localized, until at a disorder of about $\Delta/C = 3$, all three output light patterns converge to a narrow distribution centered around the input waveguide. Note that after the localization has taken place, more than half of the input photons are found at the output of the waveguide through which the input fields were sent. This characteristic property is essentially the reason for the enhanced radiation-matter interaction using localized modes [35].

Next, we present results for the variance in the mean intensity. Figure 14.7 shows the variance in the mean intensity of the output light at the 50th waveguide for a Gaussian distribution for the disorder. It is seen that the variance in the intensity increases with disorder, which is consistent with known results that fluctuations are enhanced near *Anderson localization*. The variance is the largest for squeezed light, and least for a coherent field, which reflects the different variances of the input fields. Note that the magnitude of the variance in Fig. 14.7 is sensitive to the mean number of input photons; however, the qualitative trends in the variance are similar.

To gain some additional insight on the fluctuation behavior of AL, we show the quantity, $g^{*(2)}$ (ratio of variance to square of the mean; $g^{*(2)} = g^{(2)} - 1$) at the 50th waveguide for different input *photon statistics* in Fig. 14.8. The mean photon number for all three inputs is 100. An interesting feature here is the enhancement of $g^{*(2)}$ by the disorder of the medium. Previously, a *suppression of fluctuations* was reported in multiple scattering of nonclassical light [15, 16].

Let us consider the case of single-mode, squeezed light at the input. The normalized variance for the input light is 2, whereas it increases to more than 6 for a very small disorder. For higher disorders, i.e. after complete localization, $g^{*(2)}$ is still higher than for zero disorder. A similar *enhancement of fluctuations* is seen with thermal light, and to a smaller extent with coherent light input.

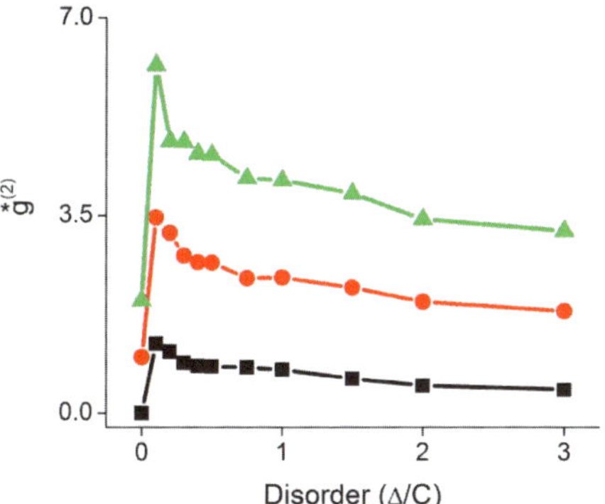

Fig. 14.8 Normalized variance vs. disorder at 50th waveguide for a Gaussian disorder. Mean photon number for all three input fields is 100. Curves shown are for coherent fields (*black*), *thermal fields* (*red*) and *squeezed fields* (*green*)

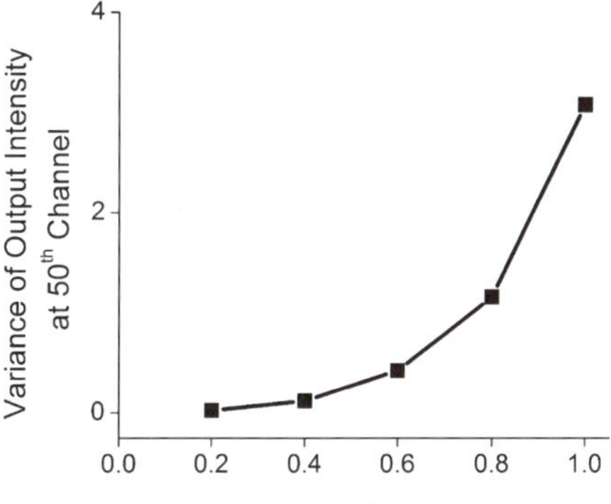

Fig. 14.9 Variance in output intensity at 50th waveguide vs. squeezing parameter, for a Gaussian disorder

We now turn to the details of the case where the input light is a single-mode, *squeezed field*. Figure 14.9 is the variance in the mean intensity at the 50th waveguide as a function of the squeezing parameter for a disorder, $\Delta/C \sim 3$. For small values of r, a situation that is similar to having two photons, the variance is very small. However, with an increase in r, there is a rapid increase in the variance at the waveguide at which the AL occurs.

The next observable we present is the *site-to-site correlations*, i.e. the quantity, $\langle I_l I_p \rangle = \langle |G_{l0}|^2 |G_{p0}|^2 \rangle \langle a_0^{\dagger 2} a_0^2 \rangle$, discussed earlier, for the case where the input light is a single-mode, *squeezed field*. For small r and no disorder in the medium, Fig. 14.10(a), we find that the magnitudes of the correlations are small, and it is apparent that in the absence of localization, there is a good probability for the output photons to be in waveguides from 10 to 90. With an increase in the disorder (see Fig. 14.10(b)), there is a *superbunching* of the photons into the waveguide into which the input photons were launched, and a diminishing probability for the photons to be found in adjacent waveguides. Of course, the magnitudes of the correlations are still small, due to the small value of the squeezing parameter. In Fig. 14.10(c) is the shown the *site-to-site correlations* for $r = 1$ and a disorder, $\Delta/C - 3$. The evidence for *superbunching* of the output photons is quite pronounced, and there is negligible probability of the photons spreading more than

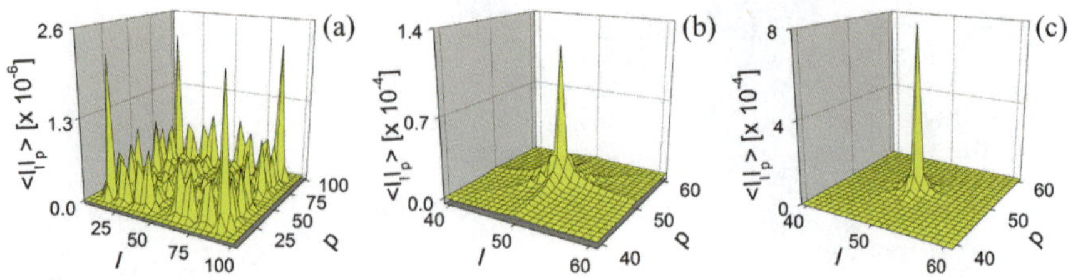

Fig. 14.10 Site-to-site correlation functions for (**a**) $r = 0.2$ and $\Delta/C = 0$, (**b**) $r = 0.2$ and $\Delta/C = 1$, and (**c**) $r = 1$ and $\Delta/C = 3$

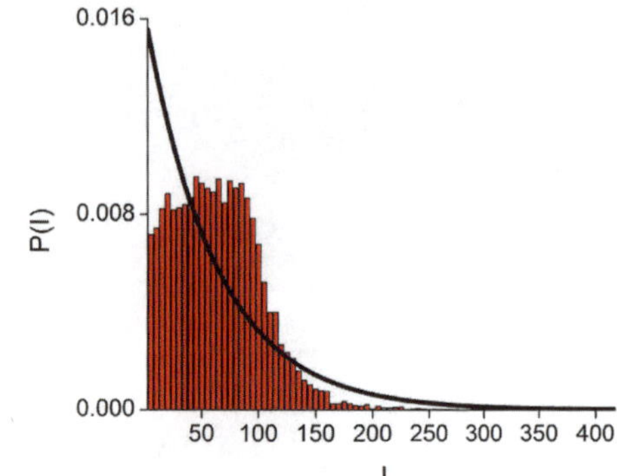

Fig. 14.11 Histogram shows the probability distribution for the output intensity at the 50th waveguide when the input field is coherent with a mean photon of 100. $\Delta/C = 3$ and the medium's disorder is Gaussian. *Black curve* is the exponential distribution for the corresponding mean intensity

about five waveguides on either side of the 50th waveguide. This *superbunching* may have possible utility in enhancing nonlinearities in the interactions between radiation and matter since we know from Mollow's work [26] that two-photon absorption in a *thermal field* is enhanced over that in a coherent field.

To examine the statical aspects of the output light even further, probability of the output light exiting the initial waveguide is investigated. As Fig. 14.11 clearly shows, the probability distribution of output light averaged over 10,000 realizations with a disorder of $\Delta/C = 3$ does not follow Gaussian statistics. The probability distribution for the output intensity is $P(I) = \frac{1}{\langle I \rangle} e^{(-I/\langle I \rangle)}$ if the light follows Gaussian statistics. This is a result of the multiplicative noise that is in Eq. (14.1) as the seen in the first term, $\beta_j a_j$. This departure from Gaussian statistics has been experimentally observed for randomly positioned scatters in waveguides [38]. For further details, see [39].

14.5 Summary

In summary, we have described all-optical, quantum physics inspired photonic transport in a system of evanescently coupled waveguide array. Starting from the tight-binding Hamiltonian that is used to describe such systems, we have derived an equation of motion that is reminiscent of the *kicked rotor model*. However, unlike that model where the *periodic potential* is usually in coordinate space, here the *periodic potential* is in Fourier space. This requires one to study the *photonic transport* in the site-space, which is fortunate because all waveguide structures are usually studied in site-space. The key element in our model for obtaining *photonic transport* is the use of coherent amplitudes which are displaced in phase.

In addition, we have also shown that modifying the tunneling function in a *tight-binding Hamiltonian*, which can be realized by an array of coupled waveguides, produces a wide range of wavepacket evolutions that are not seen in traditional models. The tunneling function $C_\alpha(k)$ affects the wavepacket properties through the bandwidth Δ_α and energy level spacings, both of which are dependent on α.

For waveguides with a fixed length, we have shown that there are qualitative differences in the wavepacket *time-evolution* depending on whether α is positive or negative. For example, when $\alpha = 1$, the equidistant energy levels lead to periodic behaviors such as wavepacket reconstruction [21]; when $\alpha < 0$, a wavepacket near the edge remains localized due to edge eigenstates.

The tunneling function has also been shown to modify the *quantum correlations* in a nontrivial manner. For example, when $\alpha = 1$, the size and the shape of bosonic and *fermionic correlations* can be tuned by the choice of initial waveguides; the periodicity of these correlations follows from the equidistant energy spectrum. For the same initial conditions, when $\alpha = 2$, we find that the correlations, including the no-coincidence region for bosons and fermions, are dramatically different.

Lastly, we focused on the *Anderson localization* of nonclassical light in an array of waveguides, which is a medium in which the disorder can be varied in a controlled manner. To put the results for nonclassical light in proper perspective, we also present corresponding results for localized light when the input light has photons statistics of coherent light, and thermal light. By numerically solving the Heisenberg equation for the field operators, we have calculated relevant quantum statistical observables, such as the variance in the intensity fluctuations of localized light, *site-to-site correlations* and the Glauber $g^{(2)}$ function. We have also reported on a comparison of the effect of the statistics associated with the disorder of the medium on *Anderson localization* and the associated *quantum statistics*.

By calculating the variance in the intensity fluctuations at the waveguide into which light is localized, we have shown that the fluctuations increase with disorder. The $g^{*(2)}$ function has a maximum for a finite disorder, before it tapers off for higher disorders to a value that is still greater than that for zero disorder.

The *site-to-site correlations* show that the probability of finding photons in waveguides that are adjacent to the one into which the input light is coupled diminishes with increasing disorder. For sufficiently high disorder, we find a *superbunching* of light into the waveguide in which localization occurs and the output light will not be Gaussian even if the input light is coherent due to the multiplicative noise in the Heisenberg equation.

Acknowledgements Parts of this work are based on work done with Professor Y.N. Joglekar and Professor G.S. Agarwal, which appeared in Refs. [11, 39, 40].

References

1. P.W. Anderson, Absence of diffusion in certain random lattices. Phys. Rev. **109**(5), 1492 (1958)
2. F. Bloch, Über die Quantenmechanik der Elektronen in Kristallgittern. Z. Phys. **52**(7), 555 (1929)
3. G.E.P. Box, M.E. Muller, A note on the generation of random normal deviates. Ann. Math. Stat. **29**(2), 610 (1958)
4. R. Brown, R. Twiss, Correlation between photons in two coherent beams of light. Nature **177**(4497), 27 (1956)
5. A.A. Chabanov, M. Stoytchev, A.Z. Genack, Statistical signatures of photon localization. Nature **404**, 850 (2000)
6. N. Chiodo, G. Della Valle, R. Osellame, S. Longhi, G. Cerullo, R. Ramponi, P. Laporta, U. Morgner, Imaging of Bloch oscillations in erbium-doped curved waveguide arrays. Opt. Lett. **31**(11), 1651 (2006)
7. D. Christodoulides, F. Lederer, Y. Silberberg et al., Discretizing light behaviour in linear and nonlinear waveguide lattices. Nature **424**(6950), 817 (2003)
8. H.S. Eisenberg, Y. Silberberg, R. Morandotti, A.R. Boyd, J.S. Aitchison, Discrete spatial optical solitons in waveguide arrays. Phys. Rev. Lett. **81**, 3383 (1998)
9. Y.N. Joglekar, Mapping between hamiltonians with attractive and repulsive potentials on a lattice. Phys. Rev. A **82**(4), 044101 (2010)
10. Y.N. Joglekar, A. Saxena, Robust PT-symmetric chain and properties of its hermitian counterpart. Phys. Rev. A **83**(5), 050101 (2011)
11. Y.N. Joglekar, C. Thompson, G. Vemuri, Tunable waveguide lattices with nonuniform parity-symmetric tunneling. Phys. Rev. A **83**(6), 063817 (2011)
12. Y. Lahini, A. Avidan, F. Pozzi, M. Sorel, R. Morandotti, D.N. Christodoulides, Y. Silberberg, Anderson localization and nonlinearity in one-dimensional disordered photonic lattices. Phys. Rev. Lett. **100**(1), 013906 (2008)
13. Y. Lahini, Y. Bromberg, D.N. Christodoulides, Y. Silberberg, Quantum correlations in two-particle Anderson localization. Phys. Rev. Lett. **105**(16), 163905 (2010)
14. G. Lenz, I. Talanina, C.M. de Sterke, Bloch oscillations in an array of curved optical waveguides. Phys. Rev. Lett. **83**(5), 963 (1999)
15. P. Lodahl, A. Lagendijk, Transport of quantum noise through random media. Phys. Rev. Lett. **94**(15), 153905 (2005)
16. P. Lodahl, A.P. Mosk, A. Lagendijk, Spatial quantum correlations in multiple scattered light. Phys. Rev. Lett. **95**(17), 173901 (2005)
17. S. Longhi, Optical realization of multilevel adiabatic population transfer in curved waveguide arrays. Phys. Lett. A **359**(2), 166 (2006)
18. S. Longhi, Photonic transport via chirped adiabatic passage in optical waveguides. J. Phys. B, At. Mol. Opt. Phys. **40**, F189 (2007)
19. S. Longhi, Bloch oscillations in complex crystals with pt symmetry. Phys. Rev. Lett. **103**(12), 123601 (2009)
20. S. Longhi, Bloch oscillations in tight-binding lattices with defects. Phys. Rev. B **81**(19), 195118 (2010)
21. S. Longhi, Periodic wave packet reconstruction in truncated tight-binding lattices. Phys. Rev. B **82**(4), 041106 (2010)
22. S. Longhi, Pt-symmetric laser absorber. Phys. Rev. A **82**(3), 031801 (2010)
23. S. Longhi, Quantum interference in photonic lattices with defects. Phys. Rev. A **83**(3), 033821 (2011)
24. S. Longhi, M. Marangoni, M. Lobino, R. Ramponi, P. Laporta, E. Cianci, V. Foglietti, Observation of dynamic localization in periodically curved waveguide arrays. Phys. Rev. Lett. **96**, 243901 (2006)
25. R. Micallef, J. Love, Y. Kivshar, Nonlinear bent single-mode waveguide as a simple all-optical switch. Opt. Commun. **147**, 259 (1998)

26. B.R. Mollow, Two-photon absorption and field correlation functions. Phys. Rev. **175**(5), 1555 (1968)
27. H.B. Perets, Y. Lahini, F. Pozzi, M. Sorel, R. Morandotti, Y. Silberberg, Realization of quantum walks with negligible decoherence in waveguide lattices. Phys. Rev. Lett. **100**(17), 170506 (2008)
28. A. Perez-Leija, H. Moya-Cessa, A. Szameit, D.N. Christodoulides, Glauber-Fock photonic lattices. Opt. Lett. **35**(14), 2409 (2010)
29. T. Pertsch, P. Dannberg, W. Elflein, A. Bräuer, F. Lederer, Optical Bloch oscillations in temperature tuned waveguide arrays. Phys. Rev. Lett. **83**(23), 4752 (1999)
30. A. Peruzzo, M. Lobino, J.C.F. Matthews, N. Matsuda, A. Politi, K. Poulios, X.Q. Zhou, Y. Lahini, N. Ismail, K. Wörhoff, Y. Bromberg, Y. Silberberg, M.G. Thompson, J.L. OBrien, Quantum walks of correlated photons. Science **329**(5998), 1500 (2010)
31. U. Peschel, T. Pertsch, F. Lederer, Optical Bloch oscillations in waveguide arrays. Opt. Lett. **23**(21), 1701 (1998)
32. A. Politi, M.J. Cryan, J.G. Rarity, S. Yu, J.L. O'Brien, Silica-on-silicon waveguide quantum circuits. Science **320**(5876), 646 (2008)
33. P. Reimann, M. Grifoni, P. Hänggi, Quantum ratchets. Phys. Rev. Lett. **79**(1), 10 (1997)
34. C.E. Rüter, K.G. Makris, R. El-Ganainy, D.N. Christodoulides, M. Segev, D. Kip, Observation of parity-time symmetry in optics. Nat. Phys. **6**(3), 192 (2010)
35. L. Sapienza, H. Thyrrestrup, S. Stobbe, P.D. Garcia, S. Smolka, P. Lodahl, Cavity quantum electrodynamics with Anderson-localized modes. Science **327**(5971), 1352 (2010)
36. D.D. Scott, Y.N. Joglekar, Degrees and signatures of broken $\mathscr{PT}$ symmetry in nonuniform lattices. Phys. Rev. A **83**, 050102 (2011)
37. M. Scully, M. Zubairy, *Quantum Optics* (Cambridge University Press, Cambridge, 1997)
38. M. Stoytchev, A.Z. Genack, Measurement of the probability distribution of total transmission in random waveguides. Phys. Rev. Lett. **79**, 309 (1997)
39. C. Thompson, G. Vemuri, G.S. Agarwal, Anderson localization with second quantized fields in a coupled array of waveguides. Phys. Rev. A **82**(5), 053805 (2010)
40. C. Thompson, G. Vemuri, G.S. Agarwal, Quantum physics inspired optical effects in tight-binding lattices: Phase-controlled photonic transport. Phys. Rev. B **84**(21), 214302 (2011)

Chapter 15
Astronomy and Cancer Research: X-Rays and Nanotechnology from Black Holes to Cancer Therapy

Anil K. Pradhan and Sultana N. Nahar

Abstract It seems highly unlikely that any connection is to be found between astronomy and medicine. But then it also appears to be obvious: X-rays. However, that is quite superficial because the nature of X-rays in the two disciplines is quite different. Nevertheless, we describe recent research on exactly that kind of link. Furthermore, the linkage lies in atomic physics, and via spectroscopy which is a vital tool in astronomy and may also be equally valuable in biomedical research. This review begins with the physics of black hole environments as viewed through X-ray spectroscopy. It is then shown that similar physics can be applied to spectroscopic imaging and therapeutics using heavy-element (high-Z) moieties designed to target cancerous tumors. X-ray irradiation of high-Z nanomaterials as radiosensitizing agents should be extremely efficient for therapy and diagnostics (theranostics). However, broadband radiation from conventional X-ray sources (such as CT scanners) results in vast and unnecessary radiation exposure. Monochromatic X-ray sources are expected to be considerably more efficient. We have developed a new and comprehensive methodology—*Resonant Nano-Plasma Theranostics* (*RNPT*)—that encompasses the use of monochromatic X-ray sources and high-Z nanoparticles. Ongoing research entails theoretical computations, numerical simulations, and *in vitro* and *in vivo* biomedical experiments. Stemming from basic theoretical studies of Kα resonant photoabsorption and fluorescence in all elements of the Periodic Table, we have established a comprehensive multi-disciplinary program involving researchers from physics, chemistry, astronomy, pathology, radiation oncology and radiology. Large-scale calculations necessary for theory and modeling are done at a variety of computational platforms at the Ohio Supercomputer Center. The final goal is the implementation of RNPT for clinical applications.

A.K. Pradhan (✉) · S.N. Nahar
Dept. of Astronomy, The Ohio State University, Columbus, OH 43210, USA
e-mail: pradhan@astronomy.ohio-state.edu

A.K. Pradhan
Biophysics Graduate Program, The Ohio State University, Columbus, OH 43210, USA

A.K. Pradhan
Chemical Physics Program, The Ohio State University, Columbus, OH 43210, USA

M. Mohan (ed.), *New Trends in Atomic and Molecular Physics*,
Springer Series on Atomic, Optical, and Plasma Physics 76,
DOI 10.1007/978-3-642-38167-6_15, © Springer-Verlag Berlin Heidelberg 2013

As an example, we present Monte Carlo numerical simulations for platinum ($Z = 78$) as radiosensitizing agent for killing cancerous cells via increased linear-energy-transfer (LET) and dose enhancement. Radiation therapy in clinical environments is generally administered from high-energy Linear Accelerators (LINAC) emitting broadband 6–15 MeV X-rays with high intensity. Monte Carlo simulations for X-ray energy absorption and dose deposition in tissues were carried out using the Geant4 code for 100 kV, 170 kV and 6 MV broadband X-ray sources. It is found that only X-ray energies $\sim$100 keV are efficient in achieving the required dose enhancement. We confirm previous results for gold ($Z = 79$) that it is the low-energy component around 100 keV from the 6 MV LINAC that is effective in dose-enhanced cell killing. We also describe a simple device for broadband-to-monochromatic (B2M) conversion. A broadband 100 kV source is used to produce monochromatic fluxes in the K-alpha lines from a zirconium ($Z = 40$) target. Monochromatic X-rays such as obtained from B2M conversion should be most effective for theranostics, provided they can be tuned to resonant energies in the targeted material. That is also a quest for X-ray sources such as synchrotrons, high-intensity laser produced plasmas, and accelerator generated free-electron lasers. The project described in this report highlights the fact that interdisciplinary research can be immensely valuable in addressing major problems of heretofore intractable complexity, such as cancer treatment. Approximately 100 news items have appeared on this research in the media; selected links to some articles are given at www.astronomy.ohio-state.edu/~pradhan.

15.1 Introduction

Atomic physics provides the link between astronomy and medicine. This review begins with the simplest possible atomic transition 1s–2p. In hydrogen it gives rise to the Lyman-alpha line. In all other atomic systems it is referred to as the K-alpha transition. But in non-hydrogenic atoms and ions there are a large number of Kα transitions associated with the 1s–2p transition arrays (for a detailed discussion see [7]).[1] In Sect. 15.2 we describe the most interesting observation of the K-alpha line in astronomy—as spectral signature of black holes. From Sect. 15.3 onwards we discuss in detail the rather unexpected application of the atomic physics of Kα transitions in building and using monochromatic X-ray sources for use in biomedical research, and potentially for clinical imaging and therapy. The biomedical physics and methodology is termed *Resonant Nano-Plasma Theranostics (RNPT), or Resonant Theranostics (RT)*, and is described in several previous publications [3, 5, 8].

---

[1]This textbook bridges physics and astronomy and is divided evenly between modern atomic physics and the spectroscopy of astrophysical objects. Figure 15.1b is from the chapter on Active Galactic Nuclei and Quasars. More details are given at the author's webpages: www.astronomy.ohio-state.edu/~pradhan and www.astronomy.ohio-state.edu/~nahar.

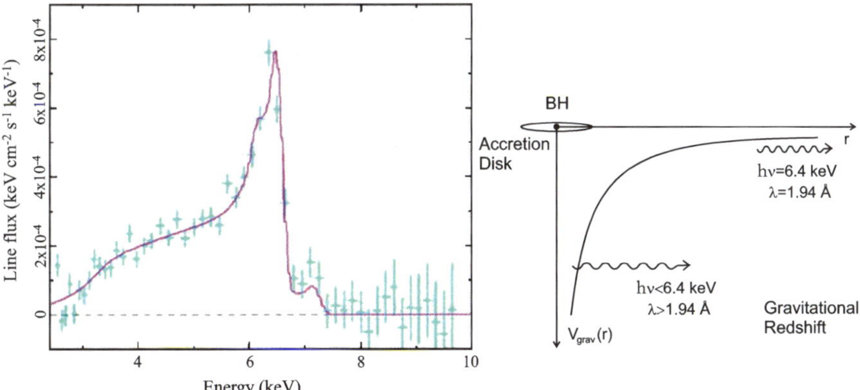

Fig. 15.1 (**a**) Relativistically broadened Iron Kα line first observed by Tanaka et al. [9]. Such asymmetric lines have since been observed from many active galactic nuclei and are regarded as evidence of iron emission in close proximity of supermassive black holes at the center of galaxies. (**b**) Schematic diagram of relativistic broadening and gravitational redshift of photon energy from regions close to the black hole and accretion disk, and farther away (reproduced from Pradhan and Nahar [7], *Atomic Astrophysics and Spectroscopy*, Cambridge University Press, 2011)

15.2 X-Rays in Astronomy

How do we know black holes exist? By definition they cannot be seen. Any evidence must necessarily be indirect. Ergo: A preponderance of evidence must be offered before concluding that they exist at all. Indeed, astronomers have by now accumulated a variety of factual observations that lead to that conclusion.

From a physics perspective—even more from the point of view of atomic physics and spectroscopy—the most convincing observation of the existence of a black hole is shown in Fig. 15.1. It is now widely accepted that at the centers of many if not all galaxies must be a supermassive black hole. The reason is that the central regions or nuclei of galaxies exhibit tremendous activity and output of energy at all wavelength ranges of the electromagnetic spectrum, from radio waves produced by relativistic electrons in jets directed perpendicular to the galactic plane, to hard **X-rays** and gamma rays from environments in the vicinity of the black hole and the accretion disk. The energy emitted is so stupendous that no physical process except gravitational infall into the extremely small region inhabited by the black hole can account for the energy budget.

But the most astonishing observation is that of a single atomic line. It is an X-ray line due to the K-alpha 2p $\rightarrow$ 1s transition in iron which corresponds to an energy of 6.4 keV. It arises due to transition(s) between the filled 2p-subshell of iron down to the 1s-subshell, following K-shell ionization by the hard X-rays originating from the immense accretion activity around the black hole. Figure 15.1a shows the line center and profile of this Fe K-alpha line. It is immediately obvious that something is very strange. First of all, the profile is asymmetric. The line center is at 6.4 keV as it should be. But then the profile is extremely broad on the low-energy side; its "red-ward" broadening extends down to about 4 keV. In other words, the line is broadened by over 3 keV. Plasma or Doppler broadening would not explain both the asymmetry and the width of the line. It is now believed that this peculiar shape is due to gravitational broadening as predicted by Einstein's General Theory of Relativity [9]. A photon emitted in close proximity of the black hole with rest-frame energy of 6.4 keV, from the inner-most stable orbits in the accretion disk, loses energy as it climbs out of the gravitational potential as shown in Fig. 15.1b. Therefore, the photon energy in the observer's frame of reference $E < 6.4$ keV. The closer the emitting iron atom is to the black hole, the bigger the energy loss and larger the broadening. The converse is also true: the Fe $K\alpha$ line emitted far away from the black hole would not experience the gravitational redshift and the peak energy would be at $E = 6.4$ keV.

We note in passing an interesting fact that may not have escaped notice by physicists. The gravitational potential shown in Fig. 15.1b is similar in shape to the coulomb atomic or molecular potential. The $V \sim 1/r$ behavior of both the gravitational and the coulomb potential implies a singularity at the nucleus as $r \to 0$ and $V \to \infty$. And, of course, the gravitational coupling constant is weaker than the electromagnetic coupling constant by a factor of 10^{39}!

15.3 X-Rays in Medicine

But what does this have to do with cancer research? The answer to that question necessitates an in-depth analysis of how X-rays are utilized in medical devices, such as ordinary X-ray machines with Roentgen X-ray tubes and Computerized Tomography (CT) scanners used for 2D and 3D imaging, and linear accelerators (LINAC) used for radiation therapy.

Conventional X-ray sources produce bremsstrahlung broadband radiation over a wide range up to the peak voltage in the X-ray tube. But X-rays at most of these energies are ineffective and result in unnecessary exposure for imaging and therapy. Despite some very low-energy filtration of X-rays with $E \sim 10$–20 keV, the remaining low-energy emission does not penetrate to sufficient depths. On the other hand, X-rays with energy more than a few hundred keV are largely Compton-scattered, which is ineffective. High-Z (HZ) radiosensitizing agents are therefore introduced to increase X-ray photoelectric absorption by heavy atoms. Among these agents are platinum compounds and gold nanoparticles [1–4]. Although HZ material embedded in tissue increases dose absorption, the lack of specificity of broadband X-ray sources still results in radiation overdose. It is therefore proposed that monochromatic X-rays would be far more effective, as their energies can be precisely matched to atomic absorption cross sections, and the dosage determined with accuracy a priori.

However, there are two problems. First, a convenient monochromatic X-ray source must be developed. Second, a procedure for efficient photoabsorption of HZ material needs to be implemented. The underlying physical process is inner-shell

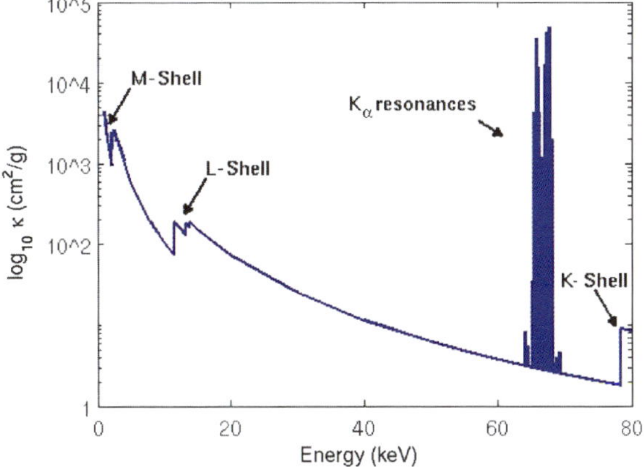

Fig. 15.2 X-ray attenuation coefficients for platinum as function of incident photon energy. In addition to the M-, L- and K-shell edges we also show the K-alpha resonances superimposed on the smooth background. The Kα resonances may be targeted following K-shell ionization and Auger cascades that result in holes in the upper electronic shells. The region between the Kα and the K-edge is therefore filled with similar resonances that enhance photoelectric absorption by orders of magnitude above the otherwise very low background cross sections [4]

ionization, leading to Auger cascades of secondary electrons. Up to 20 or more Auger electrons may be produced following a single primary K-shell photoionization of HZ elements such as Pt or Au via Coster-Kronig and Super-Coster-Kronig transitions [7]. Most of these electrons have relatively low energies of about 1 keV or less. Therefore their effective impact range is only on nanoscales comparable to the size of the cell nucleus. That localizes the linear-energy-transfer (LET) of Auger electrons resulting in cell killing without damage to surrounding cells.

While HZ compounds or nanoparticles may act as contrast agents, they cannot be effective for X-ray energies much higher than 100 keV. The reason is that after the K-shell jump, the photoabsorption cross sections decrease rapidly with energy. For example, Fig. 15.2 shows X-ray attenuation coefficients due to photoelectric absorption by platinum (Pt). The K-shell energies for the generally safe heavy elements are well below 100 keV; for example they are 81 keV and 78 keV for Au and Pt respectively. However, high-energy Linacs generating X-rays up to 6 to 18 MeV are commonly used for radiation therapy. It follows that for dose enhancement using HZ radiosensitizers only the relatively low-energy component around 100–200 keV from the high-energy Linacs is likely to be effective. Similar to earlier work on Au nanoparticles [1, 2, 4, 8], we verify this hypothesis in the present study.

Another serious problem in conventional theranostics is the use of broadband radiation that lacks specificity in its interaction with HZ material. Figure 15.1 shows the existence of features in the attenuation coefficients or cross sections for photoabsorption of X-rays by Pt. It is clear that these features can be targeted using monochromatic X-rays at resonant frequencies. In earlier works [3, 5] we have shown that monochromatic irradiation at resonant energies will allow the creation of inner-shell vacancies. In principle it is then possible to not only ionize the K-shell electrons that result in corresponding holes in all electronic shells, but also excite the $K\alpha$ resonances. This combination of resonant K-shell ionization-excitation would *accelerate* the induced Auger effect and corresponding X-ray energy deposition. Numerical simulations for Au atoms show that HZ radiosensitization using X-rays tuned to these resonances can cause an 11-fold increase in the dose absorbed by the tumor [3].

Considerable work has also been carried out using Au nanoparticles in experiments and models [9]. Simulations demonstrate that X-ray irradiation at about 100–200 keV has the highest dose enhancement factors (DEF) and is most effective in cell killing. In this study we examine the sensitivity due to Pt, which is commonly used in chemotherapy as cisPt, carboPt, and other compounds.

15.4 Resonant Enhancement of Auger Effect

In order to understand in more detail the basis for therapy based on radiosensitization of HZ atoms, it is necessary to examine the Auger process and the pathways in which it manifests itself upon irradiation by incident X-ray X-rays. The atomic structure and radiative and autoionization processes play the dominant role. Figure 15.3 illustrates the Auger processes-ejection of electron(s) upon ionization of an inner shell by X-ray photon(s).

Figure 15.3a shows the Auger effect: a transition $K \rightarrow L$ followed by a radiationless ejection of an electron from an outer shell. Special cases are: Coster-Kronig transitions, when a transition within the L-shell results in ejection of en electron from another shell (M), and Super-Coster-Kronig transitions where both the initial transition and the ejected electron are from the same shell (N).

Figure 15.3b is a diagram that highlights two main features underlying the RNPT mechanism. First, an X-ray photoionization from the K-shell can lead to multiple electron vacancies in the outer shell; the number of holes can double in each outer shell, as shown. Second, the Auger cascades may be reversed by resonant irradiation by photons exactly equal in energy to the transitions $K\alpha$, $K\beta$, etc. Employing high-intensity monochromatic X-ray sources, such resonant absorption, and hence enhancement of the "Auger cycle" may be implemented. However, it is clear that a monochromatic and precisely tuned X-ray source is required.

15.5 Resonant Nano-Plasma Theranostics (RNPT) Methodology

A basic sketch of an experimental scheme to implement RNPT [3, 5, 8] is shown in Fig. 15.4. A monochromatic X-ray source and suitable HZ material can be combined for both imaging and therapy. The former depends on fluorescence from the atoms undergoing Auger decays. The latter is the outcome of resonantly enhanced absorption that would release Auger electrons capable of killing adjacent cells.

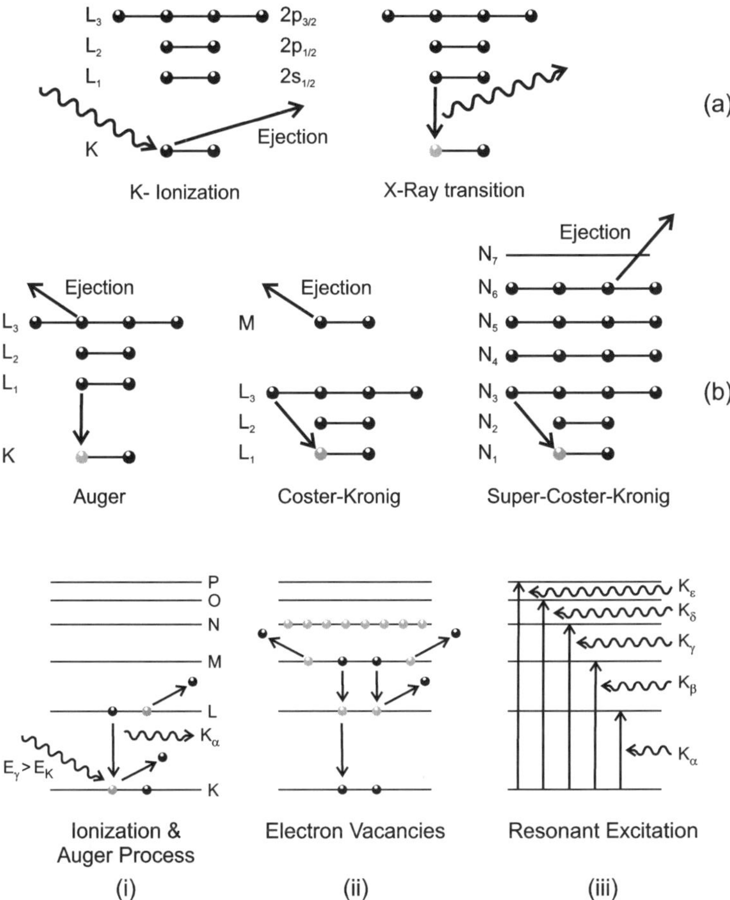

Fig. 15.3 (**a**) The Auger process and Coster-Kronig and Super-Coster-Kronig transitions (Pradhan and Nahar [7], Chap. 5, *Atomic Astrophysics and Spectroscopy*, Cambridge University Press, 2011) (**b**) (*i*) K-shell ionization followed by Auger decays, (*ii*) multiple electron vacancies propagate to outer shells, (*iii*) resonant excitation by Kα, Kβ, etc. Resonance fluorescence induced by monochromatic X-ray beams can enhance Auger decays, resulting in localized energy deposition and Auger electron-photon emission in high-Z atoms (nanoparticles) that capable of killing cancerous cells [3, 5, 8]

The **RNPT** methodology also requires a significant development in nanobiotechology: design, targeting, and delivery of nano-moieties to the cancerous cells. Considerable progress in being made in both areas, the development of high-intensity monochromatic sources, as well as HZ nanoparticles, particularly gold nanoparticles that are safe and sufficiently heavy for optimal radiosensitization. Among the X-ray sources being explored are high-powered laser produced plasmas, free-electron lasers, and synchrotrons.

Fig. 15.4 Schematic diagram for implementation of the RNPT methodology for theranostics. The detector on top perpendicular to the monochromatic beam is for imaging via fluorescence from the high-Z nanoparticles embedded in the tumor. The detector along the same direction as the X-ray beam measures the radiation dose absorbed and delivered to the tumor

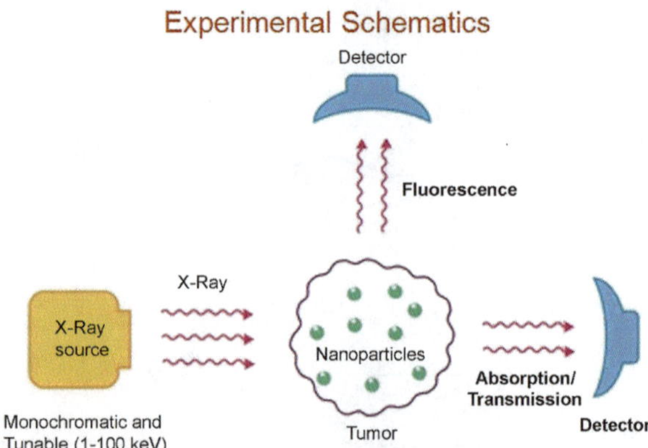

15.6 Radiosensitization Using Broadband X-Ray Sources

In this section we describe computational work using existing X-ray machines producing broadband radiation but in very different energy ranges. We consider three broadband X-ray sources in our numerical simulations: 100 kV, 170 kV and 6 MV. Their output spectra are shown in Fig. 15.5 for the former two devices and in Fig. 15.6 for a 6 MV Linac. In Fig. 15.5 the K-alpha energy and the K-shell ionization energy of Pt are marked. Each K-shell ionization in a HZ atom generally results in Kα emission, triggering Auger electron cascades that create vacancies in upper L, M, N, O and P electronic shells of the **platinum** atom. Therefore, it is possible to resonantly excite electrons from the K-shell into higher shells. The large resonant coefficients for Kα excitation are shown in Fig. 15.2. The Resonant Nano-Plasma Theranostics (RNPT) methodology has been proposed to implement the Auger trigger and *in situ* enhancement of localized X-ray energy deposition via resonant excitation using monochromatic X-rays [5, 8].

Dose Enhancement: Dose enhancement factor (DEF) is defined as the ratio of radiation dose absorbed with and without a radiosensitizing agent. In general the simulations and calculations of DEFs are quite involved. One needs a realistic model of the targeted and intervening tissue in the body. The usual approach is to assume a configuration called a "phantom". The tissue at each depth along the phantom and the interaction of X-ray radiation must then be computed at various depths as it traverses the body. Concentration of HZ moieties embedded in the targeted tissue must also be taken into account. Such simulations require a sophisticated computational program. We employed a general-purpose open software package for Monte Carlo simulations Geant4, as described in the next section.

Numerical Simulations: We adapted the Geant4 toolkit version 9.4 to simulate the deposition of X-ray energy through human tissue with the use of a 15 cm × 5 cm × 5 cm water phantom. A tumor was simulated using a 2 cm × 5 cm × 5 cm region located 10 cm into the phantom. This region was filled with a 7.0 mg/ml concentration of Pt homogeneously distributed in water to simulate the presence of Pt radiosensitizers in tumors.

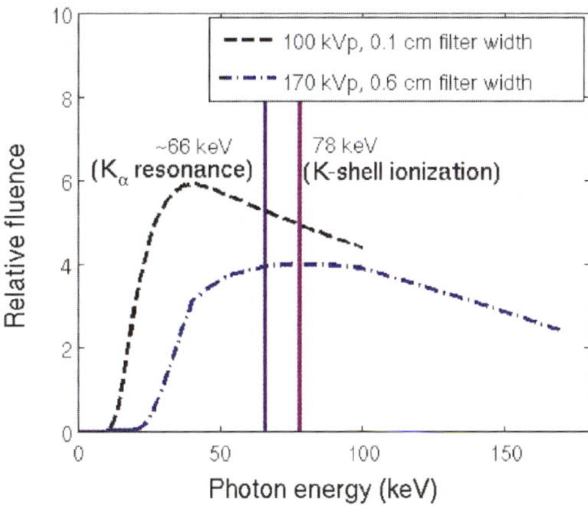

Fig. 15.5 The 100 kV and 170 kV bremsstrahlung spectra. The K-shell ionization energy and average energy of the Kα resonance complex of Pt shown in Fig. 15.2 are marked

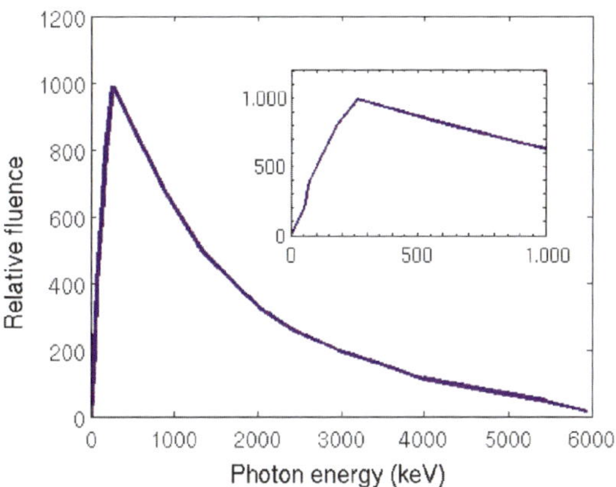

Fig. 15.6 The 6 MV LINAC bremsstrahlung spectrum. The peak around 200 keV results from filtering the lower energy X-rays

The physical interactions of Pt and water with X-rays were governed by the atomic data built into Geant4. These atomic coefficients also determine the background photoabsorption by Pt as in Fig. 15.2. Photon fluence for each broadband source was determined using the spectra shown in Figs. 15.5 and 15.6. Energy intervals of 10 keV at low energies ($<$500 keV) and 500 keV for higher energies were used.

We calculated DEFs for radiation therapy using a 6 MV Linac, compared to the relatively low energy range from the 100 kV and 170 kV. At each energy interval, the **X-rays** were tracked as they traveled through the phantom, with the total dose absorbed by the tumor model recorded. No difference in dose enhancement was seen with the tumor region located near the surface of the phantom compared to deep inside it. **Dose** deposition was determined for the tumor phantom containing either 7.0 mg/g or 0 mg/g, representing a tumor with and without radiosensitizers respectively. Dose enhancement due to Pt was then determined by dividing the dose in tumor with Pt over the dose delivered to the tumor without Pt.

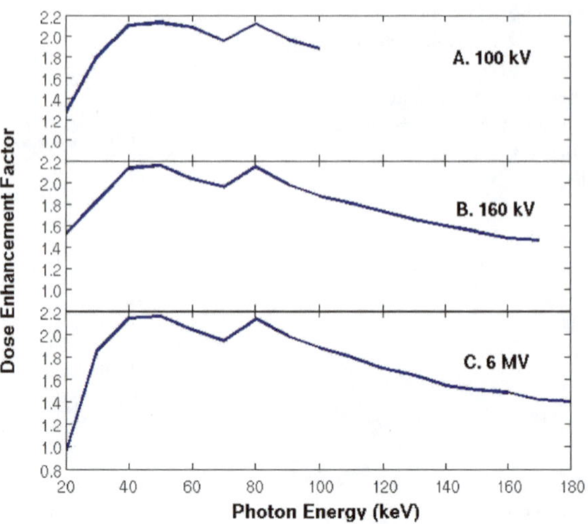

Fig. 15.7 Similarities in dose enhancement factors with platinum radiosensitization and X-ray irradiation from different broadband sources: **A.** 100 kV, **B.** 170 kV, **C.** 6 MV [3]

Figure 15.7 shows the dose enhancements achieved using a 100 kV, 170 kV and 6 MV photon sources. As can be seen, dose enhancements peak at low energies ($E < 100$ keV). The explanation for the double-peaked structure is as follows. The low energy X-rays in the otherwise bremsstrahlung spectrum are assumed to be filtered out. The combination of low-energy filter and L-shell ionization produces the first peak around 40 keV. The DEF decreases thereafter until the second peak at approximately 80 keV corresponding to the K-shell ionization of Pt (Fig. 15.1). Dose enhancements decrease monotonically after K-shell ionization since there are no other absorption features in Pt photoabsorption cross section that induce another rise in the DEF. The interesting point however is that the DEF tends to a constant value at energies above 250 keV. At this and higher energies, Compton scattering predominates instead of photoelectric absorption [4]. Since there is little difference in attenuation due to largely elastic Compton scattering for both water and Pt, DEF's are small in the high-energy range $E > 250$ keV.

These results lead to two conclusions: first, dose enhancement characteristics of HZ radiosensitizers are largely due to photoionization and subsequent Auger ejections of electrons. That causes the enhancement in the DEFs shown in Fig. 15.7. The second is that elastic scattering dominates photoelectric absorption from about 200 keV onwards, and is relatively constant up to all higher energies even into the MeV range characteristic of Linacs. Therefore no further enhancement of DEF is possible. Also, these high-energy X-rays deposit very limited amounts of their energy to the tumor—the attenuation coefficients of HZ radiosensitizers at these energies are only slightly above that of normal tissue.

Figure 15.8 illustrates these two points and the low and high energy behavior of the Pt DEF using a 6 MV Linac. Following the peak around 200 keV the DEF falls rapidly to a constant value with very little enhancement up to the maximum energy 6 MeV. The drop thereafter is due to negligible interaction of X-rays with atomic species and the final limiting value of DEF is unity.

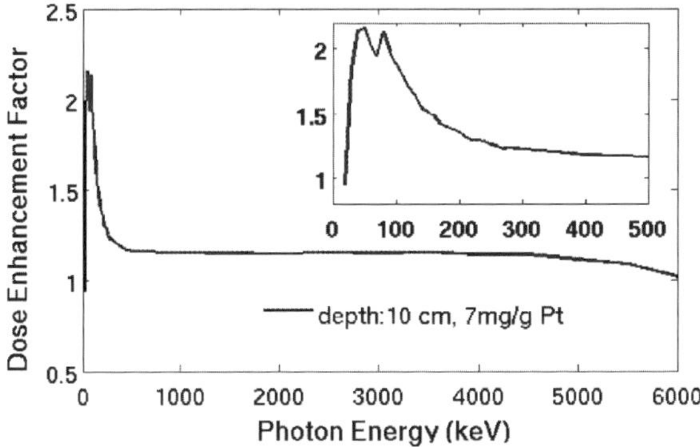

Fig. 15.8 Dose enhancement factors with a 6 MV source and Pt reagent. The DEF peaks at approximately 100 keV and then drops to low and relatively constant values up to high MeV energies. It is evident that there is little dose enhancement due to radiosensitization at energies higher than 100–200 keV, and that LINACs with MeV X-ray output are not effective throughout most of the energy range

Experimental X-Ray Irradiation: The numerical simulations presented herein will be tested experimentally. The experiments are in progress in several stages. First, specific cancer cell lines, such as F98 rat glioma (brain cancer) cells, are prepared. Second, they are treated with Pt compounds such as platinated dendrimers, cisPt or carboPt. Third, the radiosensitized cell lines are then irradiated using 168 kV and 6 MV sources. Fourth, measurements of the cell survival fractions are made to validate the models. The aim is to verify the proposition that it is the low energy component of the broadband spectrum that results in effective cell killing; the high energy component has little effect on the DEF and survival fractions.

15.7 Broadband-to-Monochromatic Conversion

Experimental results from a broadband-to-monochromatic (B2M) device are given, as well as modeling results from Geant4 simulations of DEFs using various broadband devices. We first report proof-of-principle results from a device to convert broadband radiation from a conventional X-ray source into largely monochromatic radiation at the Kα, Kβ energies [3]. This is achieved via fluorescent emission resulting from K-shell ionization of an element (Fig. 15.3a). For example, Fig. 15.9 shows experimental results using an ordinary 100 kV simulator and a zirconium target. The input spectrum is as in Fig. 15.3. The output spectrum in the top panel of Fig. 15.4 shows the Zr Kα lines above the scattered background; the isolated features after subtracting the background are shown in the bottom panel.

Using different HZ targets, it is possible to obtain monochromatic flux via Kα fluorescence at higher energies that would enable greater penetration in the human body. Figure 15.9 demonstrates that conversion of **X-rays** from a broadband

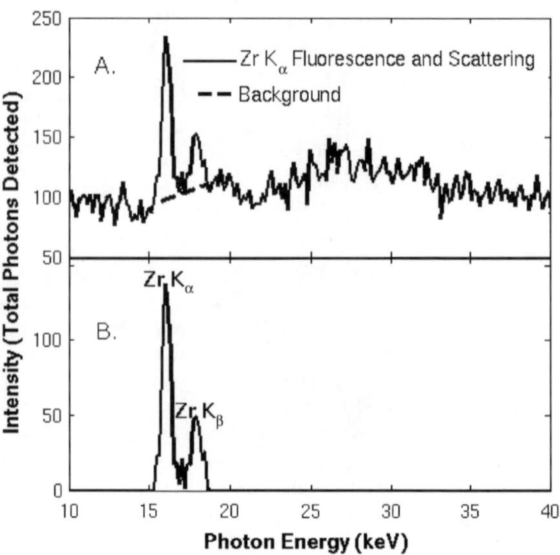

Fig. 15.9 Monochromatic Kα line fluxes detected after conversion of the bremsstrahlung output from a 100 kV broadband X-ray machine [3]. The *top panel* shows the monochromatic flux together with the Compton scattered background; the *bottom panel* shows the isolated features

source to largely monochromatic spectrum is feasible. Moreover, Kα fluxes can be produced at a range of desired energies by selecting different HZ targets. However, the intensity is still insufficient to enable imaging and therapy to the extent needed for clinical applications and more experimental work is required. The B2M conversion efficiency is estimated to be around 0.02. Much lower monochromatic flux is required to activate the targeted features in photoabsorption (Fig. 15.2) than from broadband sources such as simulated in this report. The resulting **DEFs** from monochromatic X-ray irradiation should be considerably higher than obtained herein (Fig. 15.7).

15.8 Conclusion

We conclude this review by highlighting the main features.

1. Based on atomic and astrophysical X-ray spectroscopy, a new methodology—Resonant Nano-Plasma Theranostics (RNPT) has been developed [3, 5, 8]. It aims to complement or supplant existing X-ray devices used for radiation therapy by utilizing monochromatic X-ray sources and high-Z nanomaterials.
2. Heavy elements such as Pt or Au, and HZ compounds, are subject to Auger decays of secondary electrons upon X-ray irradiation which can efficiently kill malignant cells with minimal collateral damage.
3. A general pattern is demonstrated in the behavior of the DEF with a peak of $\sim$2.2 close to the L- and K-edges of HZ material. The low energy range $E < 100$ keV is most effective, and drops substantially to a constant value of $\sim$1.2 for $E > 250$ keV.

4. The results confirm earlier studies using Au nanoparticles [4]. These studies on both Au and Pt should be of considerable interest to the biomedical community pertaining to the use of high-energy megavoltage radiation up to 6–10 MV in conjunction with HZ radiosensitization.
5. Monochromatic X-ray sources would be most efficient for imaging and therapy since they can be tuned resonantly to specific features that enhance photoabsorption, thereby enabling contrast and radiosensitization with minimum of radiation exposure. While we have presented first results from a proof-of-principle device for B2M conversion, more experimental work is needed to generate required monochromatic fluxes.

Acknowledgements   We would like to thank our close collaborators from several disciplines on the myriad developments underlying the RNPT mechanism. In particular we would like to acknowledge Yan Yu and his medical physics team in the department of Radiation Oncology at the Thomas Jefferson University Medical School in Philadelphia, Pennsylvania. Max Montenegro and Sara Lim have carried the Geant4 simulations reported herein. Biomedical experimental research at the Ohio State University is primarily due to Rolf Barth (Pathology), Erica Bell (Radiation Oncology), Enam Chowdhury (Physics), Russell Pitzer (Chemistry) and Claudia Turro (Chemistry).

References

1. R.I. Berbeco, W. Ngwa, G.M. Makrigiorgos, Localized dose enhancement to tumor blood vessel endothelial cells via megavoltage X-rays and targeted gold nanoparticles: New potential for external beam radiotherapy. Int. J. Radiat. Oncol. Biol. Phys. **1**(81), 270–276 (2011)
2. S. Jain, J.A. Coulter et al., Cell-specific radiosensitization by gold nanoparticles at megavoltage radiation energies. Int. J. Radiat. Oncol. Biol. Phys. **79**(2), 531–539 (2011)
3. S. Lim, M. Montenegro, A.K. Pradhan, S.N. Nahar, E. Chowdhury, Y. Yu, Broadband and monochromatic X-ray irradiation of platinum: Mone Carlo simulations for dose enhancement factors and resonant theranostics, in *Proceedings of the 2012 World Congress on Medical Physics and Biomedical Engineering*, Beijing, May (2012), pp. 26–31
4. M.K.K. Lueng, J.C.L. Chow et al., Irradiation of gold nanoparticles by X-rays: Monte Carlo simulation of dose enhancements and the spatial properties of the secondary electrons production. Med. Phys. **38**, 624–631 (2011)
5. M. Montenegro, S.N. Nahar et al., Monte Carlo simulations and atomic calculations for Auger processes in biomedical nanotheranostics. J. Phys. Chem. A **113**(45), 12364–12369 (2009)
6. S.N. Nahar, A.K. Pradhan, S. Lim, K-α transition probabilities for platinum and uranium ions for possible X-ray biomedical applications. Can. J. Phys. **89**, 483–494 (2011)
7. A.K. Pradhan, S.N. Nahar, *Atomic Astrophysics and Spectroscopy* (Cambridge University Presss, Cambridge, 2011)
8. A.K. Pradhan, S.N. Nahar et al., Resonant X-ray enhancement of the Auger effect in high-Z atoms, molecules, and nanoparticles: potential biomedical applications. J. Phys. Chem. A **113**(45), 12356–12363 (2009)
9. Y. Tanaka et al., Gravitationally redshifted emission implying an accretion disc and massive black hole in the active galaxy MCG-6-30-15. Nature **375**, 659 (1995)

Chapter 16
Transmission of Slow Highly Charged Ions Through Ultra-thin Carbon Nano-sheets

N. Akram, H.Q. Zhang, U. Werner, A. Beyer, and R. Schuch

Abstract Transmission properties of slow highly charged ions through nanometer thick foils are discussed. We also report on the measurement of the energy loss and the charge states of 46.2 keV Ne^{10+}-ions and 11.7 keV Ne^{3+}-ions transmitted through ultra-thin carbon nano-sheets. The sheets had a thickness of 1.2 nm (single molecular layer) and 3.6 nm (three molecular layers). The measured energy loss of the transmitted ions is considerably smaller than the calculated energy loss by SRIM but it is in agreement with energy loss calculated using the Firsov model. The majority of the transmitted ions retain their initial charge state (up-to 98 %) contrary to prediction by the classical over-the-barrier model. The results suggest that the energy loss of slow highly charged ions in such thin sheets is only due to the electronic excitations, without charge exchange inside the target.

16.1 Introduction

The interaction of energetic charged particles with matter has been a subject of large interest because of their potential applications such as material modifications, radiation detectors, biological and medical treatments, and ion-implantation. The energy and charge state distribution of transmitted ions after passing through the matter is also studied due to the intrinsic interest on the mechanisms of ion-solid interaction. The ions lose their kinetic energy while passing through the matter, as a result of collisions with target nuclei and electrons [1–5]. The energy loss of ions passing through the matter can be separated in two processes, nuclear and electronic, which are usually treated independent of each other [3]. The nuclear energy loss is due to elastic collisions between the incident ions and the target atoms and is influenced by the Coulomb field of the interacting nuclei of both projectile and target atom

N. Akram · H.Q. Zhang · R. Schuch (✉)
Atomic Physics, Fysikum, AlbaNova Physics Centre, Stockholm University, 106 91 Stockholm, Sweden
e-mail: schuch@fysik.su.se

U. Werner · A. Beyer
Fakultät für Physik, Universität Bielefeld, Universitätsstr. 25, 33615 Bielefeld, Germany

M. Mohan (ed.), *New Trends in Atomic and Molecular Physics*,
Springer Series on Atomic, Optical, and Plasma Physics 76,
DOI 10.1007/978-3-642-38167-6_16, © Springer-Verlag Berlin Heidelberg 2013

which is partially screened by the electrons. The electronic stopping inside matter happens inelastically and the electrons in projectile and target are excited, captured, and ionized. Vacancies that are created result in several secondary processes including plasmon excitation, X-ray emission and Auger electron emission. The dynamics of interaction between the projectile and the target depends strongly on the incident velocity of the projectile.

The research field of ion-solid interactions is very wide. We concentrate here on the subfield of slow ions, i.e. ions whose velocity is much smaller than the Bohr velocity of the involved bound electrons. And, on the penetration of these slow ions through nanometer thick foils. The availability of sources for highly charges ions (HCI) made it possible to study the interaction of HCI with the matter [6–8]. Different aspects of HCI-matter interaction including HCI-induced surface modifications [9], image charge acceleration of HCI before the target surface [10], neutralization and relaxation of HCI in front and below surfaces [11, 12]; the energy loss of HCI [13] and HCI-guiding through capillaries in insulating materials [14–20] have been extensively studied.

Recently, interaction of charged particles with nano-structures has opened new possibilities in the ion-solid interaction. Nano-structures in materials provide the possibility to study certain physical aspects of highly charged ions (HCI)-solid interactions, which have not been possible before, i.e., the formation of hollow atoms above the surfaces during the passage of HCI through metallic nanocapillaries and the pre-equilibrium energy loss in ultra-thin nano-sheets due to a very short interaction time of HCI.

For HCI as a projectile, the energy loss in matter may depend on the charge state and can be a result of increased momentum transfer to the target atoms due to reduced screening of the nuclear charge [21]. For a slow HCI, the electron capture into high Rydberg states from target atoms starts even before the HCI enters the target given by the classical over-the-barrier (COB) model [22]. During the passage of HCI through matter, its charge state varies as a result of electron capture and loss due to collisions with the target atoms. In fact, the condition of ions in matter is complex. The electrons are not in their ground state, but excited and frequently lost by auto-ionization and ionization in subsequent collisions. Energy and charge state-dependent energy loss of HCI have been measured [21, 23] after transmission, as well as, in the scattering experiments [24–26].

When the electron capture rate and the electron loss rate compensate each other for an ion inside the target, an equilibrium of the charge distribution is reached. The equilibrium distribution of charges is achieved after certain number of collisions with the target atoms, depending on the target material, its thickness and velocity of incident ions [27, 28].

Information of charge state equilibration time for a projectile inside matter provides better understanding of microscopic de-excitation processes. Several experiments have been performed to evaluate the equilibrium charge state of ions and the corresponding time to reach charge state equilibrium [29–33]. Also, in the scattering experiments, energy dependent neutralization of slow HCI was found along with the higher probability of multiple charge states after interaction. For transmission experiments, it was reported that the typical time to reach charge state equilibrium inside solids is several fs e.g. it is $\sim$7 fs for Xe^{44+} and Au^{68+} [30]. In an experiment that measured equilibrium of inner shell electrons, the time constants of 22 fs for Pb^{58+} and 68 fs for Pb^{53+} was found [33].

Ultra-thin sheets with thickness of a molecular layer offer an interaction time of few fs with keV HCI in the transmission geometry. Such a short interaction time, $\sim$1–4 fs, is in the range or even shorter than the equilibration times measured before. The charge state of the transmitted HCI might not be equilibrated inside the sheet and therefore provides a unique opportunity to study the pre-equilibrium characteristics of HCI including the energy loss and the exit charge states. Recently, free standing ultra-thin foils of nm thickness (single molecular layer) have been manufactured [34].

In this paper, we not only review the transmission characteristics of slow HCI through thin foils [21, 23, 29, 30, 33] but also present the preliminary results of pre-equilibrium energy loss and charge states of HCI after transmission through ultra-thin carbon nano-sheets which are of thickness of single and three molecular layers. The energy loss results are compared with the estimated results obtained from SRIM and the Firsov model. The exit charge states of transmitted ions are compared with the calculations from COB-model. In the next section, we briefly review the experimental results of slow HCI transmission through thin foils. In Sect. 16.3, the experimental set-up used to study the transmission characteristics of slow HCI through ultra-thin nano-sheets is described. The experimental results are presented in Sect. 16.4. The results are discussed in comparison with calculations from the COB-model, the SRIM and the Firsov model in Sect. 16.5.

16.2 Brief Review of Transmission of Slow Highly Charged Ions Through Thin Sheets

For decades, much interest has been devoted to theoretical and experimental work in order to obtain information on energy and charge state distribution of transmitted ions through the solid. In this section, we try to provide a brief review of recent experimental work performed on the aforementioned subject focusing on slow HCI transmission through thin foils of different thicknesses of nm dimension.

Hermann et al. [29] investigated transmission of 576 keV $Ar^{8,12,16+}$-ions through 31 nm thick carbon foil by measuring projectiles energy loss and final charge-state distributions. They have not observed any dependence of initial charge states on transmission characteristics. From the experimental data, it was estimated that the ions reach their charge-state equilibrium after passing through one atomic layer in the target. The approximated short equilibration length would require extremely fast charge exchange processes between the projectile and the target.

Later, Schenkel et al. [23] observed energy loss of projectiles with charge-states ranging from $q = 3$ for oxygen to $q = 69$ for gold ions after passing through 10.4 nm thick carbon foil. They observed charge-state dependent energy loss enhancement for Xe^{q+} and Au^{q+} ions with $q > 40$. The charge state-dependent energy loss increase indicated strong pre-equilibrium contribution and an upper limit of ~ 21 fs for charge-state equilibrium inside the target was estimated.

Schenkel et al. [21] also observed energy loss HCI in 10.4 nm thick carbon foil as a function of impact velocity. Velocity dependent energy loss for Xe^{44+}-ions and Au^{69+}-ions was seen. The pre-equilibrium contributions to the observed energy loss were found due to finite de-excitation time of HCI in target.

Later, Hattass et al. [30] investigated non-equilibrium charge state distributions of slow HCI with charge-states ranging from $q = 33$–75 of different ions after transmission through carbon foils of thickness of 5 nm and 10.4 nm. The charge equilibration times were determined from dependence of exit charge states on projectile velocity and foil thickness. A typical time of ~ 7 fs was found for Xe^{44+}-ions and Au^{68+}-ions to be equilibrated inside the solid. For such a quick equilibration of HCI inside the solid, it was suggested that very fast multiple Auger-transition cascades must take place in order to fill the inner shell vacancies.

Pešić et al. [33] studied the relaxation of slow HCI, $8.5q$ keV and $23.5q$ keV Pb^{q+}-ions ($q = 53$–58) in 65 nm and 456 nm thick Ta foil by measuring the X-ray emission from the ions inside the target. The mean relaxation times of M-shell vacancies were found to be in the range of 22 fs to 68 fs for different charge states of the projectile. This relaxation time is much longer than previously determined by Hattass et al. [30] using a different method for carbon foils. The mean relaxation depth was found to be between 50 to 100 atomic layers contrary to equilibrium length of single atomic layer as estimated by Hermann et al. [29]. It was also found by Pešić et al. [33] that the mean relaxation depth decreases with increasing charge state of the projectile. These results in [33] were explained by an Auger cascade and electron capture model.

16.3 Set-up of Present Experiment

The experiments were performed using 46.2 keV Ne^{10+}-ion beam and 11.7 keV Ne^{3+}-ion beam in Stockholm Electron Beam Ion Trap (S-EBIT) laboratory at Stockholm University, Stockholm. Ne^{10+}-ions were produced in S-EBIT [35] and extracted at energy of 46.2 keV while Ne^{3+}-ions were produced using Penning Ion Gauge Source and were extracted at 11.7 keV. The ion beam was collimated to a size of 1×1 mm^2 using a pair of four-jaw slits. A pressure of 10^{-8} mbar was maintained in the beam line and the experimental chamber. The schematic view of the experimental set-up is shown in Fig. 16.1-d.

The carbon nano-sheets are free standing ultra-thin sheets which consist of cross-linked self assembled mono-layers (SAMs) of biphenyl molecules $((C_6H_5)_2)$. These nano-sheets are manufactured at Bielefeld University, Bielefeld, Germany. The detailed description of formation the nano-sheets can be found elsewhere [34]. The typical thickness of a single molecular layer biphenyl nano-sheet is about 1.2 nm.

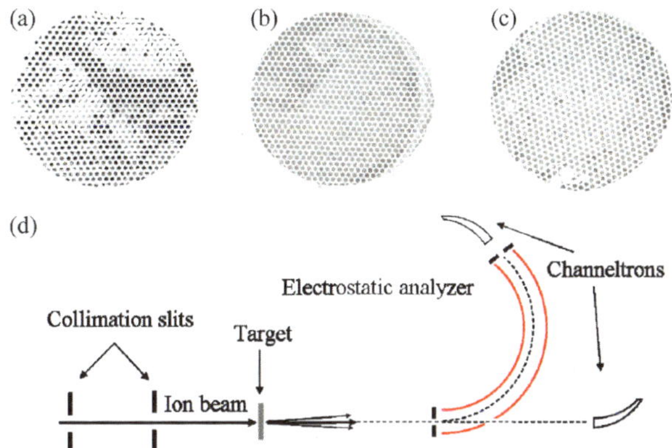

Fig. 16.1 TEM images of the carbon nano-sheets placed on the TEM-grid. (**a**) Three molecular layer ($\sim$45 % covered), (**b**) three molecular layer (75 % covered) and (**c**) single molecular layer (85 % covered). The *white patches on the grid* show the nano-sheets. (**d**) Schematic of the experimental set-up, showing target, detectors and electrostatic analyzer

The nano-sheets are placed on tunnelling electron microscope (TEM) grid to perform the energy loss experiments in the transmission geometry. TEM images of single layer and three layer nano-sheets used in the experiment are shown in Fig. 16.1-a, 16.1-b, 16.1-c. The holder containing carbon nano-sheet (TEM) grids was connected to a goniometer. The goniometer with five degree of freedom, three translations and two rotations, allows to precisely align the nano-sheets perpendicular to the incident beam.

The energy and the charge state of the transmitted beam through nano-sheets were measured using an electrostatic analyzer. The electrostatic analyzer is placed at 20 cm after the nano-sheets and has an entrance aperture of 3 mm which allows the acceptance angle from ions scattered in the target to be $\sim$0.9°. The analyzed ions were collected using a channeltron detector. A LabView programme was used to apply uniform sweeping voltage across the plates and to count the ions for each voltage. The electrostatic potential applied to the analyzer corresponds to a specific kinetic energy of the analyzed ion. In the direction of the normal incidence, the outer plate of the analyzer has an aperture of 3 mm, which facilitates the measurement of neutral atoms or the direct beam with the grounded plates by using another channeltron detector.

16.4 Results

The energy distributions of the primary beam, 46.2 keV Ne^{10+}-ions, and of the transmitted ions through three molecular layer (45 % covered) carbon nano-sheet is shown in Fig. 16.2-a. As seen in the figure, the energy of the transmitted ions is shifted towards lower energy by an amount of $\sim$40 eV. The energy difference is calculated by comparing the centroids of both peaks as well as by considering the energy distribution of ions as Gaussian. Both methods give a similar value of $\sim$40 eV. As the nano-sheet is covering only 45 % of the grid (Fig. 16.1-a) which implies that almost 55 % of the transmitted beam did not pass through the nano-sheet and herein contains primary beam component. The spectrometer resolution is not

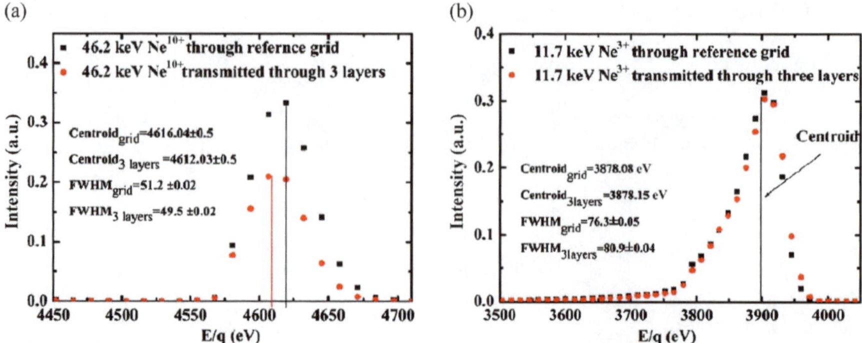

Fig. 16.2 The energy distribution of transmitted ions through (a) three layer nano-sheet (45 % covered, Fig. 16.1-a) and the primary beam through a reference grid for 46.2 keV Ne^{10+}-ions and (b) for 11.7 keV Ne^{3+}-ions through a 75 % covered three molecular layer carbon nano-sheet (Fig. 16.1-b) and the primary beam through reference grid. *Vertical line* shows the centroid position

Fig. 16.3 The charge state distribution of 46.2 keV Ne^{10+}-ion beam transmitted through three molecular layers thick carbon nano-sheet (45 % covered, Fig. 16.1-a). *Red lines* are the expected positions of charge states from Ne^{9+} to Ne^{6+}

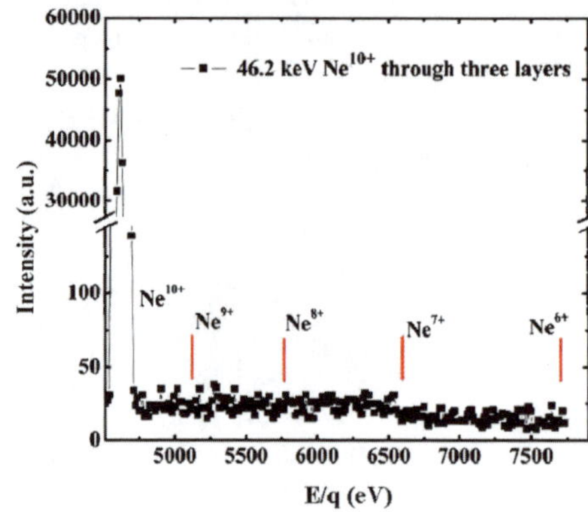

high enough to resolve these components. Thus, the energy loss through the three layer sheet by removing 55 % primary beam component is ~89 eV for 46.2 keV Ne^{10+}-ions.

A charge state analysis of 46.2 keV Ne^{10+}-ion beam transmitted through the three molecular layer sheet was performed. The charge state distribution of transmitted ions from Ne^{9+} to Ne^{6+} is shown in Fig. 16.3. The ions of other charge states were hardly seen at expected positions as indicated in Fig. 16.3.

In order to determine the energy loss in the lower charge state, we used 11.7 keV Ne^{3+}-ions. The energy distribution of the incident ions, 11.7 keV Ne^{3+}-ions, and the ions transmitted through three molecular layer sheet (75 % covered) is shown in Fig. 16.2-b. As seen in the figure, we did not see any energy loss of transmitted Ne^{3+}-ions through three layer nano-sheet, as well as ions of other charge states, i.e., Ne^{2+} and Ne^{1+}, were not seen. Also for single molecular layer carbon nano-sheets covering 85 % of the grid, the transmitted Ne^{10+}-ions retained their incident energy, 52 keV in this case. All the Ne^{10+} and Ne^{3+} ions transmitted through three and single layer nano-sheets kept their initial charge state up-to $\sim$98 %. In order to see the neutrals if any, we used the channeltron detector placed at $0°$. However, we did not observe any neutralized ions by transmission through the nano-sheets.

To see sputtering effect on the nano-sheets by the incident ions or a damage by installation in the set-up, the three layer nano-sheet ($\sim$45 % covered, Fig. 16.1-a) was examined by TEM after the experiments. It was found that the sheet was fully retained.

16.5 Discussions

The classical over-the-barrier model has been extensively used to explain charge transfer to HCI in slow atomic collisions and solid surfaces [22]. In our case the COB-model predicts an effective distance, Rc, for resonant charge transfer of $\sim$5.8 Å for Ne^{10+} and $\sim$3.2 Å for Ne^{3+}. The inter-atomic distance between carbon atoms in biphenyl is about $\sim$1.5 Å which is less than effective distance, Rc, for Ne^{10+} as well as Ne^{3+}. The cross-section for the resonant charge transfer is 1.027×10^{-14} cm^2 for Ne^{10+} and 0.31×10^{-14} cm^2 for Ne^{3+}. Through a single molecular layer carbon nano-sheet, the probability of resonant charge transfer is $\sim$4.7 for Ne^{10+} and $\sim$1.4 for Ne^{3+}. Thus, we expect charge transfer probability of 100 %.

We also estimated the scattering fraction of transmitted ions beyond the acceptance angle of the detector. For single layer nano-sheet, $\sim$3 % of transmitted ions should be scattered beyond the acceptance angle of the detector.

The experimental results of the energy loss in the sheets are compared with calculations performed using SRIM [36] and the Firsov model [37]. The SRIM (Stopping and Range of Ions in Matter) is a Monte Carlo based computer program widely used to calculate the stopping and range of ions in matter. In SRIM, the charge state of incident ions can not be set as the experimental one as it only uses neutral atoms as input. Mono-energetic beams of energy 42.6 keV and 11.7 keV for Ne are used as input. For the target, biphenyl SAMs ((C_6H_5)$_2$), a relative density, 0.953 g/cm^3 is used. For 42.6 keV Ne (simulating Ne^{10+}) transmitted through three layers, the energy distribution are shown in Fig. 16.4-a. As seen in the figure, the calculated energy loss is 529 eV through a three layer sheet.

For 11.7 keV Ne (simulating Ne^{3+}) transmitted through three layers, the energy distribution are shown in Fig. 16.4-b. As seen, the calculated energy loss is 338 eV in a three layers thick nano-sheet.

From Fig. 16.4, it is obvious that the transmitted ions have higher straggling (energy spread) and higher energy loss for 42.6 keV Ne (simulating Ne^{10+}) as compared to 11.7 keV Ne (simulating Ne^{3+}). However, when we checked the different factors contributing to the energy loss, it was found that almost 50 % of the energy loss for 42.6 keV Ne is coming from ionization of the projectiles inside the target. This contribution can not be present in the experiment of Ne^{10+}-ions. Similarly for 11.7 keV Ne, a contribution of $\sim$27 % comes from ionization of projectiles inside

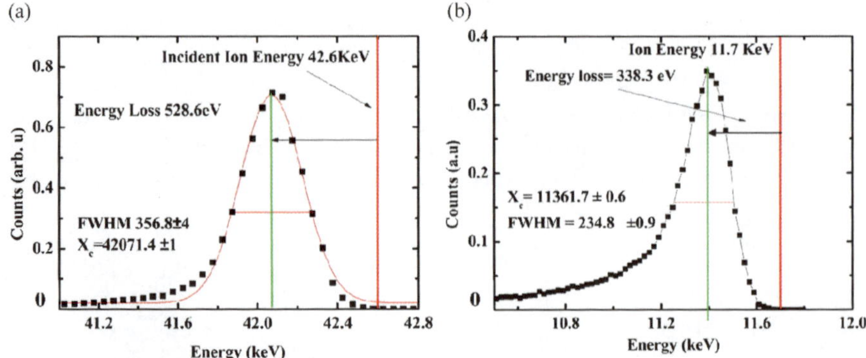

Fig. 16.4 The energy distribution of (**a**) 42.6 keV Ne (simulating Ne^{10+}) transmitted through three layers sheet and (**b**) 11.7 keV Ne (simulating Ne^{3+}) transmitted through three layers sheet, using SRIM 2008.04

the target. Again this contribution is absent in our case where the incident charge state is already ionized to Ne^{3+}. Removing these channels of the energy loss would lead to a lower value from SRIM that is close to our experimental results.

Furthermore, no charge exchange observed in the experiments suggests that ionization and electron capture do not occur during the interaction of HCI with the nano-sheets.

Given the situation of the present experiment, it seems important to calculate inelastic energy loss due to electronic excitations. The Firsov model has been used to calculate the inelastic energy loss in absence of electron capture and loss [37] and was quite successful to explain energy loss measurements with HCI [24, 25]. Here we apply this model to our case. In the Firsov model, the solid is assumed to be consisting of free atoms, where the energy loss is calculated by summing the binary collisions with individual target atoms. For each individual interaction, the inelastic energy loss comes from electron flux at the given impact parameter. A dividing surface between the interacting nuclei is introduced to calculate the electron flux. The surface, S, intersects the internuclear axis of the two colliding atoms where the electron density has a minimum which means that S separates the potential attraction of the colliding atoms. It is assumed that an electron of one atom will lose its initial momentum temporarily and acquires new momentum centered around the velocity of other atom when it passes through S and thus loses its energy. If the nuclear charges of the interacting atoms differ by not more than four times, the energy loss can be given by [37]:

$$E = \frac{(Z_a + Z_b)^{5/3} 4.3 \times 10^{-8} u}{[1 + 3.1(Z_a + Z_b)^{1/3} 10^7 R_0]^5},$$

where u is the velocity of the projectile expressed in cm/sec and R_0 is the distance of closest approach given in cm. Z_a and Z_b are the atomic numbers of projectile and the target atoms respectively.

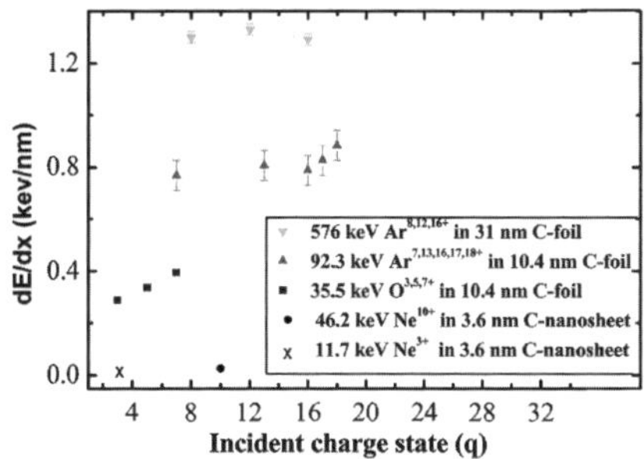

Fig. 16.5 Energy loss per unit length in carbon foils as a function of different charge states, 576 keV Ar$^{8,12,16+}$ through 31 nm carbon foil [29]; 92.3 keV Ar$^{7,13,16,17,18+}$ and 35.5 keV O$^{3,5,7+}$ through 10.4 nm carbon foil [23]; 46.2 keV Ne^{10+} and 11.7 keV Ne^{3+} through 3.6 nm carbon nano-sheet (our work)

Using the atomic numbers of the two interacting nuclei in our case and the distance of closest approach corresponding to the acceptance angle of the detector, the energy loss is $\sim$30 eV for 46.2 keV Ne^{10+}-ions through three layers, $\sim$10 eV for 52 keV Ne^{10+}-ions through a single layer and $\sim$15 eV for 11.7 keV Ne^{3+}-ions through three layers.

In the Firsov model, the energy loss is given by the interaction of electrons of the colliding partners without considering the charge state of the projectile. In case of Ne^{3+}-ions, the projectile has electrons which can interact with the target electrons to provide the inelastic energy loss which agrees rather well for the observed energy loss. For 52 keV Ne^{10+}-ions through single molecular layer sheet, the observed energy loss agrees with the results obtained from the Firsov model even though the projectile is a bare nuclei. But for 46.2 keV Ne^{10+}-ions through three layers sheet, we see a discrepancy between the observed energy loss and the value calculated from the Firsov model.

For a comparison of the presently measured energy loss of HCI in carbon nano-sheet, we scaled the energy loss per unit length with the earlier work discussed in review section [23, 29] for various ions and charge states through carbon foils of different thicknesses, which is shown in Fig. 16.5.

16.6 Summary and Conclusions

Equilibration of the charge of slow HCI and their energy loss in thin carbon sheets and other materials was considered. The results obtained, including those presented here, reveal many open questions. In experiments reported here the pre-equilibrium charge state and energy loss was studied for keV Ne^{q+}-ions with $q = 10, 3$ through ultra-thin carbon nano-sheets. No observable energy loss was seen for 52 keV Ne^{10+} transmitted through single layer sheet and 11.7 keV Ne^{3+}-ions through three layer sheet. A small energy loss of 89 eV is observed for 46.2 keV Ne^{10+}-ions transmitted through three molecular layer carbon nano-sheet. A surprising result is that the transmitted ions retain their initial charge states up-to 98 % in all the cases. The results agree with the calculated inelastic energy loss due to the electron excitations using the Firsov model.

Acknowledgements This work was funded by the European network ITS-LEIF, the K&A Wallenberg foundation and the Swedish Research Council (VR). NA also acknowledges the financial support of Higher Education of Pakistan (HEC) via Swedish Institute (SI).

References

1. H.A. Bethe, Theory of the passage of fast corpuscular rays through matter. Ann. Phys. **5**, 325 (1930)
2. F. Bloch, Zur Bremsung rasch bewegter Teilchen beim Durchgang durch die Materie. Ann. Phys. **16**, 285 (1933)
3. N. Bohr, Kgl. Danske Videnskab. Selskab. Mat.-fys. Medd. **18**(8) (1948)
4. J. Lindhard, M. Scharff, Energy dissipation by ions in the keV region. Phys. Rev. **124**, 128 (1961)
5. P. Sigmund, Low-speed limit of Bohr's stopping-power formula. Phys. Rev. A **54**, 3113 (1996)
6. E.D. Donets, Historical review of electron beam ion sources (invited). Rev. Sci. Instrum. **69**, 614–619 (1998)
7. R. Geller, *Electron Cyclotron Resonance Ion Source and ECR Plasma* (Institute of Physics, Bristol, 1996)
8. D. Schneider, M.W. Clark, B.M. Penetrante, J. McDonald, D. DeWitt, J.N. Bardsley, Production of high-charge-state thorium and uranium ions in an electron-beam ion trap. Phys. Rev. A **44**(5), 3119–3124 (1991)
9. F. Aumayr, S. Facsko, A.S. El-Said, C. Trautmann, M. Schleberger, Topical review: Single ion induced surface nanostructures—a comparison between slow highly charged and swift heavy ions. J. Phys. Condens. Matter **23**, 393001 (2011)
10. H. Winter, C. Auth, R. Schuch, E. Beebe, Image acceleration of highly charged xenon ions in front of a metal surface. Phys. Rev. Lett. **71**, 1939 (1993)
11. W. Huang, H. Lebius, R. Schuch, M. Grether, N. Stolterfoht, Neutralization rates of slow highly charged Ar ions in single and double scattering from a Au(111) surface. Phys. Scr. T **80**, 228 (1999)
12. Z.D. Pešić, G. Vikor, H. Hanafy, A. Enulescu, R. Schuch, Two-photon coincidence studies of highly-charged ion relaxation in solids. Eur. Phys. J. D **54**, 623–628 (2009)
13. Z.D. Pešić, J. Anton, J.H. Bremer, V. Hoffmann, N. Stolterfoht, G. Vikor, R. Schuch, Inelastic energy loss in large angle scattering of Ar^{9+} ions from Au(111) crystal. Nucl. Instrum. Methods Phys. Res., Sect. B, Beam Interact. Mater. Atoms **203**, 96 (2003)
14. G. Vikor, R.T.R. Kumar, Z.D. Pesic, N. Stolterfoht, R. Schuch, Guiding of slow highly charged ions by nanocapillaries in PET. Nucl. Instrum. Methods Phys. Res., Sect. B, Beam Interact. Mater. Atoms **233**, 218 (2005)
15. M.B. Sahana, P. Skog, G. Vikor, R.T. Rajendra Kumar, R. Schuch, Guiding of highly charged ions by highly ordered SiO_2 nano-capillaries. Phys. Rev. A **73**, 040901(R) (2006). Rapid Communication
16. P. Skog, I.L. Soroka, A. Johansson, R. Schuch, Guiding of highly charged ions through Al_2O_3 nano-capillaries. Nucl. Instrum. Methods Phys. Res., Sect. B, Beam Interact. Mater. Atoms **258**, 145–149 (2007)
17. R. Schuch, A. Johansson, A. Rajendra Kumar, M.B. Sahana, P. Skog, I.L. Soroka, G. Vikor, H.Q. Zhang, Guiding of highly charged ions through insulating nano-capillaries. Can. J. Phys. **86**, 327–330 (2008)

18. P. Skog, H.Q. Zhang, R. Schuch, Evidence of sequentially formed charge patches guiding ions through nano-capillaries. Phys. Rev. Lett. **101**, 223202 (2008)

19. H.Q. Zhang, P. Skog, R. Schuch, Guiding of slow highly charged ions through insulating nano-capillaries. J. Phys. Conf. Ser. **163**, 012092 (2009)

20. H.Q. Zhang, P. Skog, R. Schuch, Dynamics of guiding of highly charged ions through Si_2O nanocapillaries. Phys. Rev. A **82**, 052901 (2010)

21. T. Schenkel, A.V. Hamza, A.V. Barnes, D.H. Schneider, Energy loss of slow, highly charged ions in solids. Phys. Rev. A **56**, R1701 (1997)

22. J. Burgdörfer, P. Lerner, F.W. Meyer, Above-surface neutralization of highly charged ions: The classical over-the-barrier model. Phys. Rev. A **44**, 5674 (1991)

23. T. Schenkel, M.A. Briere, A.V. Barnes, A.V. Hamza, K. Bethge, H. Schmidt-Böcking, D.H. Schneider, Charge state dependent energy loss of slow heavy ions in solids. Phys. Rev. Lett. **79**, 2030 (1997)

24. S. Winecki, M.P. Stöckli, C.L. Cocke, Energy loss of highly charged argon ions at grazing incidence on a graphite surface. Phys. Rev. A **55**, 4310 (1997)

25. W. Huang, H. Lebius, R. Schuch, M. Grether, N. Stolterfoht, Energy loss in large-angle scattering of slow, highly charged Ar ions from a Au surface. Phys. Rev. A **58**, 2962 (1998)

26. R. Schuch, W. Huang, H. Lebius, Z.D. Pešić, N. Stolterfoht, G. Víkor, Angular dependence of energy loss in scattering of slow highly charged Ar ions from a Au surface. Nucl. Instrum. Methods Phys. Res., Sect. B, Beam Interact. Mater. Atoms **157**, 309 (1999)

27. A.B. Wittkower, G. Ryding, Equilibrium charge-state distributions of heavy ions (1–14 MeV). Phys. Rev. A **4**, 226 (1971)

28. K. Shima, T. Ishihara, T. Miyoshi, T. Mikumo, Equilibrium charge-state distributions of 35–146-MeV Cu ions behind carbon foils. Phys. Rev. A **28**, 2162 (1983)

29. R. Herrmann, C.L. Cocke, J. Ullrich, S. Hagmann, M. Stoeckli, H. Schmidt-Boecking, Charge-state equilibration length of a highly charged ion inside a carbon solid. Phys. Rev. A **50**, 1435 (1994)

30. M. Hattass, T. Schenkel, A.V. Hamza, A.V. Barnes, M.W. Newman, J.W. McDonald, Charge equilibration time of slow, highly charged ions in solids. Phys. Rev. Lett. **82**, 4795 (1999)

31. Z.D. Pešić, H. Lebius, R. Schuch, G. Víkor, V. Hoffman, D. Niemann, N. Stolterfoht, Energy dependence of neutralization in scattering of slow highly charged Ar ions from an Au surface. Nucl. Instrum. Methods Phys. Res., Sect. B, Beam Interact. Mater. Atoms **164–165**, 511 (2000)

32. Z.D. Pešić, V. Hoffman, D. Niemann, N. Stolterfoht, G. Víkor, R. Schuch, Degree of the neutralization of highly charged Ar ions scattered at different angles from a Au surface. Phys. Scr. T **92**, 237 (2001)

33. Z.D. Pešić et al., Relaxation of slow highly charged ions hitting thin metallic foils. Phys. Rev. A **75**, 012903 (2007)

34. W. Eck, A. Küller, M. Grunze, B. Völkel, A. Gölzhäuser, Freestanding nanosheets from crosslinked biphenyl self-assembled monolayers. Adv. Mater. **17**, 2583 (2005)

35. R. Schuch, S. Tashenov, I. Orban, M. Hobein, S. Mahmood, O. Kamalou, N. Akram, A. Safdar, P. Skog, A. Solders, H. Zhang, The new Stockholm Electron Beam Ion Trap (S-EBIT). J. Instrum. **5**, C12018 (2010)

36. J.F. Ziegler, http://www.srim.org/SRIM/SRIM%2008.pdf

37. O.B. Firsov, A qualitative interpretation of the mean electron excitation energy in atomic collisions. Sov. Phys. JEPT **36**, 1076 (1959)

Chapter 17
Relevance of Electron-Molecule Collision Data for Engineering Purposes

Gorur Govinda Raju

Abstract Innumerable applications have resulted from the application of gaseous electronics to engineering purposes, from the mundane tube lights and neon signs to its rejuvenated version of compact fluorescent bulbs, gas lasers, plasma TV among others. Research data, both experimental and theoretical, from this area continue to be used for engineering purposes. Engineers often look for qualitative similarities in the various properties of interest as a function of electron energy or some other parameters which are easy to measure and relate to practical situations. These aspects are dealt with in the paper.

17.1 Introduction

With constant emergence of new research and application possibilities, gaseous electronics has become more important than ever in various disciplines of engineering, electrical, mechanical, electronics, environmental, aeronautical, and communication, just to name a few. Some recent research areas in gaseous electronics include environmentally efficient and protective lighting devices. Plasma research, both from application to power generation and disturbances in communication circuits in space applications has become important. Research into uses of plasma devices in medical applications, involving skin treatment and healing has emerged as a frontier area. In view if such diverse applications, the presentation here is restricted to outline the broad principles involved and indicate the status of the research rather than dwell into a single topic with much greater depth.

17.2 Dielectric Strength

Every gas has a critical electric field strength which it can safely withstand, and which, if exceeded, results in failure as insulating medium. Traditionally, com-

G.G. Raju (✉)
Department of Electrical and Computer Engineering, University of Windsor, Windsor, Ontario, Canada N9B 3P4
e-mail: raju@uwindsor.ca

M. Mohan (ed.), *New Trends in Atomic and Molecular Physics*,
Springer Series on Atomic, Optical, and Plasma Physics 76,
DOI 10.1007/978-3-642-38167-6_17, © Springer-Verlag Berlin Heidelberg 2013

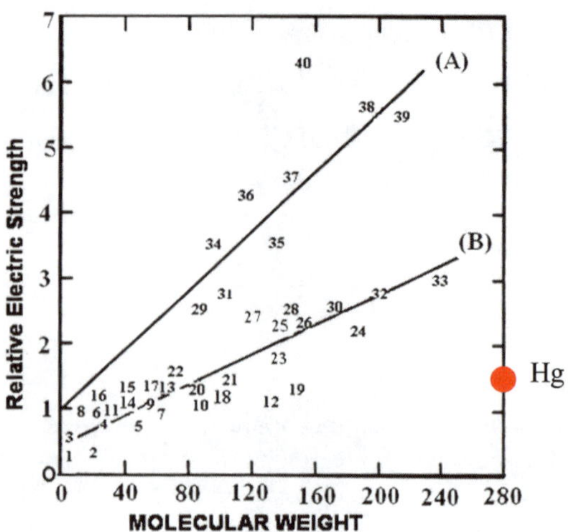

Fig. 17.1 Relative electric strength of molecular gases as a function of molecular weight. For limitation of space individual gases are not identified. *1*—helium, *6*—N_2, *27*—Freon, *28*—SF_6, *40*—CCl_4. Mercury is in vapor phase. Gases that belong to branch *A* have a—CN group or—SO fragment and do not contain hydrogen [14]

pressed air, more preferentially nitrogen and sulfur hexafluoride (SF_6) have been used in these gas insulated systems (GIS). Some experimental installations have employed mixtures of N_2 and CO_2. The gas pressure used is 0.1–1.5 MPa (1 atmospheric pressure = 0.1 MPa) for nitrogen and 0.4–0.5 MPa for sulfur hexafluoride. The dielectric strength of SF_6 is 90 kV/cm, three times as large as that of nitrogen or air.

The environmental concerns of SF_6 molecule and its dissociation is due to the fact that the residence time life time of the particles are long, (800–3200) years [2] and the green house effect in the environment is estimated to have a potential of approximately 24 000 times that of CO_2 over a period of hundred years. The search for a replacement gas, most probably man made, has interested engineers to seek and learn from the fundamental aspects of electron motion in gases, in place of the traditional hit and miss method of measuring the dielectric strength after making the molecule and compare it with SF_6 for dielectric strength.

As a starting exercise let us look at the dielectric strength as a function of molecular weight. The relative dielectric strength of about 40 molecules are shown Fig. 17.1 [12, 18]. For the sake of brevity the entire list of numbers identifying the molecules is not given, and only representative gases are identified. Two branches *A* and *B* are found. It is interesting that many of the gases we deal with are in the lower branch *B*, (nitrogen—6, carbon dioxide—5, SF_6—28, CCl_2F_2—27) whereas the upper branch shows higher dielectric strength, but not suitable for engineering purposes due to other reasons such as toxicity (carbon tetrachloride, CCl_4—40) or corrosion issues (Perfluoropropane C_3F_8—39). A general observation is that gases with—CN group, or—SO fragment are in the higher branch *A* and they do not contain H atom.

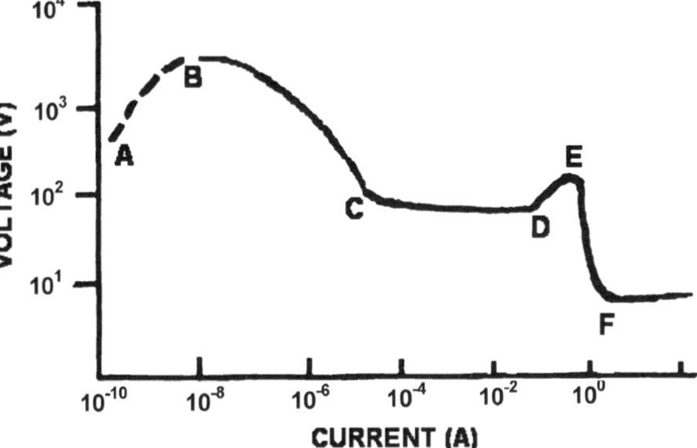

Fig. 17.2 Current voltage relationship in a parallel plane gap. *A–B*, townsend dark discharge; *B–C*, self sustained discharge; *C–D*, normal glow; *D–E*, abnormal glow; *EF*-arc [14]

17.3 Ionization Coefficients

As expected a simple relationship with molecular weight does not provide sufficient insight into the physics of the problem and the next level of complexity is to consider the current growth coefficients. If we measure the current as a function of the applied voltage the current the relationship takes the form shown in Fig. 17.2. In the region AB, which is the earliest stage of discharge the current growth is given by:

$$i = i_o \frac{\left\{ \overbrace{\frac{\alpha}{\alpha-\eta}}^{\text{ionization coefficient}} \exp(\alpha - \eta)d - \overbrace{\frac{\eta}{\alpha-\eta}}^{\text{attachment coefficient}} \right\}}{1 - \left\{ \underbrace{\frac{\overbrace{\gamma}{}}{(\alpha-\eta)}}_{\text{secondary processes}} \alpha \left[\exp(\alpha \underbrace{d}_{\text{gap lenght}}) - 1 \right] \right\}} \tag{17.1}$$

A typical experimental arrangement for determination of these coefficients is shown in Fig. 17.3.

Both steady state and temporal methods may be employed. Equation (17.1) is the steady state method and a similar expression obtains for electron and ion currents if temporal methods are employed. The steady state UV lamp is replaced by a flash UV lamp with 2–3 ns width output.

Measurement of ionization and attachment coefficients, α/N and η/N respectively, where N is the gas number density have been accomplished in a large number of gases under high vacuum conditions and spectroscopic pure gases where possible. Figures 17.4 and 17.5 show a set of results of hydrocarbons. A general formulation for α/N was observed in terms of the parameter E/N where E is the applied electric field. The relationship is of the form:

$$\frac{\alpha}{N} = F \exp\left(-\frac{GN}{E}\right) \tag{17.2}$$

Where F and G are constants that depend on the gas.

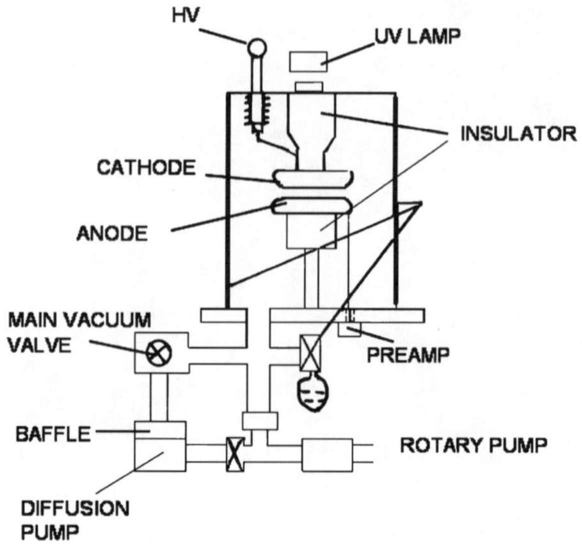

Fig. 17.3 Experimental arrangement for measurement of growth coefficients in gases [14]

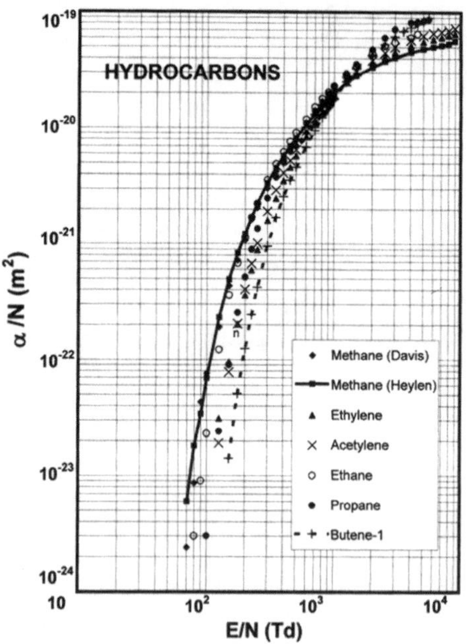

Fig. 17.4 Reduced ionization coefficients in hydrocarbons. The general dependence of α/N is according to Eq. (17.2) [14]

Equation (17.2) has remained unsatisfactory because it is semi-empirical, does not cover the entire range of E/N for which data on α/N are acquired. Attempts to relate F and G to fundamental properties of the gas have not succeeded. The constant G is dependent on the ionization potential of the gas in a complex manner.

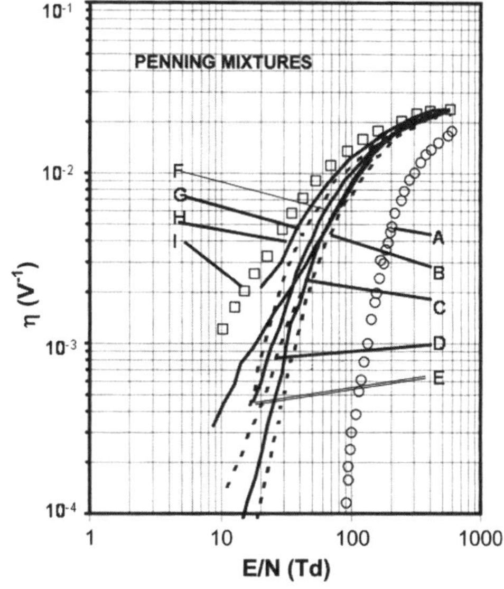

Fig. 17.5 Ionization per volt in gas mixtures of argon, ethane (C_2H_6) and mixtures thereof, showing the Penning effect. (*A*) 100 % C_2H_6; (*B*) 0.005 %; (*C*) 3×10^{-5} %; (*D*) 3×10^{-4} %; (*E*) 0.5 %; (*F*) 3 %; (*G*) 0.3 %; (*H*) 0.03 %; (*I*) 0.003 %. *A–D* 99.995 % Ar; *E–I* 99.95 % Ar [14]

17.4 Electron Attachment

17.4.1 Attachment Processes

Electron attachment may be a two body or three body process. In the two body process, the formation of negative ions by collision of an electron with a diatomic molecule may be explained by considering the formation of a molecule. Consider two isolated atoms that are situated so far apart that they do not influence each other. As they come closer together the outer electrons repel each other, and the electron of one atom is attracted to the nucleus of the other. If they are too close, of course their nuclei repel each other. At a certain critical distance between their nuclei the energy is a minimum and the molecule will be at ground state with minimum energy at the particular bond length. For shorter or longer bond lengths the energy of the molecule is not minimum and the molecule will not have a stable configuration, say X_2 (Fig. 17.6).

When an electron with a finite energy impacts with the molecule and a negative ion is formed (X_2^-) the energy of the electron must be accounted for. The binding energy of the extra electron to the molecule is known as the electron affinity. If the negative ion is stable then the electron affinity is defined as the energy difference between the ground state of the neutral and the lowest state of the corresponding negative ion.

Higher the electron affinity of the atom greater is the attachment. For example, halogen atoms have electron affinities in the range 3.1–3.5 eV compared to 0.08 eV for helium which is not attaching [14]. Oxygen atom has an electron affinity of 1.46 eV and it is moderately attaching. Atoms and molecules have different elec-

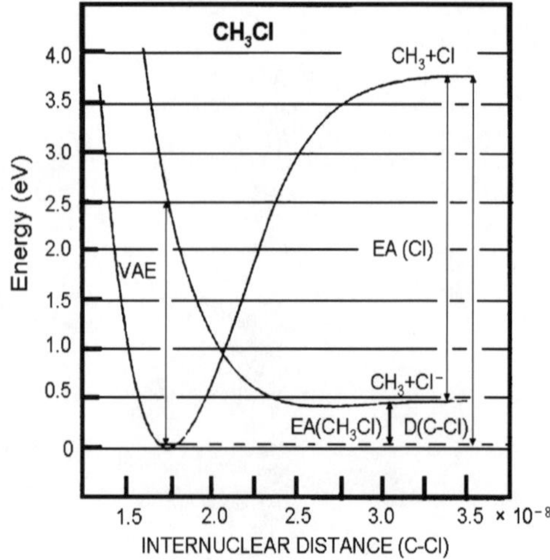

Fig. 17.6 Energy scheme for dissociative attachment of chloromethane. Curves for other species not shown for clarity (G.G. Raju)

tron affinities. For example, O_2 and O_3 have electron affinities of 0.44 and 2.10 eV respectively.

The minimum energy of the negative ion is higher and the bond length has a higher value. If we assume that the life time of the negative ion is short compared to the time of the nuclear motion the X_2-ion dissociates into X and X^- in time $\sim 10^{-13}$ s. The incoming electron must have precisely the exact energy for the single-bond distance of the molecule. This requirement translates into a peak in the cross section-energy $(Q_a - \varepsilon)$ curve and a resonance is said to occur. The resonance energy for a large number of molecules are in the 1–4 eV range.

Direct attachment is represented by

$$AB + e \rightarrow AB^- \tag{17.3}$$

The dissociative attachment is usually represented by

$$e + AB \rightarrow \left(AB^-\right) \rightarrow A + B^- \tag{17.4}$$

Ion pair production is another type of two body process that results in electron attachment. The reaction is represented by

$$e + AB \rightarrow \left(AB^*\right) + e \rightarrow A^+ + B^- + e \tag{17.5}$$

The three body processes for electron attachment require the presence of a third particle that may or may not be the same as the parent atoms. These reactions are:

$$e + AB \rightarrow \left(AB^*\right)^-; \qquad \left(AB^*\right)^- + M \rightarrow AB^- + M \tag{17.6}$$

$$e + AB + M \rightarrow AB^- + M \tag{17.7}$$

Electron affinity and total attachment cross section peaks of selected molecules are shown in Table 17.1. The cross section shown is for the combined dissociative attachment and ion pair production [reactions (17.4) and (17.5)]. The cross section for SF_6 is at least two orders of magnitude higher and the electron affinity is very

Table 17.1 Electron affinity (ε_A), attachment peak energy (ε_p) and total cross section at peak for negative ion formation in molecular gases [12]

Gas	ε_A	ε_p (eV)	Q_a (10^{-20} m^2)
O_2	0.44	6.5	1.407×10^{-2}
CO		9.9	2.023×10^{-3}
NO	0.024	8.15	1.117×10^{-2}
CO_2		8.1	4.283×10^{-3}
N_2O	~1.465	2.2	8.602×10^{-2}
SF_6	0.46	~0.1	2.146

Table 17.2 Rate constant (10^{-16} m^3 s^{-1}) of dissociative attachment to the molecule in an electric field. a (b) means $a \times 10^b$

E/N (Td)	Gas										
	SF_6	CF_4	C_2F_6	C_3F_8	C_4F_{10}	CCl_2F_2	H_2O	N_2O	HCl	HBr	HI
2								6.5 (−6)	1.6	34	53
5								1.2 (−4)	2.6		26
10								0.001			
50	6.3	0.39					0.017				
100	7	0.68					0.8				
150	7.8	0.68					1.0				
200	7.3	0.64					0.78				
250	6.8	0.5	2.7	3.0	3.3						
300	6.8	0.46	2.6	2.4	2.3	4.6					
400	6.7		1.5	1.8	1.0	4.6					
500	6.5		0.65	1.0	0.6	3.9					
600	5.8					2.5					

low, possibly the resonance peak is very narrow and the width of the electron beam partially contributes to the observed energy [12].

The experimental cross sections [12] shown in Table 17.2 were obtained using mono-energetic beam of electrons. Under swarm conditions the independent variable is the parameter E/N related to the mean energy and the rate constant for dissociative attachment is given in Table 17.3 [12]. Again note the large difference in the rate between molecules with halogen atoms and those without. The rate constant of dissociative attachment in gases as a function of mean energy can be found in [12, 18].

The usefulness of this table is demonstrated by substituting typical values for SF_6 (say): $E/N = 100$ Td, $K_a = 7 \times 10^{-16}$ m^3 s^{-1}, $W_e = 10^5$ m s^{-1}, giving reduced attachment coefficient (η/N) of 70×10^{-22} m^2.

Table 17.3 Attachment processes (SF$_6$): Peak energy and cross section

Ion	Threshold (eV)	Peak energy (eV)	Q_a (peak) (10^{-20} m^2)	Reaction
SF$_6^-$	0.00	0.00	7617	(17.11)
SF$_5^-$	0.10	0.30	4.23	(17.12)
SF$_4^-$	2.8	5.0	4.57×10^{-3}	(17.13), (17.14)
SF$_3^-$	0.7	11.0	750×10^{-6}	(17.15)
F$_2^-$		2.00	8×10^{-5}	
		4.5	7.7×10^{-4}	
		11.5	9.54×10^{-4}	(17.18)
F$^-$	0.2	5.0	0.0463	(17.16)
F$^-$	1.64			(17.17)

17.4.1.1 Oxygen

Dissociative and three body attachment processes occur according to

$$e + O_2 \rightarrow O^- + O \tag{17.8}$$

$$e + O_2 \rightarrow \left(O_2^-\right)^*; \qquad \left(O_2^-\right)^* + M \rightarrow O_2^- + M \tag{17.9}$$

The ion-pair production occurs according to

$$e + O_2 \rightarrow O^+ + O^- + e \tag{17.10}$$

A very small cross section $\sim 1 \times 10^{-28}$ m^2 has been estimated from photodetachment of O$_2^-$ [12]; in view of the small cross section we shall not discuss it further.

17.4.1.2 Sulfur Hexafluoride

Selected attachment processes are:

$$e + SF_6 \rightarrow SF_6^- \tag{17.11}$$

$$e + SF_6 \rightarrow SF_5^- + F \tag{17.12}$$

$$e + SF_6 \rightarrow SF_4^- + 2F \tag{17.13}$$

$$e + SF_6 \rightarrow SF_4^- + F_2 \tag{17.14}$$

$$e + SF_4 \rightarrow SF_3^- + F \tag{17.15}$$

$$e + SF_4 \rightarrow SF_3 + F^- \tag{17.16}$$

$$e + F_2 \rightarrow F + F^- \tag{17.17}$$

$$e + SF_6 \rightarrow SF_6^- * (5.4 \, eV) \rightarrow SF_4 + F_2^- \tag{17.18}$$

Reaction (17.11) is direct attachment, reactions (17.12) through (17.17) are dissociative attachments and reaction (17.18) is attachment through excitation.

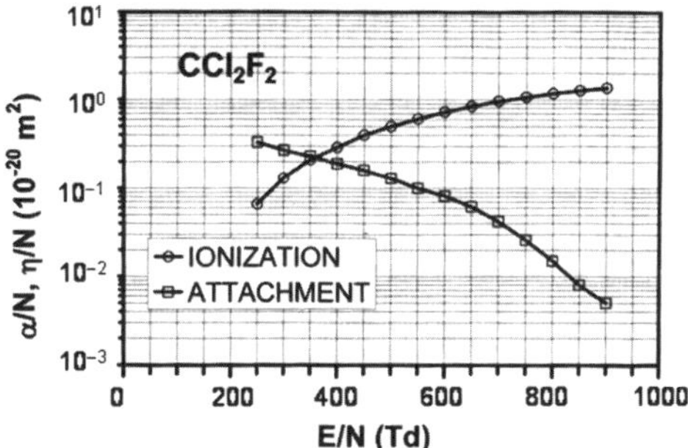

Fig. 17.7 Ionization and attachment cross sections in CCl_2F_2 (dichloro difluoro methane or Freon-12). The intersection point defines the insulation strength

17.4.2 Attachment Coefficients

Figure 17.7 shows the attachment coefficients and measurements in large number of gases show that even a simple analytical expression such as Eq. (17.2) has not been possible. But, in terms of fundamentals of collision process the objective is clearer because a shift of the intersection point, if possible, will increase the insulation strength.

17.5 Energy Distribution Functions

It has been realized that the goal of making a new gas for higher dielectric strength can not be achieved without a fundamental insight into the electron-molecule collision phenomena and this brings us to a discussion of electron energy distribution of electrons in a swarm. The energy distribution function may be determined by two alternative methods. (1) Solving Boltzmann equation which, under steady state conditions, is formulated by all the energy gain processes and energy loss processes. For the time being we do not consider the time dependent solution. (2) Monte Carlo method.

Evolution fast and super computers have influenced the search for solution numerically.

17.5.1 Boltzmann Equation

A concise form of Boltzmann equation is

$$\frac{\partial F}{\partial t} = V \bullet \nabla_r F_i + a \bullet \nabla_v F_i = \left(\frac{\partial F}{\partial t} \right)_{coll} \tag{17.19}$$

where F_i is the energy distribution for particle species i which is a function of (r, v, t) where r is the variable in the configuration space (x, y, z), V is the variable in the velocity space (V_x, V_y, V_z) and t is the time. $F_i(r, V, t)d^3r d^3V$ is the number of particles of species i in six dimensional phase space in the differential volume $dx\,dy\,dz\,dVx\,dVy\,dVz$. The mathematical difficulty of finding the solution of Eq. (17.19) is surmountable and the technique of separating the variables (t, r) is adopted. For time independent solution the solution neglects the dependence on t.

For practical application the equation that is employed is [6] where

E = electric field
i, j = species or levels
k = Boltzmann constant
m = electron mass
N = number density
M = molecular mass
Q_M = momentum transfer cross section
T = temperature
ε = electron energy

$$\frac{d}{d\varepsilon}\left(\frac{e^2 E^2 \varepsilon}{3N Q_M}\frac{df}{d\varepsilon}\right) + \frac{2m}{M}\frac{d}{d\varepsilon}\left(\varepsilon^2 N Q_M\left[f + kT\frac{df}{d\varepsilon}\right]\right)$$
$$+ \sum_i \left[(\varepsilon \pm \varepsilon_j)f(\varepsilon \pm \varepsilon_j)N Q_{\pm j}(\varepsilon \pm \varepsilon_j) - \varepsilon f(\varepsilon)N Q_{\pm j}(\varepsilon)\right] = 0$$

$$(17.20)$$

The last term in Eq. (17.20) is in fact two terms condensed into one for space limitation.

Equation (17.20) is solved by numerical methods using spherical harmonics expansion or [6] or adopting Fourier expansion [4]. The solution for $f(\varepsilon)$ is then used to determine the swarm coefficients according to the definitions:

$$W = -\frac{1}{3}\frac{E}{N}\left(\frac{2e}{m}\right)^{1/2}\int_0^\infty \frac{\varepsilon}{Q_M(\varepsilon)}\frac{d}{d\varepsilon}\left(\frac{f(\varepsilon)}{\varepsilon^{1/2}}\right)d\varepsilon \qquad (17.21)$$

$$D = \frac{1}{3N}\left(\frac{2}{m}\right)^{1/2}\int_0^\infty \frac{\varepsilon^{1/2}}{3N Q_M}f(\varepsilon)d\varepsilon \qquad (17.22)$$

$$\frac{\alpha}{N} = \left(\frac{2e}{m}\right)^{1/2}\frac{1}{W}\int_{\varepsilon_i}^\infty \varepsilon^{1/2}Q_i f(\varepsilon)d\varepsilon \qquad (17.23)$$

$$\frac{\eta}{N} = \left(\frac{2e}{m}\right)^{1/2}\frac{1}{W}\int_{\varepsilon_a}^\infty \varepsilon^{1/2}Q_a f(\varepsilon)d\varepsilon \qquad (17.24)$$

where

D = diffusion coefficient
W = drift velocity
Q_a = attachment cross section
Q_i = ionization cross section

$\alpha =$ ionization coefficient

$\eta =$ attachment coefficient

The cross sections needed for evaluating the swarm coefficients are: total scattering cross section, momentum transfer cross section, cross section for each inelastic process depending upon the degree of accuracy required—total excitation cross section, ionization and attachment cross section are more frequently used.

17.5.2 Monte Carlo Methods

In essence the Monte Carlo method for collision phenomena is an efficient book keeping exercising requiring enormous memory storage and large computation times. A certain number of electrons are released from the cathode, between 100–10,000 depending upon the storage capability of the computer and each electron is followed as it moves from cathode and anode. The energy gain is calculated at each step and a determination made, whether a collision has occurred, if yes, the nature of collision, the energy loss and angle of scattering. Running averages of swarm properties are maintained and continually updated as the electron moves through the gas. Several techniques such as the mean free path method, mean collision time method, null collision techniques have been developed [12].

17.5.3 Continuity Methods

The continuity equations for particle generation and decay are given by the equations

$$\left.\begin{aligned}
\frac{\partial N_e}{\partial t} &= N_e \alpha W_e - N_e \eta W_e - N_e N_p \beta - \frac{\partial (N_e W_e)}{\partial x} + \frac{\partial^2 (D N_e)}{dx} \\
\frac{\partial N_p}{\partial t} &= N_e \alpha W_e - N_e N_p \beta - N_n N_p \beta \frac{\partial (N_p W_p)}{\partial x} \\
\frac{\partial N_n}{\partial t} &= N_e \eta W_e - N_n N_p \beta \frac{\partial (N_n W_n)}{\partial x}
\end{aligned}\right\} \quad (17.25)$$

where the subscripts e, p, and n denote the electron, positive ion and neutral species, α, β and η are the corresponding coefficients.

17.6 Energy Loss Mechanisms

Inelastic collisions result in energy loss of the electron as it moves through the gas and each inelastic process has a cross section as a function of energy. Figure 17.8 shows the importance of the type of collision in different ranges of the parameter

Fig. 17.8 Energy loss mechanism in low energy inelastic collisions. Rotational excitations dominate in the range $0.1 \leq E/N \leq 2$ Td and vibrational excitations dominate in the range $2 \leq E/N \leq 30$ Td. The electronic excitation and ionization energy losses occur for $E/N \geq 75$ Td [12]

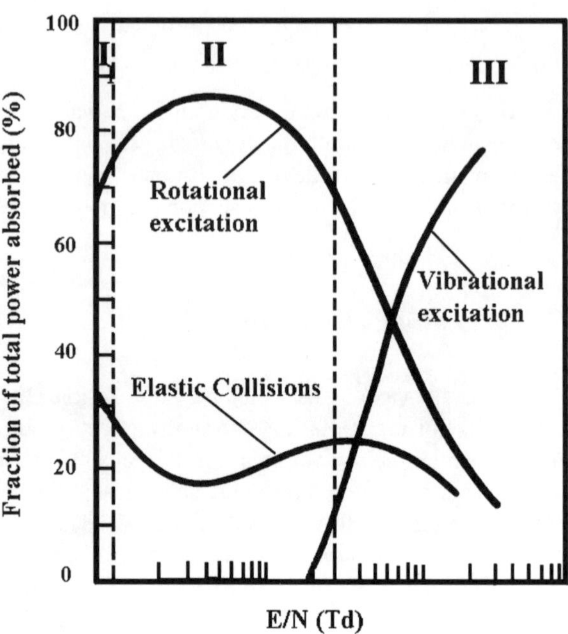

Fig. 17.9 Addition of ∼5 % of H_2 results in a lower mean energy by a factor of nearly 20 [8]

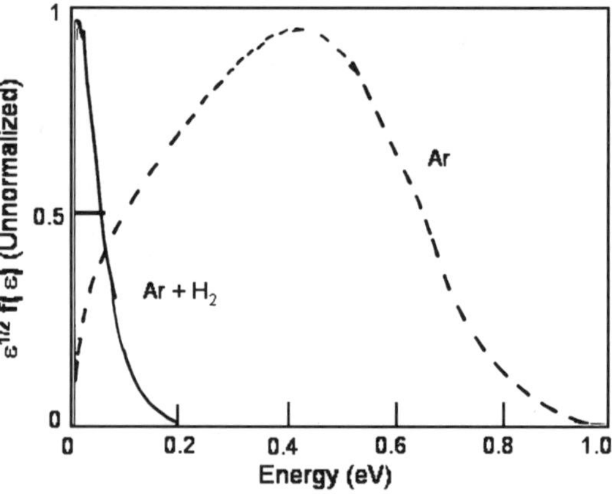

E/N. The influence of low energy losses can be dramatic as shown by Fig. 17.9 where addition of hydrogen to argon reduces the mean energy of the electron swarm by a factor of 20. Figure 17.10 shows the relative losses over a wide range of energy in CO_2 [10].

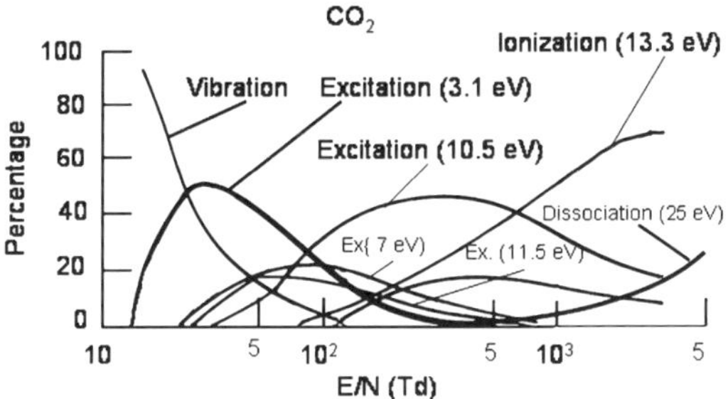

Fig. 17.10 Energy loss in collisions in CO_2. At low energy the vibrational cross sections dominate whereas at high energy ionization phenomena becomes the major loss process [14]

17.7 Cross Sections Derived from Swarm Coefficients

Equations (17.21)–(17.24) demonstrate that the measured swarm coefficients may be employed to evaluate collision cross sections. This is advantageous in situations where swarm measurements may be easier and economical to make compared with beam measurements. In a perfect world both techniques, namely swarm and beam techniques yield collision cross sections which agree with each other, but this is not so in many cases. Even in a relatively simple gas such as H_2 considerable disagreement exists [19]. Notwithstanding these limitations the technique is a useful one and we make brief comments about the techniques available.

The swarm coefficients method is subject to the restriction that the accuracy decreases if the function to be used has maxima or minima since small changes in the measured quantity can lead to rapid changes in $Q_M(\varepsilon)$ resulting in instability of the solution [11]. The method adopted is to change $Q(\varepsilon)$ and calculate the swarm coefficient used, in an iterative algorithms till satisfactory agreement between experimental and calculated values are achieved. It is not unusual to vary measured $Q_M(\varepsilon)$ and other cross sections by ±30 % to obtain agreement. In particular, low energy inelastic scattering cross sections and electronic excitation cross sections are changed by 100 % or more to achieve agreement.

Another method employed to evaluate $Q_M(\varepsilon)$ is to measure the conductivity with or without a crossed magnetic field. The latter case is dealt with in Sect. 17.8.

The DC conductivity of a plasma may be expressed as

$$\sigma = \frac{N_e}{\nu}\frac{e^2}{m} \tag{17.26}$$

where N_e is the electron density, ν the electron-neutral collision frequency, e and m the electronic charge and mass respectively. The conductivity may also be defined by an integral similar to those in Eqs. (17.21) and (17.22) involving energy and cross section. To obtain Eq. (17.26) a Maxwellian distribution is assumed, but use of other distributions are also possible in principle if numerical techniques are employed. A number different experimental combinations are possible for the measurement of conductivity, such as high pressures, DC and RF plasma etc. As we have not used plasma for these studies no further is discussion is given here.

17.8 Generalized Collision Cross Sections

In search of gaseous dielectrics for engineering purposes one looks for possible generalizations of various collision cross sections as a function of electron energy. The dependence of dielectric strength on molecular weight has already commented upon as not having the In order to determine such relationships the variation of cross section with energy should have the following specific characteristics. (1) The cross section must be a smooth function of energy. (2) The variation of the cross section must be similar as shown in Fig. 17.4. That is, a monotonic increase or decrease. (3) The functional dependence should have as few as possible, the fundamental quantities involved in the collision process, such as the electronic polarizability, dipole moment, ionization potential of the target molecule. (4) The relationship between the cross section and energy should be as simple as possible.

Ionization cross sections as a function electron energy satisfy these requirements and we shall consider it first.

17.8.1 Generalization of Atomic Ionization Cross Sections

Ionization cross sections of atoms have the potential of being used as building blocks for a general expression for molecules. Szłińska et al. [15] have proposed a simple formula:

$$\frac{Q_i^A(\frac{\varepsilon}{\varepsilon_i^A})}{(Q_i^A)_{MAX}} = \frac{Q_i^B(\frac{\varepsilon}{\varepsilon_i^B})}{(Q_i^B)_{MAX}} \tag{17.27}$$

where A and B are two species being considered, ε_i the ionization threshold, Q_i the ionization cross section, $(Q_i)_{MAX}$ the maximum of ionization cross section for each species and ε the electron energy. Figure 17.11 shows selected cross sections plotted according to Eq. (17.27).

Deutsch et al. [17] have proposed a scheme to use the ionization cross sections of constituent atoms on the basis of an additivity rule. The additivity rule incorporates weighting factors for the contribution from constituent atoms which depends explicitly on the atomic radii and the effective number of atomic electrons. The atomic radius is not easily found and the effective number of contributing electrons from each atom is empirically determined. For example, the molecule SF_4 is calculated to have contributions from 7 (out of 9) electrons from the fluorine atom and four electrons (out of 14) from silicon atom.

17.8.2 Generalization of Molecular Ionization Cross Sections

In a series of publications Harland and his group [3, 5] have proposed a scheme that relates the electronic polarizability and the maximum in ionization cross section according to

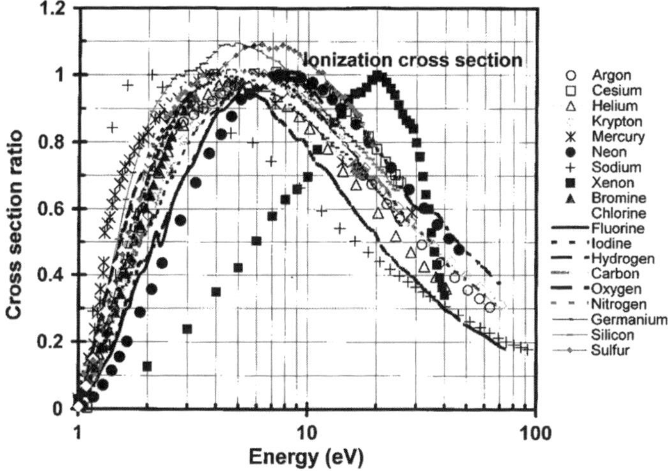

Fig. 17.11 Measured ionization cross sections of atoms plotted according to Eq. (17.27), Raju [19]. Elements belonging to the group of halogens, noble gases, columns IV, V and VI reasonably satisfy the equation except xenon. Mercury was not plotted in their publication, and satisfies the equation reasonably well

$$Q_{i(MAX)} = \frac{e}{4\pi\varepsilon_0}\left(\frac{\alpha_e}{\varepsilon_i}\right)^{1/2} \tag{17.28}$$

where e = electron charge (C), ε_0 = permittivity of free space (F/m), α_e = electronic polarizability (F m^2), ε_i = ionization potential (eV) and Q_i = the ionization cross section (m^2). Collecting the constants Eq. (17.28) may be expressed as

$$Q_{i(MAX)} = K\left(\frac{\alpha_e}{\varepsilon_i}\right)^{1/2} \tag{17.29}$$

where K is a constant. The maximum ionization cross section in most gases as a function of electron energy reaches a plateau extending over several eV. Since the magnitude is involved and not the electron energy at this maximum Eq. (17.29) has the advantage of being more accurate.

Harland and Vallance [15] have demonstrated that Eq. (17.29) is applicable to atomic gases, though good correlation was also observed between $Q_{i(MAX)}$ and α_e. They have also measured ionization cross sections for a number of molecules, over 80 in number, many for the first time, and demonstrated the general applicability of Eq. (17.29).

Figure 17.12 shows a plot of Eq. (17.29) for several gases [3, 5, 11, 15, 17] updated from recent data [12]. The electronic polarizability of certain molecules, for example, bromomethane (CH_3Br) is quoted in the literature as 6.18, 6.53 and 6.71, all in units of 10^{-40} F m^2, and an average value of 6.47 has been plotted. A special comment about the molecule, ethyl alcohol (C_2H_6O) is appropriate here. Of the three quantities, $Q_{i(MAX)}$, α_e and ε_i, if two quantities are known then the third quantity may be evaluated, within reasonable accuracy, because of the linear

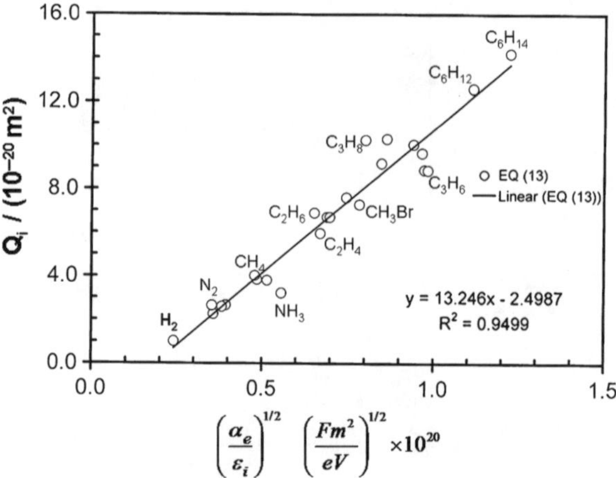

Fig. 17.12 Plot of experimentally measured $Q_{i(MAX)}$ as a function of the parameter shown for over thirty gases, see Eq. (17.29). For clarity not all molecules are identified

relation in Eq. (17.29). Ionization cross section data for HBr are not available though the value of the parameter (α_e/ε_i) is available as 0.587. Figure 17.14 yields the maximum in ionization cross section as 5.37×10^{-20} m^2. Similarly one gets 4.48×10^{-20} m^2 for $Q_{i(MAX)}$ for HCl.

17.8.3 Relationship Between Q_T and Q_i

A relationship between total scattering cross section and total ionization cross section has been proposed by Kwitnewski et al. [7] using three parameters:

$$\frac{Q_i}{Q_T} = A\left[1 - \exp\left(\frac{B - \varepsilon}{C}\right)\right] \tag{17.30}$$

in which A shows the contribution of all ionization processes to overall scattering at very high impact energies and B is related to the ionization potential. Kwitnewski et al. [7] have evaluated the constants for a number of hydrocarbons and perfluorocarbons. The constants are different for each gas limiting the general applicability of Eq. (17.30).

17.9 Scaling Formula for Total Scattering Cross Sections

The energy dependence of total cross section for gases shows peaks or resonance processes that are operative in the collision process, except possibly in helium. The polarization of the neutral species and the dipole moment of the molecule play a role in determining the energy at which resonance occurs. The energy dependence of the total scattering cross section has complex pattern as shown in Fig. 17.13 for a number of large molecules.

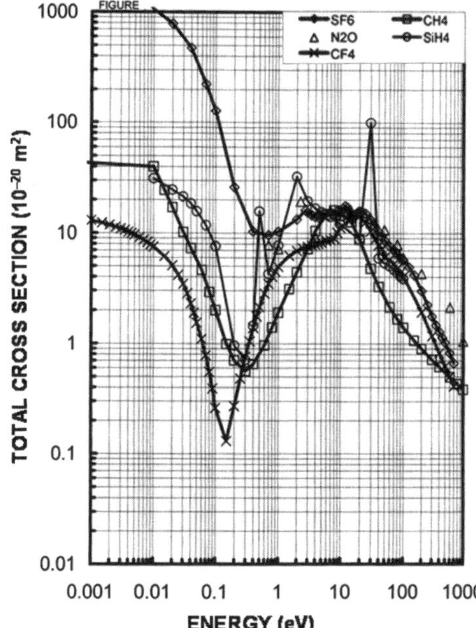

Fig. 17.13 Total scattering cross section as a function of impact energy for molecular gases. $(-\Diamond-)$ SF$_6$; (Δ) N$_2$O; $(-\times-)$ CF$_4$; $(-\Box-)$ CH$_4$; $(-\circ-)$ SiH$_4$

Raju [9] has classified the behavior for a number of gases (over 100 gases) according to a scheme shown in Fig. 17.16. In order to understand the meaning of resonance we consider the total scattering cross section as a function of electron energy. To make the description easier to understand we include all the characteristics that are observed in various species of the target particle, namely, atoms, polar molecules, non-polar molecules, electron attaching neutrals and non-attaching neutrals. A schematic variation of the total scattering cross section as a function of electron energy observed in gases is shown in Fig. 17.14. Arbitrary scales are to make the description as general as possible.

It is stressed that the figure is schematic but not fictional. The following characteristics with increasing electron energy are noted:

(1) Segment A_1B: The total cross section decreases rapidly by several orders of magnitude due to attachment of electrons having zero energy. Sulfur hexafluoride (SF$_6$) is a good example for this segment. The absolute total scattering cross section is as high as 350×10^{-20} m^2 at 0.03 eV.

(2) Segment A_2B: The total cross section decreases with increasing energy, though not as steeply as in the previous case. This behavior is observed for polar molecules. Ozone (O$_3$) which has a moderate dipole moment of 0.53 D is a good example for this segment. The total cross section is 120×10^{-20} m^2 at 0.01 eV.

(3) Segment A_3B: The molecule may not have a dipole moment or electron attaching. The total cross section tends to decline very slowly or even remain constant. Methane (CH$_4$) is a good example for this segment. The total cross section decreases from a relatively low value of 4×10^{-20} m^2 at 0.1 eV.

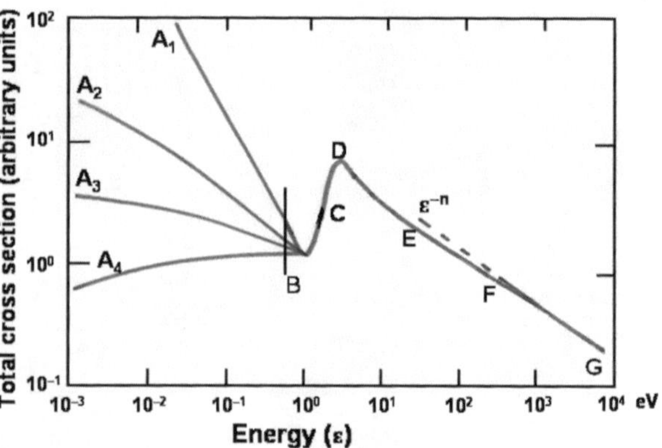

Fig. 17.14 Schematic diagram of the variation of the total scattering cross section as a function of electron energy. The nature of the target particle determines which segments of the curves are exhibited

(4) Segment A_4B: The total cross section increases slowly from moderate value of $\sim$1–10 $\times$ 10^{-20} m². Atomic helium (He) and diatomic hydrogen (H$_2$) are good examples. Scattering theory is thought to explain this behavior well.

(5) Segment BC: This segment is the well known Ramsauer-Townsend effect. Several hydrocarbon and rare gases exhibit this characteristic in the energy range 0.1–1.0 eV. Quantum mechanical scattering theory explains this behavior very well.

(6) Segment CDE: This segment is exhibited by most target particles, that is, atoms, electron attaching and non attaching molecules, polar or non polar molecules. The cross section raises to a peak in the energy range 2–10 eV and declines for higher energies. The mechanism is thought to be resonance between the velocity of the incoming electron and outer electrons of the target particle.

(7) Segment EF: A monotonic decrease of total scattering cross section with increasing electron energy without any structure (peaks, valleys, oscillations) is the principle feature. Almost all target particles exhibit this behavior. Inelastic cross sections make substantial contribution to the total.

(8) Segment FG: A further monotonic decrease in the total scattering cross section is observed. The essential difference with the previous segment (EF) is that the total cross section varies according to ε^{-n} where ε is the electron energy in eV and n is a constant, in the energy region 500–10 000 eV. Only this high energy segment has been successfully scaled for many gases, with different value of n for each gas.

A relationship between the peak in the scattering cross section at resonance, the polarizability and the dipole moment of the molecule has been identified by the present author [13] as:

$$Q_{T,\text{res}} = f\left(\alpha_e + \frac{\mu^2}{\varepsilon_p}\right) \qquad (17.31)$$

The functional dependence is:

$$Q_{T,\text{res}} = Q_o\left(\frac{\alpha_e}{10^{-40}} + \frac{\mu^2}{\varepsilon_p}\right)^n \qquad (17.32)$$

where $Q_{T,\text{res}}$ is the cross section at resonance peak (in units of 10^{-20} m²), α_e is the electronic polarizability in F m² units, μ the dipole moment in Debye units, ε_p

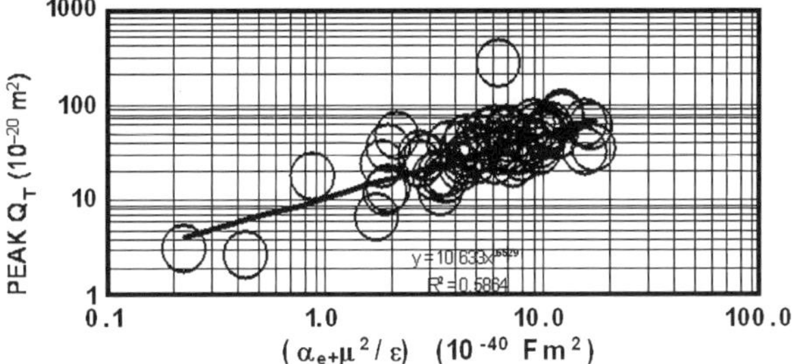

Fig. 17.15 Graphical representation of Eq. (17.32). Each circle denotes a target particle. Individual gases are not identified for the sake of clarity

the energy at resonance (eV), Q_0 (units of 10^{-20} F^{-1} m^2) and n are constants. The equation applies to resonance at D identified in Fig. 17.16. The energy range almost invariably is 1–10 eV for most gases.

Equation (17.32) shows that the total cross section increases with higher dipole moment and lower resonance energy. This is consistent with the concept that a higher dipole moment leads to intensified interaction between the target particle and electron. To explain the increase of total cross section due to lower resonance energy one has to resort to quantum theory [19] which is beyond the scope of the present paper.

Figure 17.15 shows the plot of the quantities on either side of Eq. (17.32). Individual particles, ranging from acetylene (C_2H_2) to xenon (Xe), 62 total, is composed of all species, namely atomic, polyatomic, polar and non-polar, electron attaching and non-attaching. A reasonably satisfactory agreement is obtained yielding $Q_0 = 10.66 \times 10^{-20}$ F^{-1} m^2 and $n = 0.653$. A brief comment about units employed is appropriate here. If the dipole moment is expressed in units of C m instead of Debye, (1 Debye $= 3.33 \times 10^{-30}$ C m) and the energy in Joules in place of eV (1 eV $= 1.602 \times 10^{-19}$ J) Eq. (17.32) is still applicable with marginally different values of Q_0 and n resulting, $Q_0 = 10.63$ $m^2 \times 10^{-20}$, $n = 0.654$.

Since a large number of gases have been included in the analysis the evaluated values of Q_0 and n acquire the status of, however approximately determined, universal characteristics. The values of Q_0 and n are independent of the species of the target molecule which is the distinct feature when compared with previously available formulas for other regions.

17.10 Attachment Rate Constant

Electron attachment to neutrals is a deionizing process and it is of obvious interest in engineering applications in which gaseous dielectrics are employed, either for insulation purposes or arc quenching properties. Attachment rate constants depend both on electron energy and temperature of the medium. There are a large number of gases in which the attachment cross section increases with decreasing energy, reaching a very high value for thermal electrons (C_6H_6, CCl_4, SF_6, CF_4 etc.). Christophorou and Hadjiantoniou [13] suggest that such large cross section is

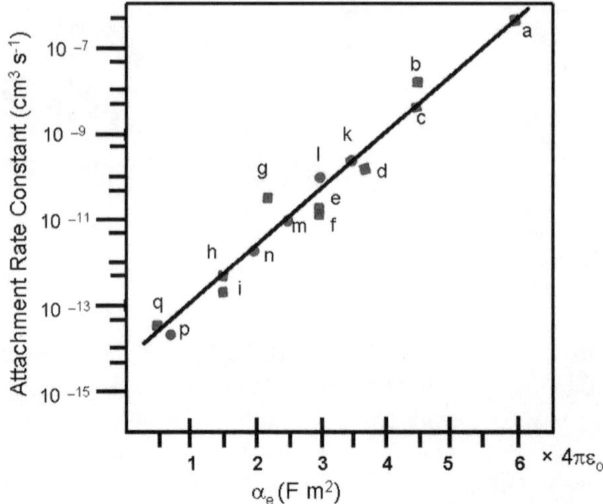

Fig. 17.16 Attachment rate constants for thermal electrons chlorine and fluorine compounds as a function of calculated electronic polarizability of accepting centers [16]. ■-Chlorine containing aliphatic compounds. ●-Fluorine containing—aliphatic compounds. a—CCl_4; b—$1,1,1$-$C_2H_3Cl_3$; c—$CHCl_3$; d—$1,1,2$-$C_2H_3Cl_3$; e—$1,1$-$C_2H_4Cl_2$; f—CH_2Cl_2; g—$1,2$-$C_2H_4Cl_2$; h—C_2H_5Cl; i—CH_3Cl; k—C_6F_{14}; l—C_5F_{12}; m—C_4F_{10}; n—C_3F_8; p—CHF_3; q—CH_2F_2

indicative of toxicity in human body. Szamrej [1] has shown that the relationship between the attachment rate constant for several chlorine and fluorine compounds increases linearly with the polarizability of the accepting center, defined as the atomic polarizability of the electronegative atoms (Br, Cl, I, F) as well as double or triple bonds, and aliphatic or aromatic rings, as shown in Fig. 17.16.

The characteristic may be explained on the basis that the electronic polarizability increases with increasing molecular radius and therefore larger the influence of the accepting center, larger is the attachment cross section resulting in a higher rate constant.

17.11 Conclusions

The need for fundamental collision cross sections in understanding and evaluating gaseous dielectrics has been summarized in the paper. General formulas for representing ionization coefficients, ionization cross sections, momentum transfer cross sections and attachment rate constants have been presented and discussed.

References

1. L.G. Christophorou, D. Hadjiantoniou, Chem. Phys. Lett. **419**, 405 (2006)

2. L.G. Christophorou, J.K. Olthoff, *Fundamental Electron Interactions with Plasma Processing Gases* (Kluwer Academic/Plenum, New York, 2004)
3. H. Deutsch, K. Becker, T.D. Märk, Int. J. Mass Spectrom. Ion Process. **167/168**, 503 (1967)
4. A.G. Engelhardt, A.V. Phelps, Phys. Rev. **131**, 2115 (1962)
5. P.W. Harland, C. Vallance, Int. J. Mass Spectrom. Ion Process. **171**, 173 (1997)
6. J.B. Hasted, *Physics of Atomic Collisions* (Elsevier, New York, 1972), p. 451
7. J.E. Hudson, C. Vallance, M. Bart, P.W. Harland, J. Phys. B, At. Mol. Opt. Phys. **34**, 3025 (2001)
8. H.N. Kücükarpaci, J. Lucas, J. Phys. D, Appl. Phys. **12**, 2123 (1979)
9. S. Kwitnewski, E. Ptasińska-Denga, C. Szmytkowski, Radiat. Phys. Chem. **68**, 169 (2003)
10. J. Liu, G.G. Raju, Can. J. Phys. **70**, 216 (1992)
11. Z.Lj. Petrovic, T.F. O'Malley, R.W. Crompton, J. Phys. B, At. Mol. Opt. Phys. **28**, 3309 (1995)
12. G.G. Raju, *Gaseous Electronics: Theory and Practice* (Taylor & Francis, Boca Raton, 2006)
13. G.G. Raju, IEEE Trans. Dielectr. Electr. Insul. **16**, 1199 (2009)
14. G.G. Raju, *Gaseous Electronics: Tables, Atoms and Molecules* (Taylor & Francis, Boca Raton, 2011)
15. B. Stefanov, Phys. Rev. A **22**, 427 (1980)
16. I. Szamrej, in *Gaseous Dielectrics VIII.* ed by L.G. Christophorou, J.K. Olthoff (Kluwer Academic/Plenum, New York, 1998), p. 63
17. M. Szlińska, P. Van Reeth, G. Laricchia, J. Phys. B, At. Mol. Opt. Phys. **35**, 4059 (2002)
18. K.K. Vijh, IEEE Trans. Electr. Insul. **EI-12**, 313 (1977)
19. A. Zecca, G.P. Karwasz, R.S. Brusa, Phys. Rev. A **46**, 3877–3882 (1992)

Chapter 18
Data-Intensive Profile for the VAMDC

C. Mendoza and VAMDC Collaboration

Abstract The Virtual Atomic and Molecular Data Centre (VAMDC) is a multi-national project launched in July 2009 involving 24 research teams from several countries of the European Union (Austria, France, Germany, Italy, Sweden and the United Kingdom), Russia, Serbia and Venezuela. It intends to upgrade and integrate at least 21 atomic and molecular databases in order to implement an interoperable cyber-infrastructure for efficient data exchange in several user communities. In the present report we review the main features of data-intensive science (e-science), and within the context of the VAMDC, we discuss relevant concepts such as virtual databases; the deployment of database-centric applications as cloud web services; the workflow and script web-service integrators; data curation models; the social network as the new end user; the XSAMS XML schema and the structure of the VAMDC node.

18.1 Introduction

Atomic and molecular (A&M) data are required in a wide variety of scientific fields (astrophysics, fusion plasmas, atmospheric physics and chemistry, environmental studies and quantum optics) and technological applications (lighting, semiconductors, nanotechnology and molecular biology). Since the mid 1980s, our experience has been mainly in the production of atomic data for astrophysical applications, in particular in international, long-term consortia to compute astrophysical opacities (The Opacity Project [1]) and radiative and collisional data for iron-group ions (The Iron Project [2]). We have also been concerned with different aspects of data diffusion, in particular with the development and maintenance of online atomic databases (`TOPbase` [3, 4]) and derived data web services (`OPserver` [5]).

C. Mendoza (✉) · VAMDC Collaboration
Centro de Física, Instituto Venezolano de Investigaciones Científicas (IVIC), PO Box 20632, Caracas 1020A, Venezuela
e-mail: claudio@ivic.gob.ve

C. Mendoza · VAMDC Collaboration
CeCalCULA, Corporación Parque Tecnológico de Mérida, Mérida 5101, Venezuela

M. Mohan (ed.), *New Trends in Atomic and Molecular Physics*,
Springer Series on Atomic, Optical, and Plasma Physics 76,
DOI 10.1007/978-3-642-38167-6_18, © Springer-Verlag Berlin Heidelberg 2013

A&M data activities have often been overlooked in the past, poorly funded and mainly the result of domestic initiatives. However, with the advent of a scientific revolution where data are the new global currency, usually referred to as *e-science* [6], such activities are rapidly coming of age. The Virtual Atomic and Molecular Data Centre (VAMDC[1] [7, 8]), the main topic of the present report, is one of the first well-funded attempts to elevate A&M data dissemination to the new order. In the next sections, we describe some of the features of data-intensive science and how the VAMDC is organizing itself to comply with them.

18.2 Data-Intensive Science

Scientific research is becoming very data intensive and mostly driven by virtual communities on the Internet. New dynamics are transforming the scale, lifecycles and economics of the scientific endeavor where both experts and newcomers are finding difficulties in remaining competitive or, at least, capable of taking advantage of the unavoidable data deluge. Many fields of science are experiencing thousandfold increases in data volume generated by large-scale instrumentation such as accelerators, astronomical telescopes and supercomputers, or by networks (e.g. of wireless sensors or genome sequencers) with almost exponential data growths [9]. Astronomy, in particular, has been one of the leaders in this data revolution, as illustrated by the Sloan Digital Sky Survey (SDSS[2]) and the International Virtual Observatory Alliance (IVOA[3]).

As shown in Fig. 18.1, data-intensive science is being proposed as a new scientific paradigm that will nucleate and consolidate the other three well-established methods, namely experiment, theory and simulation [10]. As a consequence, high-performance computing is progressively becoming database centric; scientific applications will be globally accessed as cloud web services; data storage is moving to shared virtual network disks and distributed repositories; user interfaces are being managed with scripts (e.g. `Python`[4]) and workflow systems (e.g. `Taverna`[5]); and a Web 2.0 is providing the infrastructure for community and user-group developments by means of blogs, wikis, mashups, feeds and social network environments (e.g. `myExperiment`[6]).

Key data issues are then: *access* which must be open and involving complete collections that must be preserved in time; *manipulation* where provenance and reuse

---

[1] http://www.vamdc.eu/.

[2] http://www.sdss.org/.

[3] http://www.ivoa.net/.

[4] http://www.python.org/.

[5] http://www.taverna.org.uk/.

[6] http://www.myexperiment.org/.

Fig. 18.1 The four paradigms of 21st-century science

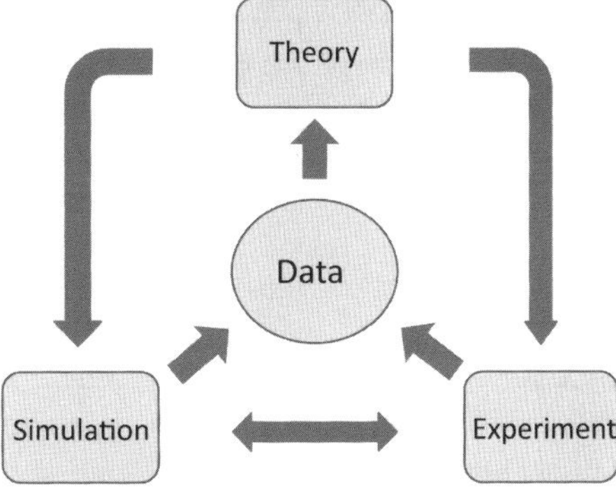

are ensured with extensive metadata; *flux* via innovative curation models and publication schemes; and *sharing* by means of open standards, XML schemata and ontologies aimed at fast knowledge creation through versatile data mining.

Furthermore, an advanced and scalable computing infrastructure for data-intensive science is definitely a distributed one, where computing power, resources and services can be obtained by the user on demand. In the past ten years, much effort has been devoted by scientific institutions and funding agencies worldwide to promote grids (e.g. TeraGrid,[7] EGEE/EGI,[8] EELA/GISELA,[9] GARUDA[10] and EchoGRID[11]). However, due to their complexity, awkwardness and low service quality, new computing environments (e.g. clouds), which foster research productivity, multidisciplinary collaborations, evolution and easy-to-use tools, are being explored (see, for instance, XSEDE[12]). A sustainable future will most probably include both grids and clouds where the driving forces are believed to be similar to those of electrical energy [11]; that is, computing will be globally managed by a mixture of local clusters and large remote facilities that can be configured to provide integrated virtual solutions to the exact requirements of end users.

18.3 Virtual Databases

Most databases containing scientific information, e.g. atomic databases,[13] have been developed by individuals or research groups using very diverse data models. Some

---

[7] https://www.teragrid.org/.

[8] http://www.egi.eu/.

[9] http://www.gisela-grid.eu/.

[10] http://www.garudaindia.in/.

[11] http://www.echogrid.ercim.eu/.

[12] https://www.xsede.org/.

[13] http://plasma-gate.weizmann.ac.il/directories/databases/.

employ relational database management systems, but many rely on flat files or simple *ad hoc* database engines. As a result of the current data explosion, such heterogeneous scientific databases are bound to proliferate causing havoc in prospective users. Endemic problems include lack of interoperability, a variety of query schemes, informal data provenance and peer-to-peer data exchanges, web user interfaces with poor functionalities and a general indifference to data preservation.

A *virtual database*, also referred to as a federated database, is a way of integrating a set of distributed, heterogeneous databases such that they can be queried by a user as a single entity. Each component keeps its geographical site, independence and functionality, but the user can navigate seamlessly across them. When large volumes of data are involved, it is impractical to impose a universal data model to all sites or to maintain a common web portal (see, for instance, GENIE[14]), and therefore, a middleware layer must be developed in order to achieve the desired level of integration.

Following the success of the XML-based Open Document Format OASIS Standard,[15] a promising approach is to implement a mapping layer with an XML schema independent of the storage models which can also be employed to normalize data exchanges. In the case of atomic, molecules and solids, a multi-institutional group under the auspices of the International Atomic Energy Agency (IAEA) has proposed the XSAMS schema [12] which is described in Sect. 18.4.

As previously discussed [13], the management of XML-based mapping layers presents some difficulties regarding data storage, access and publishing. Although many databases are implemented with efficient, unstructured relational models that are manipulated by means of the well-known SQL query language, data are now being published for exchange purposes in XML (see Fig. 18.2). Conversely, hierarchically structured XML documents must be mapped to a relational model for storage purposes while XML-based queries (e.g. XQuery) are yet to be standardized and optimized, apart from being practically unknown in most scientific circles. This back-and-forth translational process can be complex, difficult to maintain and may hamper data flow in high throughput setups. For such reasons hybrid relational/XML databases are being designed [14], but they are yet to become the option of choice.

18.4 XSAMS Schema

XSAMS[16] is a standard XML schema for the exchange of atomic, molecular and particle–solid interaction data. It was first proposed at an IAEA Data Center Network meeting in October 2003, and it has been developed since by a group from

---

[14]http://www-amdis.iaea.org/GENIE/.

[15]http://opendocument.xml.org/.

[16]http://www-amdis.iaea.org/xsams/.

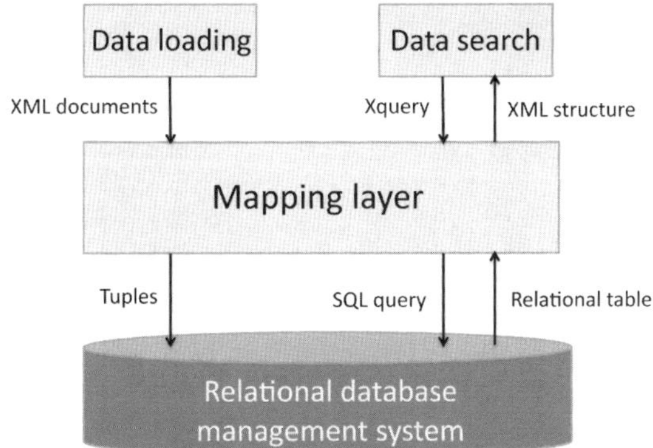

Fig. 18.2 Storing, querying and retrieving XML data sets in a relational database. Figure adapted from Fig. 3 of Ref. [13]

several international scientific institutions. Version v0.1.1 was released in January 2011.

In XSAMS, particular emphasis is given to data provenance and the methods used; physical processes are specified in terms of states in different levels of detail and coupling schemes; transitions between states are itemized by process type and by reference to the initial and final states; and numerical data for a process, e.g. a cross section, can be either tabulated or parameterized. For instance, the structure to denote neutral atomic hydrogen in XSAMS is

```
<ChemicalElement>
<NuclearCharge>1</NuclearCharge>
<ElementSymbol>H</ElementSymbol>
</ChemicalElement>
<IonCharge>0</IonCharge>
<IsoelectronicSequence>H</IsoelectronicSequence>
```

where its ionic charge (0) and isoelectronic sequence (H) are also included. In Appendix A we give the structure to identify the hydrogen 1s ^{2}S$_{1/2}$ ground state (labeled S.0101.001) which lists several attributes such as its energy in cm^{-1}. In Appendix B, we show the structure for the attributes (wavelength and A-value) of the radiative transition between the 2p ^{2}P$^o_{1/2}$ and 1s ^{2}S$_{1/2}$ states in hydrogen, identified respectively with the labels S.0101.002 and S.0101.001.

On the other hand, the working data model of the VAMDC relies on a modified version (v0.2) of XSAMS referred to as VAMDC-XSAMS[17] which introduces some improvements and extensions. Considerable work has been carried out on the molecular part while an independent schema is being introduced for solid spectroscopy whereby interoperability is achieved at the keyword level. For example, the `MolecularState` structure has been replaced with a case-by-case representation, and new structures for line broadening/shifting environments and absorption cross sections have also been added.

---

[17]http://www.vamdc.eu/documents/standards/dataModel/vamdcxsams/.

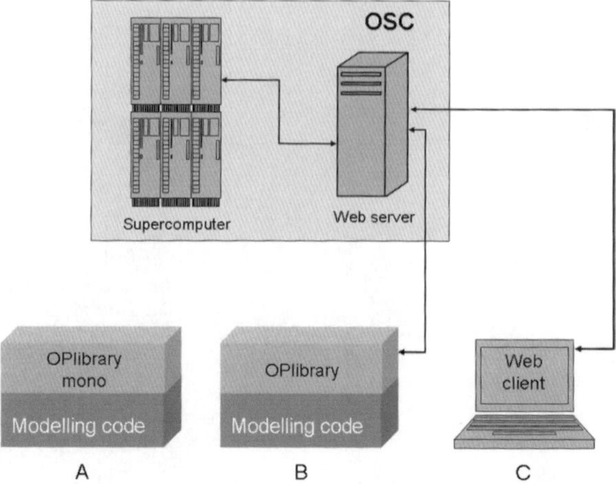

Fig. 18.3 `OPserver` enterprise showing the web-server–supercomputer tandem at the Ohio Supercomputer Center (OSC) and the three available data access modes. Reproduced from Fig. 2 of Ref. [5] with permission of ©John Wiley & Sons, Inc.

18.5 SaaS and DBaaS

With respect to database web services, we show in Fig. 18.3 the three modes for obtaining derived data (astrophysical opacities) from the `OPserver` [5] located at the Ohio Supercomputer Center (OSC), USA. In mode A, the complete database (`mono`) and API (`OPlibrary`) are downloaded locally and interfaced with the user's modeling code. In mode B, only the API is downloaded, and views of the database are accessed remotely from the central facility upon request by the modeling code at runtime. Finally, data sets can also be selected from the OSC via a web browser in mode C. In spite of all the advances that are being brought about by e-science, we expect that these three data access modes will remain functional, but mode B is definitely the most suitable for implementing services involving data or database-centric applications in the distributed environments pertaining grids and clouds.

Therefore, channels for accessing software as a service (SaaS) and databases as a service (DBaaS) are extremely important in cloud database-centric computing, particularly in collaborative networks, in sharing data repositories and in relation to efficient delivery models for scientific applications. In Fig. 18.4, for instance, we show the typical invocation procedure in a SOAP (Simple Object Access Protocol) web service, which can be used to wrap command-line `Fortran` applications such as those popular in laboratory astrophysics. It may be appreciated that communication in its client–server architecture is XML-based (Web Services Description Language): the client requests a service from a registry (server A) and is directed to a service provider (server B); an exact description of the service is then requested from the latter before it is invoked with one of its native SOAP procedures; and, finally, the output is received from server B.

In this convenient SaaS approach, the application is centrally hosted with a single configuration that facilitates software maintenance and delivery, avoiding recurrent software upgrading by users. Furthermore, once the application or database is deployed as a web service, it allows programmatic access from a remote computer

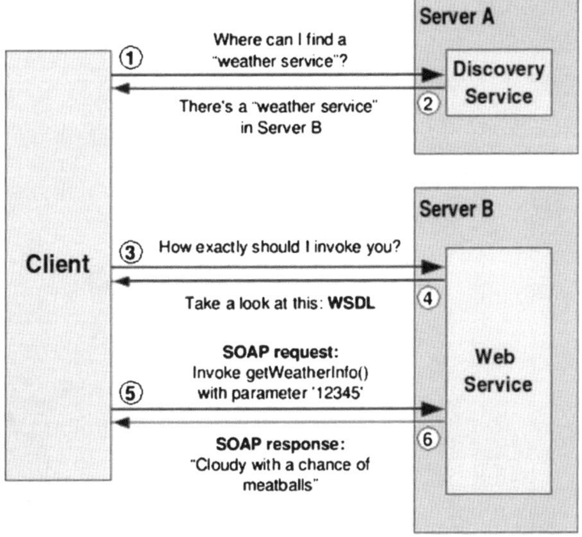

Fig. 18.4 A typical web service invocation scheme. Reproduced from Fig. 1.5 of the Globus Toolkit 4 Programmer's Tutorial

with different types of clients: a web page, a command-line client, a `Python` script or a `Taverna` workflow.

18.6 Service Integration

Since computing—whether infrastructure, platform, software or data repositories—will be mainly offered as a service, the issue of service integration is of importance to the end user. In a similar fashion to the virtual database (see Sect. 18.3), service integration is being managed by virtual integrators rather than by web portals. Thus, software tools and data services, such as the web services described in Sect. 18.5, are becoming available in registered repositories from where they can be embedded in the end-user's workflows [15, 16], the latter then being managed by scripts or workflow tools (see Fig. 18.5).

This approach suggests, on the one hand, that high-performance computing can be carried out by a scientist from a relatively modest frontend, and that scripting, as a service integrator, will be exploited to the extreme in software development, prototyping and production. On the other, since such scripts and workflows can be run from anywhere, they are bound to become the blueprints of a research process by integrating every step: collaborations, verification, replication, extension, refereeing and publication. Moreover, taking into consideration the potential of social network systems in scientific research, scripts and workflows are becoming the exchange items of choice within research communities. This is indeed the principle behind the `myExperiment` scientific social network environment which promotes collaborations between groups of researchers based on workflow interchange and the monitoring of the ensuing social transactions and developments.

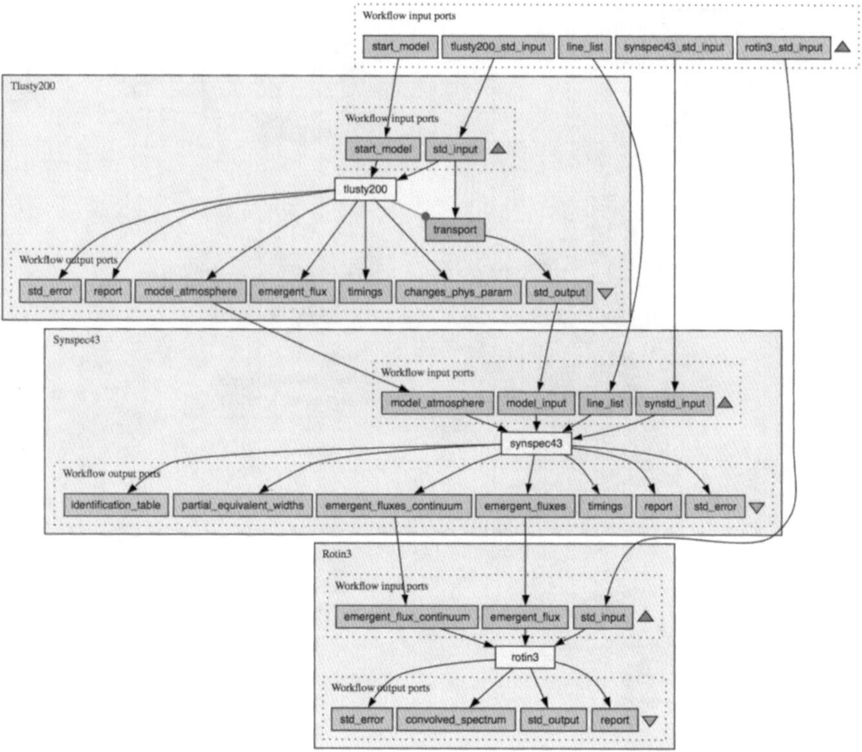

Fig. 18.5 `Taverna` workflow for the suite of web services `tlusty200-synspec43-rotin33` [17–19]. Input files can be local files, URLs or from previous stages of the workflow

18.7 Data Curation

Data curation concerns the management of data from the moment they are generated and throughout their lifecycle validating their accuracy, completeness and physical and logical integrity. It promotes and facilitates data access, publication and use, and guarantees their maintenance and preservation for reuse in future research. For data-intensive projects to prosper, data openness, trustworthiness and service quality are essential, and thus, data curation is rapidly becoming a key strategic field in its own right.

In the modified research cycle brought about by e-science, a new data curation model appears as an integral part of the cycle in addition to those of data creation and publication, where the traditional data archives rapidly evolve to become active, interoperable data repositories that can be mined by different disciplines to generate knowledge. In this context, important issues are then data provenance through extensive metadata [20] and data conceptualization through discipline specific ontologies [21], both representing quite a challenge and effort to basic scientists.

An ontology is essentially a semantic model with the terms and concepts associated with a research community which allows intelligent querying and the inte-

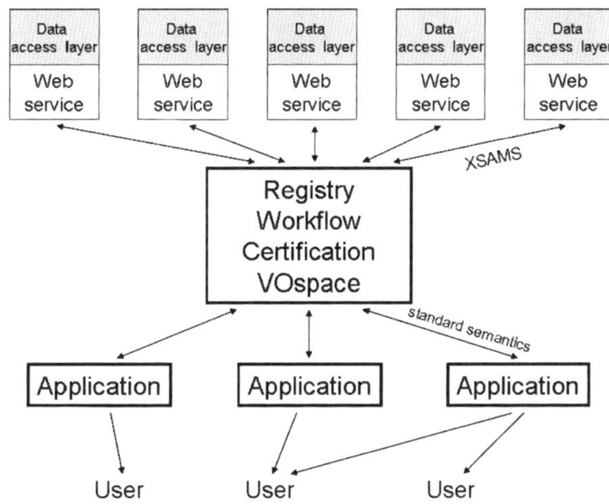

Fig. 18.6 Distributed infrastructure of the VAMDC cyber-warehouse

gration of heterogeneous information in data mining. Ready implementation and acceptance of ontologies have an impact on the processes related to knowledge inferring from data and to publishing papers linked to voluminous online resources [22].

18.8 VAMDC

The VAMDC project was launched in July 2009 by several countries of the European Union (Austria, France, Germany, Italy, Sweden and the United Kingdom), Russia, Serbia and Venezuela to upscale and integrate several (more than 21) A&M databases in a virtual interoperable cyber-warehouse. It embraces 24 research teams with ample expertise in the production and applications of A&M data in different user communities and in e-infrastructure development for data-intensive environments. VAMDC is basically a sophisticated virtual database such as those described in Sect. 18.3. Furthermore, it is also an on-going project concerned with the definitions of XML schemata, protocols and standards; the deployment of registries, database-centric web services and data-access layers within the context of the IVOA (see Fig. 18.6); and with the organization of networking activities to consolidate the communities of A&M data producers, users and database providers. Standards and software versions are released periodically in order to facilitate user upgrades.

18.8.1 VAMDC Node

The general structure of a VAMDC node is shown in Fig. 18.7. A node provides a data service using the standards and protocols prescribed by the VAMDC, namely

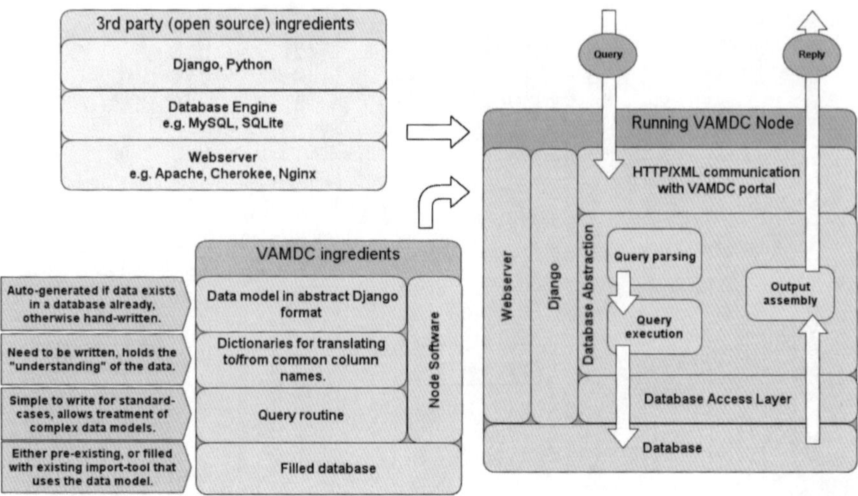

Fig. 18.7 Structure of a VAMDC node

RESTful HTTP services. The node software (v11.5rl[18]) is open source and the data sets must reside in a relational database, the node being flexible regarding data structures in as much as it can be installed coping with its arbitrary internal data formats. Moreover, the VAMDC protocols and standards do not depend on the requirements of a specific node. The *data model* contains the database layout in terms of $\mathtt{Python}$ classes ($\mathtt{Java}$ classes may also be employed due to open standards), one for each table where the class members are fields that represent the table columns. This code representation of the database allows the automatic recreation of the database independently of its engine. The VAMDC nodes that can be queried are tabulated in a central web service referred to as the *registry*.

The *dictionary*[19] lists a set of names that identifies a data type, each name being unique and uniquely linked to a description, a data type, a unit (when applicable) and restriction (if necessary). Each node also contains: *returnables* associated with its data deliverables that map the global VAMDC names to those specific to each node; *restrictables* which are key–value pairs, the keys being the global names and the values the corresponding place in the data model that express query constraints; and *requestables* for selecting subsets of the database.

---

[18] http://www.vamdc.org/documents/nodesoftware/.

[19] http://dictionary.vamdc.org/.

18.8.2 VAMDC-TAP Services

VAMDC nodes offer their data through a TAP-like interface where TAP[20] stands for the IVOA *Table Access Protocol*. In fact, VAMDC only uses a subset of the TAP standard referred to as VAMDC-TAP (v11.05[21]) which provides a common table model complying with the VAMDC-XSAMS (v0.2) schema and accepts queries in a restricted form of SQL (see Sect. 18.8.3 below). VAMDC-TAP generates virtual data which may be then retrieved via a semi-permanent URL that contains the results of a query, i.e. the URL is built from a combination of the data selection criteria of the query and the address of the archived data set to which it is sent. VAMDC-TAP provides a standard view of the relational database by means of a synchronous query, and provides output data in VAMDC-XSAMS format.

18.8.3 VSS1/VSS2 Query Languages

The query language is known as the VAMDC SQL Subset 1 (VSS1[22]), and uses an SQL-like string that does not require knowledge of the structure of the node database being queried, i.e. all nodes understand the same queries. For example, a VSS1 query could be

```
SELECT ALL WHERE AtomNuclearCharge = 25 AND AtomIonCharge < 2
```

A VSS1 query satisfies the syntactic rules of the SQL92 language, it is basically a `SELECT` statement and must not contain the `JOIN` keyword. Since it does not alter the database it is addressing, the SQL92 keywords `ALTER`, `CREATE`, `DELETE`, `DROP`, `INSERT`, `REPLACE` and `UPDATE` must also be avoided.

Many of the limitations of VSS1 are being addressed by an improved VSS2 query language to be published in September 2011. It allows more powerful searches and adds flexibility in terms of multi-species process selection and data output restriction, defining the desired branches of VAMDC-XSAMS to be returned.

18.8.4 Application Web Services

In VAMDC we have set up experimental registries of web services with two SOAP servers: Astrogrid Common Execution Architecture (CEA[23]) and Soaplab2.[24] They

---

[20]http://www.ivoa.net/Documents/TAP/.

[21]http://vamdc.org/documents/standards/dataAccessProtocol/.

[22]http://vamdc.org/documents/standards/queryLanguage/.

[23]http://www.astrogrid.org/maven/docs/HEAD/applications/.

[24]http://soaplab.sourceforge.net/soaplab2/.

offer A&M database-centric astrophysical applications such as xstar [23], the
suite of codes tlusty200-synspec43-rotin33 [17–19], OPserver [5]
and the VALD[25] (the Vienna Atomic Line Data Base) tool select3 [24], but
atomic structure codes (e.g. autostructure [25]) have also been considered.
Once the web services are deployed, they can be remotely run from several clients,
e.g. the registry web portal[26] itself, a Taverna workflow and a SUDS[27] Python
script. For instance, in Fig. 18.5 we show the Taverna workflow for running the
web service offering tlusty200-synspec43-rotin33. The input files may
be taken from the user's local disk, URLs or linked from previous stages of the calculation while the output files are always assigned URLs from where they can be
downloaded. In Appendix C a Python script is shown for running asynchronously
the select3 application with a SUDS client. The service is invoked by assigning
its URL to the Client module

```
url = 'http://caoba.ivic.gob.ve:8180/soaplab2-axis/services/\
      testing.select3?wsdl'
client = Client(url)
```

Details about the service native procedures and specs may be obtained with the
command

```
print client
```

and, after preparing the input file, it is run with the asynchronous command

```
jobid = client.service.createAndRun(input)
```

The job may be monitored with a loop of the type

```
while (client.service.getStatus(jobid) == 'RUNNING'):
   pass
print client.service.getStatus(jobid)
```

and, once it is finished, the output files and/or their URLs can be downloaded with
the command

```
result = client.service.getResults(jobid)
```

The power behind this simple script within the context of data-intensive science
is that the user can submit simultaneously a large number of jobs to the remote web
service, which are then monitored or retrieved by means of the jobid identifier.
This suggests that a web-service portfolio must rely on a fairly capable multipro-
cessor infrastructure, a key aspect which is currently being negotiated by VAMDC
with several supercomputer centers since the grid does not offer at present reliable
service quality. Furthermore, if large data transfers are involved between the application and the database, the web service must then be housed as close as possible to
the latter in order to reduce network bottlenecks.

---

[25] http://www.astro.uu.se/~vald/php/vald.php.

[26] http://caoba.ivic.gob.ve:8180/soaplab2-axis/.

[27] https://fedorahosted.org/suds/.

18.9 Concluding Remarks

We have tried to convey the changing world of A&M data dissemination—that of
data-intensive science (e-science) where recent large-scale initiatives such as the
VAMDC are dictating the future direction of the field. We are of the opinion that
high-performance computing, application deployment and data storage, management and publication have entered a new age, and appear to be ready to move on
to the cloud. Therefore, integration of heterogeneous databases and application services by means of virtual databases and workflow tools are of key importance in
order to sustain the voracious data needs of the virtual research communities. For-
eign concepts to A&M database pioneers such as XML schemata, TAP protocols,
metadata and ontologies, to mention a few, are beginning to dominate the field causing some distress in how to make life easy for a user with a simple data request. We
are now in the hands of savvy computer engineers who feel at home in the virtual distributed environments of a second generation Internet. We just hope that the
original A&M data producers can live up to it.

Acknowledgements   Most of this paper was written at Woodcut, Richmond, UK, where the hos-
pitality of Sasa Marinkov and Michael Jones was greatly appreciated. VAMDC is funded under the
'Combination of Collaborative Projects and Coordination and Support Actions' Funding Scheme
of The Seventh Framework Program. Call topic: INFRA-2008-1.2.2 Scientific Data Infrastructure.
Grant Agreement number: 239108. The Venezuelan node has been supported by Corporación Parque Tecnológico de Mérida, by EELA-2 (E-science grid facility for Europe and Latin America)
and GISELA (Grid Initiatives for e-Science virtual communities in Europe and Latin America).

Appendix A

XSAMS structure for the $1s\,^2S_{1/2}$ ground state of neutral hydrogen.

```
<AtomicState stateID="S.0101.001">
<AtomicNumericalData>
<StateEnergy><Value units="1/cm">0.0000000E+00</Value>
</StateEnergy>
</AtomicNumericalData>
<AtomicQuantumNumbers>
<Parity>even</Parity>
<TotalAngularMomentum>0.5</TotalAngularMomentum>
</AtomicQuantumNumbers>
<AtomicComposition>
<Component>
<Configuration>
<Shells>
<Shell>
<PrincipalQuantumNumber>1</PrincipalQuantumNumber>
<OrbitalAngularMomentum><Value>0</Value>
</OrbitalAngularMomentum>
<NumberOfElectrons>1</NumberOfElectrons>
</Shell>
</Shells>
```

```
<ConfigurationLabel>1s_1/2</ConfigurationLabel>
</Configuration>
<Term>
<LS>
<L><Value>0</Value></L>
<S>0.5</S>
<Multiplicity>2</Multiplicity>
</LS>
</Term>
</Component>
</AtomicComposition>
</AtomicState>
```

Appendix B

XSAMS structure for the transition $2p\ ^2P^o_{1/2} \rightarrow 1s\ ^2S_{1/2}$ in hydrogen.

```
<RadiativeTransition>
<EnergyWavelength>
<Wavelength>
<Theoretical><Value units="nm">1.215674E+02</Value>
</Theoretical>
</Wavelength>
</EnergyWavelength>
<InitialStateRef>S.0101.002</InitialStateRef>
<FinalStateRef>S.0101.001</FinalStateRef>
<Probability>
<TransitionProbabilityA><Value units="1/s">6.2684E+08</Value>
</TransitionProbabilityA>
</Probability>
</RadiativeTransition>
```

Appendix C

Python script for running the `select3` Soaplab2 web service with a SUDS client.
`select3` is a tool associated with the VALD[28] atomic database [24].

```python
#!/usr/bin/env python
import os, shutil, urllib
from suds.client import Client
# Invoke the web service
url = 'http://caoba.ivic.gob.ve:8180/soaplab2-axis/services/\
      testing.select3?wsdl'
client = Client(url)
print client
# Prepare the input file
```

---

[28]http://www.astro.uu.se/~vald/php/vald.php.

```python
input_content  = \
 " 5162.5,6180.5,0.01,2.0,\n\
 'marcs5000g50t02z0.krz'\n\
 'Fe:-4.51'\n\
 'END',\n\
 'Synth',"
input = client.factory.create('ns3:Map')
input.item = {'key':'select3_input_direct_data',\
          'value':input_content}
# Run a job with the createAndRun utility
jobid = client.service.createAndRun(input)
while (client.service.getStatus(jobid) == 'RUNNING'):
   pass
print client.service.getStatus(jobid)
# Make a directory with jobid and move into it
os.mkdir(jobid)
os.chdir(jobid)
# Get files report, std_errors (empty if no errors)
result = client.service.getResults(jobid)
# Obtain output files: report, detailed_status, OUTPUT,\
          OUTPUT_url
for item in result.item:
  if '_url' in item.key:            # get remote file
    lfile = item.key.replace('_url','')
    rfile = item.value
    urllib.urlretrieve(rfile,lfile)
  else:                             # write output and error
    fout = open(item.key,'w')
    fout.write(item.value)
    fout.close()
```

References

1. The Opacity Project Team, *The Opacity Project*, vol. 1 (Institute of Physics Publications, Bristol, 1995)
2. D.G. Hummer, K.A. Berrington, W. Eissner, A.K. Pradhan, H.E. Saraph, J.A. Tully, Atomic data for the IRON Project. I. Goals and methods. Astron. Astrophys. **279**, 298–309 (1993)
3. W. Cunto, C. Mendoza, The opacity project—the TOPbase atomic database. Rev. Mex. Astron. Astrofís. **23**, 107–118 (1992)
4. W. Cunto, C. Mendoza, F. Ochsenbein, C.J. Zeippen, TOPbase at the CDS. Astron. Astrophys. **275**, L5–L8 (1993)
5. C. Mendoza, M.J. Seaton, P. Buerger et al., OPserver: interactive online computations of opacities and radiative accelerations. Mon. Not. R. Astron. Soc. **378**, 1031–1035 (2007)
6. T. Hey, A.E. Trefethen, Cyberinfrastructure for e-Science. Science **308**, 817–821 (2005)
7. C. Mendoza, L.A. Núñez, Virtual atomic and molecular data centre, in *Proceedings of the First EELA-2 Conference*, ed. by R. Mayo, H. Hoeger, H.E. Castro, L.N. Ciuffo, R. Barbera, I. Dutra, P. Gavillet, B. Marechal (CIEMAT, Madrid, 2009), pp. 219–225
8. M.L. Dubernet, V. Boudon, J.L. Culhane et al., Virtual atomic and molecular data centre. J. Quant. Spectrosc. Radiat. Transf. **111**, 2151–2159 (2010)
9. G. Bell, T. Hey, A. Szalay, Beyond the data deluge. Science **323**, 1297–1298 (2009)
10. T. Hey, S. Tansley, K. Tolle (eds.), *The Fourth Paradigm: Data-Intensive Scientific Discovery* (Microsoft Research, Redmond, 2009)

11. I. Foster, Y. Zhao, I. Raicu, S. Lu, Cloud computing and grid computing 360-degree compared, in *Grid Computing Environments Workshop (GCE'08)* (2008), pp. 60–69. doi:10.1109/GCE.2008.4738445
12. B.J. Braams (ed.), *XSAMS: XML Schema for Atoms, Molecules and Solids, INDC(NDS)-0570* (International Atomic Energy Agency, Vienna, 2010)
13. J. Freire, M. Benedikt, Managing XML data: an abridged overview. Comput. Sci. Eng. **6**, 12–19 (2004)
14. K. Beyer, R. Cochrane, M. Hvizdos et al., DB2 goes hybrid: integrating native XML and XQuery with relational data and SQL. IBM Syst. J. **45**, 271–298 (2006)
15. S. Shankar, A. Kini, D.J. DeWitt, J. Naughton, Integrating databases and workflow systems. ACM SIGMOD Rec. **34**, 5–11 (2005)
16. C. Goble, D. De Roure, The impact of workflow tools on data-centric research, in *The Fourth Paradigm: Data-Intensive Scientific Discovery*, ed. by T. Hey, S. Tansley, K. Tolle (Microsoft Research, Redmond, 2009), pp. 137–145
17. I. Hubeny, A computer program for calculating non-LTE model stellar atmospheres. Comput. Phys. Commun. **52**, 103–132 (1988)
18. I. Hubeny, T. Lanz, Non-LTE line-blanketed model atmospheres of hot stars. I. Hybrid complete linearization/accelerated lambda iteration method. Astrophys. J. **439**, 875–904 (1995)
19. T. Lanz, I. Hubeny, Atomic data in non-LTE model stellar atmospheres, in *Stellar Atmosphere Modeling*, ed. by I. Hubeny, D. Mihalas, K. Werner. ASP Conf. Ser., vol. 288, (2003), pp. 117–129
20. Y.L. Simmhan, B. Plale, D. Gannon, A survey of data provenance in e-science. ACM SIGMOD Rec. **34**, 31–36 (2005)
21. M. Cannataro, C. Comito, A data mining ontology for grid programming, in *Proceedings of the 1st International Workshop on Semantics in Peer-to-Peer and Grid Computing, in Conjunction with WWW2003*, (2003), pp. 113–134
22. J. Blake, Bio-ontologies—fast and furious. Nat. Biotechnol. **22**, 773–774 (2004)
23. M.A. Bautista, T.R. Kallman, The XSTAR atomic database. Astrophys. J. Suppl. Ser. **134**, 139–149 (2001)
24. N.E. Piskunov, F. Kupka, T.A. Ryabchikova, W.W. Weiss, C.S. Jeffery, VALD: the Vienna atomic line data base. Astron. Astrophys. Suppl. Ser. **112**, 525–535 (1995)
25. N.R. Badnell, A Breit–Pauli distorted wave implementation for AUTOSTRUCTURE. Comput. Phys. Commun. **182**, 1528–1535 (2011)

Chapter 19
Seismology of the Sun: Inference of Thermal, Dynamic and Magnetic Field Structures of the Interior

K.M. Hiremath

Abstract Recent overwhelming evidences show that the sun strongly influences the Earth's climate and environment. Moreover existence of life on this Earth mainly depends upon the sun's energy. Hence, understanding of physics of the sun, especially the thermal, dynamic and magnetic field structures of its interior, is very important. Recently, from the ground and space based observations, it is discovered that sun oscillates near 5 min periodicity in millions of modes. This discovery heralded a new era in solar physics and a separate branch called *helioseismology* or *seismology of the sun* has started. Before the advent of *helioseismology*, sun's thermal structure of the interior was understood from the evolutionary solution of stellar structure equations that mimicked the present age, mass and radius of the sun. Whereas solution of MHD equations yielded internal dynamics and magnetic field structure of the sun's interior. In this presentation, I review the thermal, dynamic and magnetic field structures of the sun's interior as inferred by the *helioseismology*.

19.1 Introduction

From the dawn of civilization, sun is revered and held as an awe inspiring celestial object in the sky. In the world, there are many stories and poems woven around the sun god in different folklores and the magnificent architectures are dedicated to the mighty sun. The flora and fauna on the earth mainly depends on the sun for their survival. Recent observational evidences are building up that even the unpredictable climates and rainfalls of the earth are excited and are maintained by the sun's influence ([67, 101, 119] and references there in; [44, 53, 58, 60, 94, 120, 125] and references there in; [1, 62, 105]). Analysis of vast stretch in time of the paleoclimatic records show that the sun's activity is imprinted in the global temperature and precipitation variabilities. The imminent influence ([67] and references there in; [62] and references there in) of the sun also can be traced in the well documented instrumental rainfall records.

K.M. Hiremath (✉)
Indian Institute of Astrophysics, Bengaluru 560034, India
e-mail: hiremath@iiap.res.in

M. Mohan (ed.), *New Trends in Atomic and Molecular Physics*,
Springer Series on Atomic, Optical, and Plasma Physics 76,
DOI 10.1007/978-3-642-38167-6_19, © Springer-Verlag Berlin Heidelberg 2013

Among all the distant stars, the sun is very closest to us and is apparently $\sim$100 billion times brighter than any other star. Thus owing to it's proximity and brilliance, the sun can be closely studied in details; like it's chemical composition, thermal and dynamic structures of the interior, etc. The sun is a gigantic cosmic laboratory where one can test the physical phenomena discovered on the earth. Hence the sun is very important astronomical object that needs to be studied carefully. If we understand the sun, we can understand the distant stars and physics of whole universe.

In the recent decade, helioseismology—a tool to probe the solar internal structure and dynamics—has changed our conventional perception of the internal structure and dynamics. Observational evidences show that sun oscillates over a wide range of periods that range from minutes to perhaps on time scales of centuries. Observations from the ground (GONG, BISON, IRIS, etc.) and from the space (like SOHO, HINODE, STEREO, etc.) provided solar oscillation frequency data with very high precision.

Before the advent of helioseismology, sun's internal thermal structure was understood from the *standard solar model*. A standard solar model is built from the evolutionary calculations of stellar structure equations (conservation of mass, momentum and energy and supplemented with known equation of state, knowledge of transfer of energy by convection in the outer part of the sun, equation of nuclear energy generation and nature of opacity of the solar internal plasma) that ultimately mimic the sun's present age, mass and radius respectively. As for the dynamical structure, earlier studies concentrated on studying the rotation rate of the solar core as this parameter is related to evolutionary history of angular momentum of the sun and its influence on the orbit of nearby planet Mercury for testing the Einstein's general theory of relativity.

Similarly, models based on the turbulent $\alpha\omega$ dynamo mechanism required that radial part of the angular velocity should increase from the surface towards the interior in order to reproduce the correct sunspot butterfly diagrams. On the other hand, results of non-linear hydrodynamical simulations [45] show that angular velocity decreases from the surface towards the interior. These simulations also yield internal rotational isocontours that have cylindrical geometry. However, it will be known from the following sections that the helioseismic inferences yield entirely different picture of the rotational profile in the solar interior. We will also know from the helioseismic inferences and modeling that the thermal structure (such as pressure, temperature, etc.) of the solar interior is almost similar to the structure obtained from the *standard solar model*, although recent surface chemical abundances and hence helioseismic inferences yield the substantial disagreement.

Plan of this presentation is as follows. In Sect. 19.2, a brief introduction to the observational aspects of the sun is presented. Recent discoveries of the sun's oscillations from the ground and space are presented in Sect. 19.3. Section 19.4 is devoted to inferences of thermal, dynamic and magnetic field structures of the sun's interior. Concluding remarks are presented in the last section. In this presentation, I mainly concentrate on the *global seismology* to infer global large scale structure of the sun's interior although one can also infer near surface structure from the *local helioseismology*.

19.2 A Brief Observational Introduction to the Sun

Important physical parameters of the sun are: (i) mass—2×10^{30} Kg, (ii) radius—7×10^8 meter, (iii) mean density—1409 $Kg\,m^{-3}$, (iv) temperature at the photosphere—5780 K and, (v) the total amount of energy radiated by the sun (i.e., luminosity) measured at one astronomical unit (i.e., distance between the earth and the sun)—3.9×10^{26} joules/sec.

When one considers the vertical cross section of the sun parallel to it's rotation axis, based on the dynamical and physical properties, solar interior mainly can be classified into three distinct zones: (i) *the radiative core* where the energy is mainly generated by the nuclear fusion of hydrogen into helium and is transfered by the radiation, (ii) *the convection zone* where the energy is transferred from the center to the surface by the convection of the plasma and, (iii) *the photosphere* where the energy is radiated into the space. Above the photosphere, the sun's atmosphere consists of the *chromosphere* and the *corona*. The temperature increases from the layer of photosphere to million degree kelvin in the corona.

If the sun were static in time with a constant output of energy, the planetary environments in general and the earth's environment in particular would have received the constant output of energy. However, the sun's energy output is variable due to spatial and temporal variability of the sun's large scale magnetic field structure, dynamics and flow of mass (both the neutral and charged particles) from the sun. The most outstanding activity of the sun is *sunspots*—cool and dark features compared to the ambient medium-on the sun's surface that modulate sun's irradiance and the galactic cosmic rays that enter the planetary environments. Variation of the occurrence of number of sunspots over the surface of the sun with an average periodicity of $\sim$11 years is termed as sunspot or solar cycle. As the sunspots are bipolar magnetic field regions with positive and negative polarities, every 11 years polarity changes and, by 22 years, sunspots reverse back to their original polarities. This constitutes the 22 year sun's magnetic cycle.

The *flares* that are mainly associated with the sunspots [59] release the vast amount of energy (upto 10^{25} Joules) with in a short span of time. The sun also ejects sporadically the mass ($\sim10^{15}$ gm) of plasma to the space and such an event is called *coronal mass ejection*. There is also a continuous flow of wind ($\sim10^{31}$ charged particles per second or 6–7 billion tons per hour) from the sun towards the space called the *solar wind*.

The sun's activity varies on time scales of few minutes to months, years to decades and perhaps more than centuries. The solar different activity indices vary on the time scales of $\sim$27 days due to solar rotation and $\sim$150 days due to the flares. Unlike the Earth, sun rotates differentially, rotating faster near the equator compared to the regions near the poles. The 1.3 year periodicity is predominant in different solar activity indices. The next prominent and ubiquitous *viz.*, 11 year solar cycle periodicity is found not only in the present day sun's activity indices but also in the past evolutionary history derived from the solar proxies (such as C^{14} and Be^{10} respectively).

Occurrence of transient events, such as either flares or coronal mass ejections, of the solar activity that are directed towards the earth create havoc in the earth's atmosphere by disrupting the global communication, reducing life time of the low-earth-orbit satellites and, cause electric power outages. Owing to sun's immense influence of space weather effects on the earth's environment and climate, it is necessary to predict [61] and know in advance different physical parameters such as amplitude and period of the future solar cycles.

19.3 Observations of the Solar Oscillations

In a gravitationally stratified and compressible medium such as sun and with vigorous convective activity in the convective envelope, different types of oscillation modes can be excited. For example, noise due to turbulence, especially in the convective envelope, perturb [46] gradients of pressure and excite the sound waves or *p modes* that alternatively consist of compressions and rarefactions and, travel with the velocity of local sound speed. Where as perturbations of the stable radiative core excite *gravity or g* modes and are due to vertical displacement of a parcel of fluid from its normal position. In principle, these two types of waves can be observed on the surface of the sun through surface displacement of velocities (by Doppler shifts of the spectrum) or intensity (temperature) variations. Oscillations of the sun's photosphere were first discovered in the early 1960s [79, 80]. Doppler shifts of the spectral lines formed in the atmosphere revealed the period of oscillations of about five minutes. Later it was discovered [35] from theoretical investigations [78, 128] that these oscillations are due to superposition of large number of individual modes each having its own characteristics frequency. Surface observations indicate the velocity fluctuations are $\sim$1000 m/sec. Where as amplitude of individual modes ranges from 0.3 m/sec to 0.01 m/sec (Fig. 19.1). Some of the groups [72, 106, 107, 111] reported the oscillations that may represent internal gravity modes. Observations of whole disk Doppler velocities revealed an oscillation with a period 160.01 minute that is believed to be a *g* mode oscillation of the sun. Thus new era in the history of solar physics *viz.*, seismology of the sun or *helioseismology* (as this term is initially coined by Severny et al. [112]) has started that is yielding rich dividends on physics of the sun's interior.

19.3.1 Surface Patterns of the Global Oscillation Modes

In spherical coordinates, one can express velocity V of oscillation in a specific normal mode as the real part of

$$V(r, \vartheta, \phi, t) = V_n(r) Y_l^m(\vartheta, \phi) e^{-2\pi \nu t}, \tag{19.1}$$

where $Y_l^m(\vartheta, \phi) = P_l^m(\cos \vartheta) e^{im\phi}$ is spherical harmonic function of degree l and azimuthal order m, ϑ is co-latitude and ϕ is longitude of the sun.

For each pair of l and m, there is a discrete spectrum of modes with distinct frequency ν. These differ in spatial structures as a function of radius. $V_n(r)$ are the

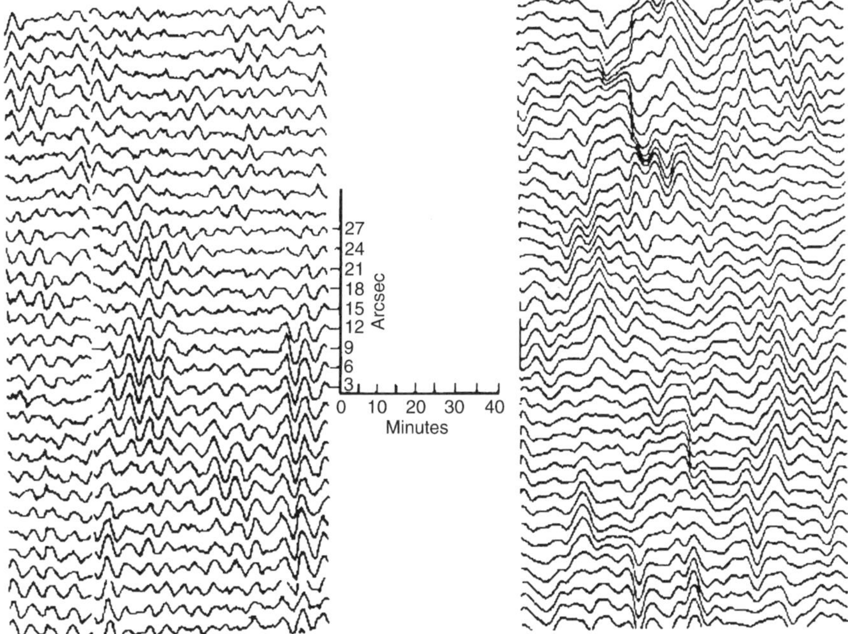

Fig. 19.1 Observations of Doppler velocities on the solar disk, when carried out in one spatial dimension and time, reveal the 5-min oscillations as apparent wave packets that last for about five or six wave periods [90]. *Right figure* illustrates the spatial intensity variations with time [91]

eigen functions which are oscillatory in space and typically possess n zeros or nodes in the radial direction. The degree l can be understood as the number of complete circles on the surface of a star where the normal component of velocity is always zero.

The angular degree l is related to the total horizontal component of the wave number, K_h as

$$K_h = \frac{[l(l+1)]^{1/2}}{R_\odot},\tag{19.2}$$

where $R_\odot$ is the radius of the star. Thus high degree l modes sense the near surface structure and very low degree l modes sense near center of the sun. Thus with different degree modes observed on the surface, one can infer the internal structure of sun. The azimuthal order m, characterizes the orientation of the mode relative to the axis of the spherical coordinate system and has $(2l+1)$ values (i.e., $-l, \ldots, l$). The zero velocity lines are like lines of latitude when $m = 0$ and are like lines of longitude when $m = \pm l$. When $m = 0$, the spherical harmonic is called as *zonal harmonics* and the harmonic with $m = l$ is called *sectoral harmonics*. *Tesseral harmonics* is obtained when $m \neq l$ (see Fig. 19.2).

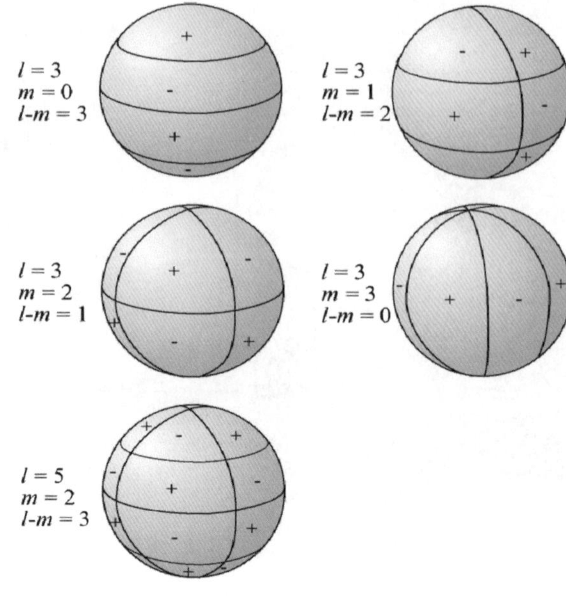

Fig. 19.2 Schematic representation of $Y_{l,m}$ on the unit sphere and its nodal lines. $Y_{l,m}$ is equal to 0 along m great circles passing through the poles, and along $l - m$ circles of equal latitude. The function changes sign each time it crosses one of these lines. This figure is adopted from Wikipedia

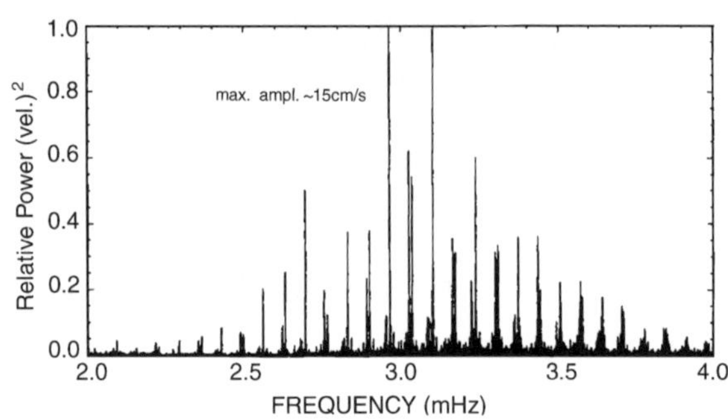

Fig. 19.3 Power spectrum of integrated low degree p modes [36]

19.3.2 Amplitudes, Frequencies and Line Widths of 5 min Oscillations

There are two ways, *viz.*, *unimaged* and *imaged* methods, of measuring the amplitudes and frequencies of 5 min oscillations [56]. In case of *unimaged* method, oscillations are due to low-l modes that are obtained from the integrated whole disk measurements. These modes have longer life times ($\sim$ of months) and hence their frequencies can be accurately measured.

Figure 19.3 illustrates an example of low-l power spectrum obtained from Birmingham integrated whole disk measurements. One can notice from Fig. 19.3 that, significant power lies between the 1–5 mHz. From the spherical Fourier analysis, it is concluded that these observed oscillations are identified as low degree ($l = 0, 1, 2$ and 3) and low radial order ($0 \leq n \leq 10$) acoustic modes that penetrate deeply the solar interior.

Fig. 19.4 Line width (*top*), power per mode (*middle*) and total energy per mode (*bottom*) as a function of frequency of p modes for the modes $l = 20$ are illustrated [81]

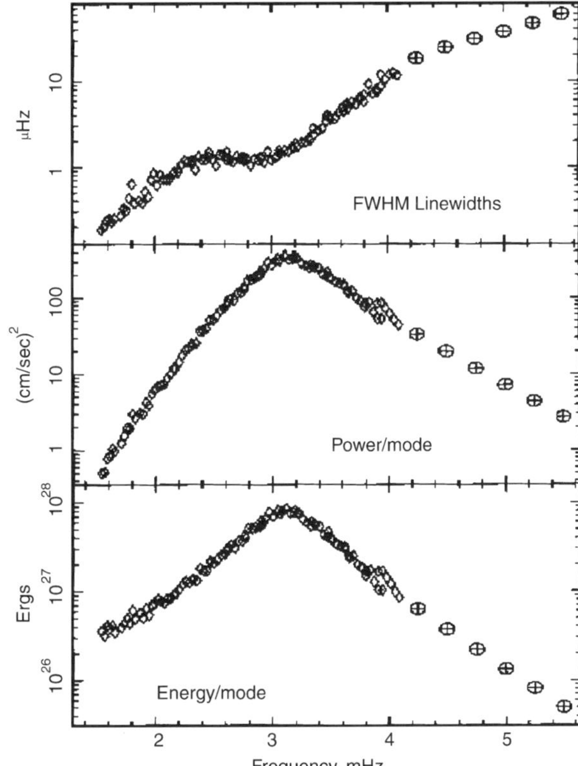

As for the measured amplitudes of oscillations in Doppler velocity, it depends upon the choice of the spectral lines (that occur at different heights) used for sensing the oscillations. For example, in potassium and sodium lines, maximum amplitudes from the power spectra are estimated to be $\sim$15 and 25 cm^{-1} respectively. In case of irradiance, amplitude of oscillations is estimated to be $\sim$2–3 $\times$ 10^{-6}. One can get the tables of amplitudes of oscillations from the observations of Doppler velocity measurements [4, 49, 93].

In case of *imaged* method, amplitudes and frequencies of the solar oscillations are computed from the spatially resolved disk of the sun. With this method, amplitudes and frequencies of many individual modes can be computed. For example, from the information of measured amplitudes, one can get the information of power and energy per mode. For the degree $l = 20$, power and energy per mode of oscillations as a function of frequency are illustrated in Fig. 19.4.

Another useful physical parameter measured from the *nonimaged* and *imaged* observations is the line width of different oscillation modes that provide vital information of excitation and damping mechanism of the solar oscillations. Measurements of line widths suggests that line width increases with both the frequency (ν) and degree (l) of the oscillations. For example, for $l = 20$, in Fig. 19.4, FWHM (Full width at half maximum) of the line widths as a function of frequency is illustrated.

When one compares the observations of line width as a function of frequency and theoretical estimates [33, 46, 75, 76], it is found that growth of the individual modes probably is constrained by the perturbations of the oscillations rather than non-linear interaction among the individual modes.

19.3.3 Observed *l–v* Diagnostic Power Spectrum of the Oscillations

For the *imaged* observations, continuous and long series of oscillation data set is subjected to spherical harmonic Fourier analysis, averaged over different m azimuthal modes that ensures the removal of effect of sun's rotation and $l–v$ diagnostic power spectrum of the solar oscillations is obtained. In Fig. 19.5, such a $l–v$ diagnostic power spectrum is illustrated. In this figure, one can notice that maximum power is concentrated near a frequency of 3 mHz that corresponds to a period of 5 minutes. Another notable interesting characteristic property of this $l–v$ diagnostic diagram is that concentration of power is not distributed randomly, rather it is systematically concentrated in the narrow curved ridges indicating that observed oscillations are due to internal standing waves (confined within the resonant cavity) whose amplitudes vary from central core to near the surface.

19.3.4 Frequency Splittings due to Rotation and Magnetic Field

In the previous section, solar power spectrum is presented by removing the effect of rotation. Presence of rotation, Coriolis force and magnetic field of the sun affect on the dynamics of the oscillations and hence on the frequencies. For example, if the sun is not rotating, non-magnetic and spherically symmetric, then the modes with same n and l, oscillation frequencies would be degenerate in azimuthal order m. However, sun is rotating and magnetic resulting in lifting the degeneracy that makes the frequencies depend upon the azimuthal order m. In case of sun, compared to magnitudes of rotation and magnetic field structures, effects due to Coriolis and centrifugal forces are negligible. From the observed data, rotational frequency splittings are computed as $\delta v_{nlm} = v_{nlm} - \bar{v}_{nl}$, where $\bar{v}_{nl}$ is average frequency for $m = 0$. In Fig. 19.6, typical power spectrum of solar oscillations, for different m with $l = 20$, obtained from the Big Bear Solar Observatory [82] is presented. Clearly one can notice the frequency splittings due to rotation in this illustration.

Traditionally frequency splittings are related to frequency v_{nlm} of a mode as follows

$$v_{nlm} = v_{nl} + \sum_{i=1}^{i_{max}} a_i(nl) P_i^{nl}(m), \tag{19.3}$$

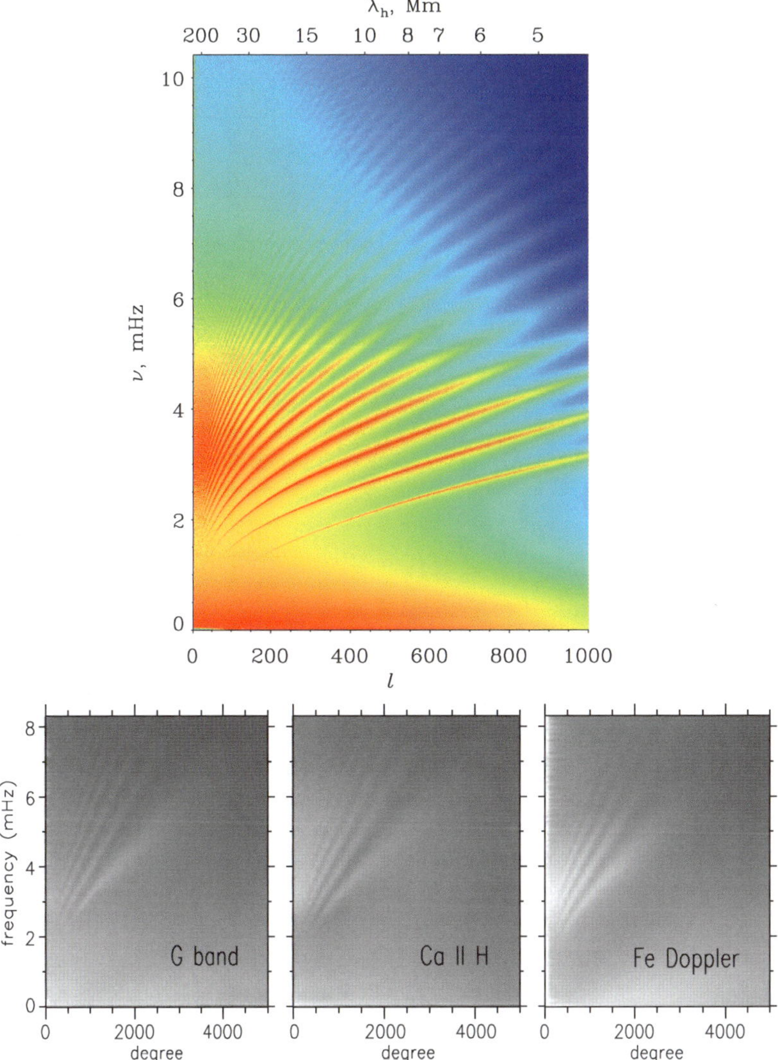

Fig. 19.5 Typical $l-\nu$ diagnostic diagram of the p modes: *top panel* from the SOHO observations (image courtesy SOHO/MDI) and, the *lower panel* from the one day HINODE observations (image courtesy HINODE/SOT, Sekii)

where a_i are the splitting coefficients and P are the polynomials related to Clebsch-Gordon coefficients. Odd degree splitting coefficients a are due to rotational effects and the combined effects (of perturbations due to magnetic field structure, structural asphericities and second order effect due to rotation) contribute to the even degree splitting coefficients.

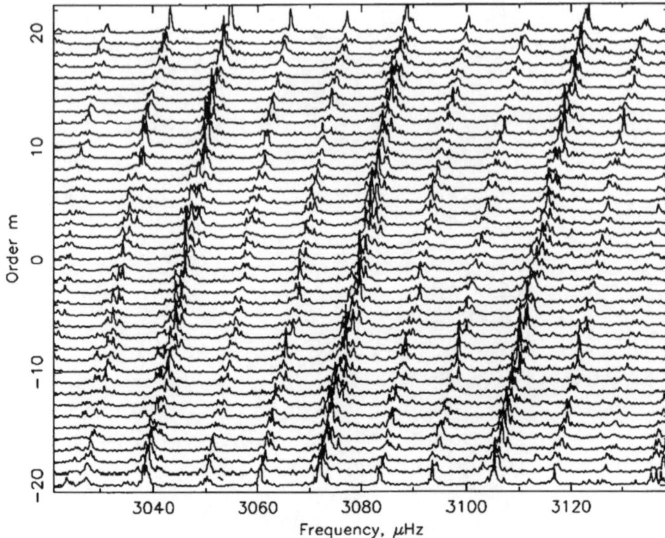

Fig. 19.6 For $l = 20$ and different m, observed frequency splittings due to rotation of the sun [82]

19.4 Helioseismic Inferences of Thermal, Dynamic and Magnetic field Structures

Since the discovery of sunspots by Galileo, scientists' quest is to understand the physics of the sun's interior. From the helioseismic observations, one can precisely estimate the frequencies and their splittings of different modes of oscillations. For example, by knowing difference between computed theoretical and observationally estimated frequencies, one can infer thermal structure such as internal sound speed and density of the solar interior. Where as the frequency splittings are used for the inference of internal dynamics (such as steady and time dependent parts of rotation and angular momentum) and magnetic field structures respectively.

19.4.1 Inference of Thermal Structure by Comparing the Observed and Computed Frequencies

Thermal structure of the solar interior can be understood in the following two ways. By using information of macro and micro physics of the solar interior, compute theoretical frequencies [31, 129] and match with the observed frequencies. Macrophysics involves standard hydrodynamic equations that are to be linearized and the perturbed variables are assumed to vary with $e^{i\omega t}$, where ω is angular frequency. Where as micro physics requires other additional details of the solar interior, *viz.*, equation of state, opacity and rate of nuclear energy generation. Using macro and micro physics, with additional knowledge of physics of the convection, solar internal structure such as pressure, temperature, density, etc., are computed.

As the observed amplitudes of oscillations are very small, equation of energy can be neglected and, linearized adiabatic equations with appropriate boundary conditions are used to compute the theoretical frequencies and are matched [31, 129] with the observed frequencies. Although most of the observed frequencies perfectly match very well for the low and intermediate degree modes, computed frequencies do not match very well with the high degree modes. Possibly this suggests our poor knowledge of physics of near surface effects.

19.4.1.1 Inference of Thermal Structure: Primary Inversions

In another method, observed frequencies and model of the solar structure are used to invert the thermal structure (for example, sound speed and density) of the solar interior in the following way. For the case of adiabatic oscillations, with additional constraint of conservation of mass, generalized form of equation [31, 47, 83, 129] of oscillation that takes into account the effects of velocity flows $\mathbf{v}$ and magnetic field structure $\mathbf{B}$ is given as follows

$$\mathcal{L}(\boldsymbol{\xi}) - \rho\omega^2\boldsymbol{\xi} + \nabla\delta p + \rho\left[\omega\mathcal{M}(\boldsymbol{\xi}) + \mathcal{N}(\boldsymbol{\xi}) + \mathcal{B}(\boldsymbol{\xi})\right] = 0, \qquad (19.4)$$

where

$$\mathcal{L}(\boldsymbol{\xi}) = \nabla\left(c^2\rho\nabla.\boldsymbol{\xi} + \nabla P.\boldsymbol{\xi}\right) - g\nabla.(\rho\boldsymbol{\xi}) - G\rho\nabla\left(\int_V \frac{\nabla.(\rho\boldsymbol{\xi})\,dV}{|r - r'|}\right), \qquad (19.5)$$

$$\mathcal{M}(\boldsymbol{\xi}) = 2i\left[\boldsymbol{\Omega}_0 \times \boldsymbol{\xi} + (\mathbf{v} \cdot \nabla)\boldsymbol{\xi}\right], \qquad (19.6)$$

$$\mathcal{N}(\boldsymbol{\xi}) = (\mathbf{v} \cdot \nabla)^2\boldsymbol{\xi} - 2\boldsymbol{\Omega}_0 \times \left[(\boldsymbol{\xi} \cdot \nabla)\mathbf{v} - (\mathbf{v} \cdot \nabla)\boldsymbol{\xi}\right] - (\boldsymbol{\xi} \cdot \nabla)(\mathbf{v} \cdot \nabla)\mathbf{v}, \qquad (19.7)$$

$$\mathcal{B}(\boldsymbol{\xi}) = (4\pi\rho)^{-1}\left[\left[\rho^{-1}(\boldsymbol{\xi} \cdot \nabla)\rho + \nabla \cdot \boldsymbol{\xi}\right]\mathbf{B} \times (\nabla \times \mathbf{B})\right.$$
$$\left. - \left[(\nabla \times \mathbf{B'}) \times \mathbf{B} + (\nabla \times \mathbf{B}) \times \mathbf{B'}\right]\right], \qquad (19.8)$$

where $\mathcal{L}$ is differential operator, $\boldsymbol{\xi}$ is eigen function of oscillations, c is speed of sound, ρ is density, P is pressure, $\mathcal{M}(\boldsymbol{\xi})$ and $\mathcal{N}(\boldsymbol{\xi})$ are effects due to velocity flows, $\mathcal{B}(\boldsymbol{\xi})$ is effect due to magnetic field structure and other symbols have usual meanings. Although effects due to flows and magnetic field structure in the solar interior can not be neglected, for simplicity, these terms are neglected and with special boundary conditions such that density and pressure perturbations completely vanish at the surface, Eq. (19.4) leads to the form $\mathcal{L}(\boldsymbol{\xi}) = \rho\omega^2\boldsymbol{\xi}$. This equation is an eigen value problem and is also Hermitian. Hence, variational principle [25]; see also [129] can be used to linearize the equation and frequency difference $\delta v_{n,l}$ between the sun and standard solar structure model can be represented ([15] and references there in) as follows

$$\frac{\delta v_{n,l}}{v_{n,l}} = \int_0^R K^{nl}_{c^2,\rho}(r)\frac{\delta c^2}{c^2}(r)dr + \int_0^R K^{nl}_{\rho,c^2}(r)\frac{\delta\rho}{\rho}(r)dr + \frac{F(v_{n,l})}{E_{nl}}, \qquad (19.9)$$

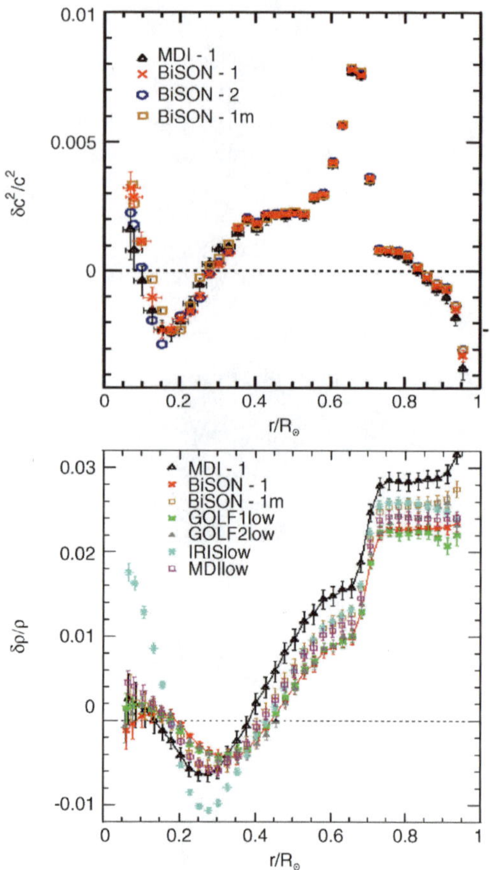

Fig. 19.7 Inference of thermal structure: (i) *top figure* illustrates the sound speed difference between the sun and the model and, (ii) *bottom figure* illustrates the density difference between the sun and the model. Image courtesy, Basu

where $K^{nl}_{c^2,\rho}(r)$ and $K^{nl}_{\rho,c^2}(r)$ are kernels that involve eigen functions of the oscillations and the known solar structure. Other terms $\frac{\delta c^2}{c^2}$ and $\frac{\delta\rho}{\rho}$ are the differences of sound speed and density between the sun and the reference solar structure model. The last term (see for details [15]) is a correction due to surface effects and does not arise due to linearization. Hence, from the observed oscillation frequencies and with the known reference solar structure model, one can invert radial profiles of $\frac{\delta c^2}{c^2}$ and $\frac{\delta\rho}{\rho}$ in the solar interior. There are many inversion techniques ([30, 48, 129] and references there in) to infer these profiles and one such inversion that yields the differences of sound speed and density (between the sun and the solar structure model) is presented [18] in Fig. 19.7. It is interesting to note that most of the models ([15] and references there in) almost match with the reference solar structure models, yet there are unexplained statistically significant sound speed differences mainly near base of the convection zone and in the radiative core. Similarly, density profile of solar structure models developed so far do not match very well with the density profile of the real sun. All of the solar structure models require the surface chemical heavy elemental abundances Z (or ratio Z/X, where X is the hydrogen abundance) that indirectly contribute to the equation of state and the opacity and hence, ultimately affect the computed frequencies, modeled sound speed and density structures. Most of these models ([15] and references there in) used the heavy elemental abundance ratio ($Z/X = 0.0229$) by Grevesse and Noels [50]. However, recent determination [2, 3, 10, 11] lowered this ratio ($Z/X = 0.0166$) leading to further deterioration of match between sound speed of the sun and different models. Hence, this conundrum of unexplained differences of thermal structures of sound speed and density is remained to be explained.

19.4.1.2 Inference of Thermal Structure: Secondary Inversions

By knowing the reference solar structure model, in the previous section, observed frequencies are used to compute the thermal structure differences of sound speed and the density. From these differences and knowing the sound speed and density structures as determined by the reference models, one can obtain sound speed and density profiles of the sun. As the sound speed and the density structures are thermodynamic quantities, in the following, using stellar structure equations (with additional information of equation of state and the opacity structures), one can uniquely obtain the thermal structure of the solar interior. There are two advantages of this solar seismic model, as in the standard solar models, (i) no assumption of history of the sun and, (ii) no need to have ad hock computation of the convective flux. In addition, chemical compositions such as hydrogen and helium abundances are obtained as part of solution. Thus, in essence, this seismic model yields a snap shot model of the present day sun.

For example, if equation of state of the solar plasma is available, inverted sound speed can be constrained to yield the ratio T/μ (where T is temperature and μ is the mean atomic weight). Following are the four set of stellar structure differential equations that need to be solved by imposing sound speed obtained by the primary inversions

$$\frac{dM_r}{dr} = 4\pi r^2 \rho, \tag{19.10}$$

$$\frac{dP}{dr} = -\frac{GM_r\rho}{r^2}, \tag{19.11}$$

$$\frac{dL_r}{dr} = 4\pi r^2 \rho\epsilon, \tag{19.12}$$

$$\frac{dT}{dr} = -\frac{3}{4ac}\frac{\kappa\rho}{T^3}\frac{L_r}{4\pi r^2} \quad \text{if radiative,}$$

$$= \left(\frac{dT}{dr}\right)_{conv} \quad \text{if convective,} \tag{19.13}$$

where the radial variables M_r, P, L_r and T are the mass, pressure, luminosity and temperature respectively. Other variables ϵ, κ and $(\frac{dT}{dr})_{conv}$ are the rate of nuclear energy generation, opacity of matter and knowledge of convective energy transport in the outer 30 % of the sun. Further axillary equations such as equation of state, equations of opacity and rate of nuclear energy generation are required. It is assumed that sun is in mechanical and thermal equilibrium such that whatever energy generated in the deep radiative core must be transported to the surface. However this latter condition can not be guaranteed.

As the sound speed is thermodynamically determined quantity, it is a function of other three variables, *viz.*, pressure, temperature and the chemical composition. As for the chemical composition, hydrogen X and helium Y are separately considered and all other elements are treated as single entity and is called as heavy elemental abundance Z.

With appropriate boundary conditions and assuming Z as constant, earlier models [7, 113–115, 123] solved above equations by considering the radiative core only. Although deduced profiles of thermal structure are almost similar to the profiles of solar standard models, there is no guarantee that these deductions satisfy the observed luminosity and the mass. By considering the inverted sound speed [123] for the whole region from center to the surface, a global seismic model [116–118] is developed that satisfies both the observed luminosity and the mass. It is found that seismic model satisfying one solar mass at the surface varies strongly on the nature of the sound speed profile near the center and a weak function of either depth of the convection zone or heavy elemental abundance Z/X. It is also found that one solar mass at the surface is satisfied for the sound speed profile which deviates at the center by $\sim$0.22 % from the sound speed profile of solar model [34], if we adopt the value of nuclear cross section factor S_{11} value of 4.07×10^{-22} keV barns. The resulting chemical abundances at base of the convective envelope are obtained to be $X = 0.755$, $Y = 0.226$ and $Z = 0.0185$ in case computed chemical abundance ratio $Z/X = 0.0245$ matches with the earlier [50] chemical abundance ratio. Neutrino fluxes of different reactions are computed and are not different than the neutrino fluxes computed from the standard solar model.

Recently, by considering the inverted sound speed kindly provided by Antia, thermal structure and neutrino fluxes are computed. One can notice from Fig. 19.8 that the deduced internal structural parameters such as mass, luminosity, pressure, density and temperature are almost similar to the structural parameters obtained by the standard solar models. Neutrino fluxes computed by using sound speed profile computed from the standard model and the seismic models with different Z values are presented in Table 19.1. In Table 19.1, nomenclatures for SSM, Seism and BCZ are standard solar model of Dalsgaard [34], seismic model and base of the convection zone respectively.

One can notice that, for the earlier [50] chemical abundance ratio $Z/X = 0.0247$, neutrino fluxes (Table 19.1) computed from the seismic model is not different from the neutrino fluxes computed from the solar standard solar model that lead to a turning point with a strong conclusion that deficiency of neutrinos emitted by the sun lies in the neutrino physics ruling out the astrophysical solutions. Even changing of uncertain physics of the interior, such as opacity, equation of state, etc., or chemical composition (especially the heavy elemental abundance Z) could not alleviate the solar neutrino problem.

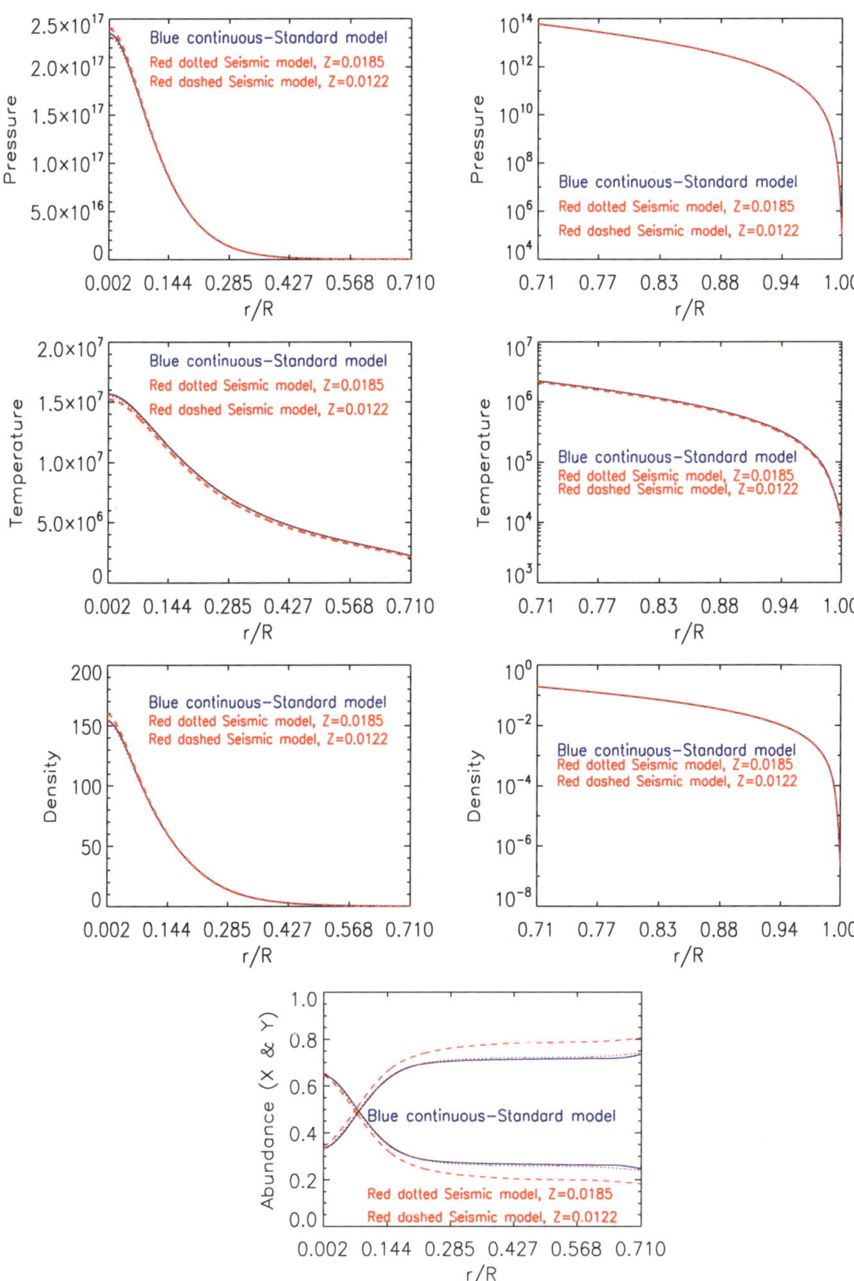

Fig. 19.8 Thermal structure such as pressure, temperature, density and hydrogen abundance X and helium abundance Y obtained by the solar seismic model for surface heavy elemental abundances $Z = 0.0185$ and $Z = 0.0122$ respectively. Units are in cgs scale

Table 19.1 Neutrino Fluxes (in the units of 10^{10} cm^{-2} sec^{-1}) estimated by the solar seismic model

Source	SSM	Seism1	Seism2	Seism3	Seism4
	$Z = 0.0185$	$Z = 0.0185$	$Z = 0.0165$	$Z = 0.015$	$Z = 0.0122$
	$Z/X = 0.0247$	$Z/X = 0.0247$	$Z/X = 0.0215$	$Z/X = 0.0188$	$Z/X = 0.0158$
	$BCZ = 0.709$	$BCZ = 0.709$	$BCZ = 0.709$	$BCZ = 0.709$	$BCZ = 0.72$
pp	6.0	6.01	6.05	6.07	6.1
pep	0.014	0.015	0.014	0.015	0.015
hep	8×10^{-7}	0.3×10^{-9}	0.3×10^{-9}	1.3×10^{-7}	1.32×10^{-7}
^{7}Be	0.47	0.44	0.42	0.41	0.379
^{8}B	5.8×10^{-4}	4.5×10^{-4}	4.03×10^{-4}	3.76×10^{-4}	3.18×10^{-4}
^{13}N	0.06	0.057	0.047	0.04	0.0287
^{15}O	0.05	0.052	0.042	0.036	0.025
^{17}F	5.2×10^{-4}	4.02×10^{-4}	3.2×10^{-4}	2.7×10^{-4}	1.92×10^{-4}

However and surprisingly, for the recently determined heavy elemental abundance ($Z/X = 0.016$; seismic model 4, Table 19.1), neutrino flux, especially for ^{8}B is substantially reduced and is almost similar to the observed neutrino fluxes, although helium abundance deduced by this seismic model at the base of the convective envelope is very low ($\sim$0.18) and is inconsistent with other cosmic helium abundances ($\sim$0.23). Recent determination [11] of heavy elemental abundances (that substantially lowered the abundances of carbon, nitrogen, oxygen and neon) has given a rebirth of astrophysical solution to the solar neutrino problem [126, 127].

19.4.2 Inference of Dynamic Structure

The term *dynamic structure* is mainly due to rotation of the whole sun, although large-scale weak meridional flow ($\sim$ few meters/sec from equator to both the northern and southern poles) and strong convective flows in the interior exist. Observations show that, with a typical linear velocity of $\sim$2 km/sec, sun rotates differentially, rotating faster at the equator and slower at the poles. As mentioned in Sect. 19.3.4, rotation of the sun lifts the degeneracy of the oscillations that leads to frequency of oscillations depend upon azimuthal order m and, similar to Zeeman effect, split the frequencies yielding a relation $\omega_m = \omega_0 + \Omega m$ (where ω_0 is frequency of the oscillations in the rotating frame of reference and Ω is the angular velocity of the sun). In this simple description, angular velocity Ω is assumed to be of rigid body rotation and hence is independent of radius and latitude respectively. Observations show that sun is rotating differentially and, by neglecting the effects

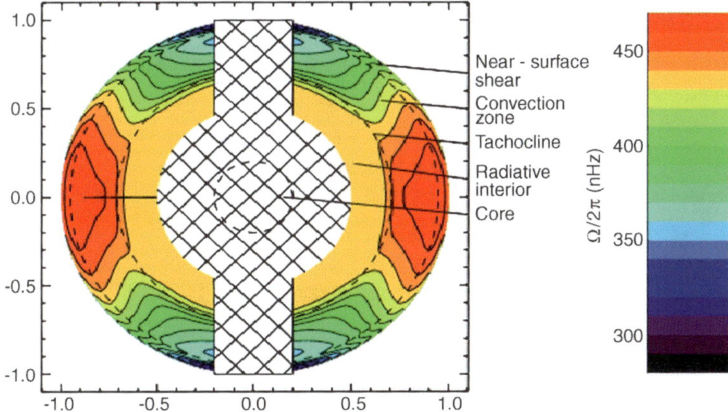

Fig. 19.9 Isorotational contours inferred from 12 years average data of MDI data. The cross hatched areas indicate the regions in which it is difficult or impossible to obtain reliable inversion results with the available data (Courtesy: Howe)

due to centrifugal force and the magnetic field structure, Eq. (19.4) can be modified [30, 31, 124, 129] as follows

$$\delta\omega_{nlm} = \omega_{nlm} - \omega_{nl0} = m \int_0^R \int_0^\pi K_{nlm}(r,\theta)\Omega(r,\theta)r\,dr\,d\theta, \qquad (19.14)$$

where the kernels [29, 54, 108] $K_{nlm}(r,\theta)$ involve the eigen functions of oscillations in the nonrotating frame. These kernels depend only upon m^2 and hence the rotational splittings $(\omega_{nlm} - \omega_{nl0})$ are odd function of m. As the kernels are also symmetrical about the equator, the rotational splittings sense only the symmetrical component of Ω. From the observed rotational splittings and the computed kernels $K_{nlm}(r,\theta)$ from the adiabatic oscillations, one can invert Eq. (19.14) and hence rotation rate $\Omega(r,\theta)$ of the whole region of the solar interior is obtained. There are different inversional techniques ([8, 32, 70, 71, 96, 97, 108, 109, 124] and references therein; [68] and references there in; [6]) and inferred rotation rate of the solar interior is presented in Fig. 19.9. One can notice from Fig. 19.9 that sun rotates differentially in the entire convective envelope and has a rigid body rotation rate in the radiative core. In addition, near the surface and the base of convection zone, two rotational shears exist. The rotational shear near base of convection zone, where the transition from differential rotation to rigid body rotation occurs, is called *tachocline* where the seat of so called *dynamo* (that is believed to be origin and maintenance of solar cycle and activity phenomena) is supposed to exist ([63, 66] and references there in). However, yet it is not completely understood how the sun attained such a rotational structure in its interior.

19.4.3 Inference of Magnetic Field Structure

Recent observations suggest that the entire universe is pervaded by a large scale weak uniform magnetic field structure ([104] and references there in) and sun is also no exception. The sun is pervaded by a large scale steady (diffusion time scale of $\sim$ billion yrs) global dipole like magnetic field structure with a strength of $\sim$1 G [122] and time dependent magnetic field structure ($\sim$ kilo gauss) that fluctuates over 22 years. If magnetic field lines vary from pole to pole, such a geometry is called *poloidal field* structure. Where as in case of *toroidal* field structure, field lines are parallel to the equator. Sun is pervaded by a combined weak ($\sim$1 G) poloidal and strong ($\sim$10^3 G) toroidal field structure. Sunspots are supposed to be formed due to Alfven perturbations of the steady toroidal magnetic field structure [66] embedded in the solar interior. Large scale magnetic field structure induces a force and distorts the geometrical figure of a star. During the early evolutionary history of the sun, magnetic field structure might have played a dominant role in transferring angular momentum to the solar system. This could be one of the main reasons why most of the angular momentum is concentrated in the solar system rather than in the sun. It is believed that 22 yr magnetic cycle is due to so called dynamo mechanism that is the result of interplay of convection and rotation on the large scale poloidal magnetic field structure although there are other alternative views [63, 66].

19.4.3.1 Inference of Magnetic Field Structure: Primary Inversions

Signatures of both the poloidal and toroidal magnetic field structures can be found in the observed even degree frequency splittings. In addition, second order rotational effect (where as first order or linear rotational effect is sensed by the odd degree frequency splittings as given in Sect. 19.4.2) and aspherical sound speed perturbations [73] also contribute to the even degree frequency splittings. Hence, by assuming that aspherical sound speed perturbations are negligible, with the internal rotational structure as inferred from the helioseismology (Sect. 19.4.2), contribution due to second order rotational effects is computed and is subtracted from the even degree frequency splittings and, resulting residual of even degree splittings are compared with the computed frequency splittings due to assumed magnetic field structure. Detailed inversion procedures can be found in the previous studies [5, 9, 13, 14, 40, 41, 48]. Recently Baldner et al. [13] came to the conclusion that the observed variation of even degree frequency splittings can be explained if the sun is pervaded by a right combination of poloidal ($\sim$100 G) and toroidal (that varies from $\sim$ 10^3 G near the surface to 10^4 G near base of the convection zone) magnetic field structure. By considering the SOHO/MDI magnetograms and from the analysis of sunspot data during their initial appearance on the surface, recently, Hiremath and Lovely [66] estimated strength of toroidal magnetic field structure of similar order confirming the previous theoretical estimates [27, 37, 38, 57].

19.4.3.2 Inversion of Poloidal Magnetic Field Structure: Secondary Inversions

Owing to large diffusion time scales due to large dimension and finite conductivity of the solar plasma, sun might have retained a large scale poloidal magnetic structure from its protostar phase even after the convective Hayashi phase [28, 77, 86, 95, 121]. Presence of such a large scale poloidal magnetic field structure, especially during the solar minimum, can be clearly discerned during the solar total eclipse. Field lines of dipole like magnetic field structure are clearly delineated in the intensity patterns of the white light pictures taken during the eclipse. If the Lorentz force due to either poloidal or toroidal magnetic field structure is very weak compared to dynamical effects such as rotation, one can show [64, 65] that weak poloidal magnetic field structure isorotates with the solar plasma. In fact, in case of the sun, especially for the poloidal magnetic field structure ($\sim$1 G), this condition is valid. That means, if Ω is angular velocity of the plasma and Φ is the magnetic flux function representing flux of the poloidal magnetic field structure, one can show that

$$\Omega = \text{function}(\Phi). \tag{19.15}$$

If one knows the internal rotational structure of the sun, by suitable combination of poloidal magnetic field structure that is obtained consistently from the solution of MHD equations, one can satisfy the above condition. To a first approximation, right hand side of equation can be linearized of the form

$$\Omega = \Omega_0 + \Omega_1(\Phi). \tag{19.16}$$

With reasonable assumptions and approximations and by using Chandrasekhar's [23, 24] MHD equations for the case of axisymmetry and incompressibility, Hiremath and Gokhale [65] have shown that poloidal component of the sun's steady magnetic field structure can be modeled as an analytical solution of magnetic diffusion equation and the magnetic flux function (Φ) for both the radiative core (RC) and the convective envelope (CE) are expressed as

$$\Phi_{RC}(x, \vartheta) = 2\pi A_0 R_c^2 x^{1/2} \sin^2 \vartheta \sum_{n=0}^{\infty} \lambda_n J_{n+3/2}(\alpha_n x) C_n^{3/2}(\mu), \tag{19.17}$$

where $x = r/R_c$, R_c is radius of the radiative core, $\mu = \cos \vartheta$, r and ϑ are radial and colatitude coordinates, n is non negative integer, $C_n^{3/2}$ is the Gegenbaur's polynomial of degree n, $J_{n+3/2}(\alpha_n x)$ is Bessel function of order $(n + 3/2)$ and α_n are the eigen values that are to be determined from the boundary conditions. Here A_0 is taken as a scale factor for the field and hence $\lambda_n = A_n/A_0$. Similarly, magnetic flux function for the case of convective envelope is

$$\Phi_{CE}(x, \vartheta) = \pi B_0 R_c^2 \sin^2 \vartheta \left[x^2 C_0^{3/2}(\mu) + \sum_{n \geq 0} \hat{\mu}_{n+1} x^{-(n+1)} C_n^{3/2}(\mu) \right], \tag{19.18}$$

where $\hat{\mu}_{n+1} = M_n/(\pi B_0 R_c^{n+3})$ are strengths of multipoles that are scaled to an asymptotically uniform magnetic field structure B_0. One can notice from the above equation that magnetic field structure approaches an asymptotically uniform field

Fig. 19.10 *Top figure* illustrates the meridional cross section of poloidal component of magnetic field structure. This figure is reproduced from, and all rights are reserved by, the Astrophysical Journal. Where as *bottom figure* illustrates radial and latitudinal variations of the magnitude of the poloidal magnetic field structure B_p normalized to an asymptotic uniform magnetic field that merges with the interplanetary field structure

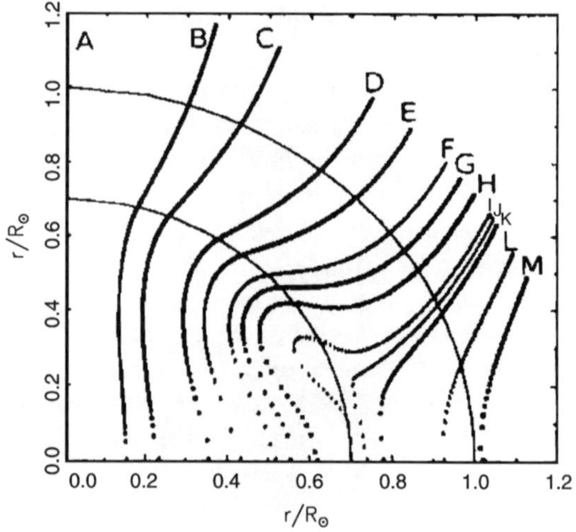

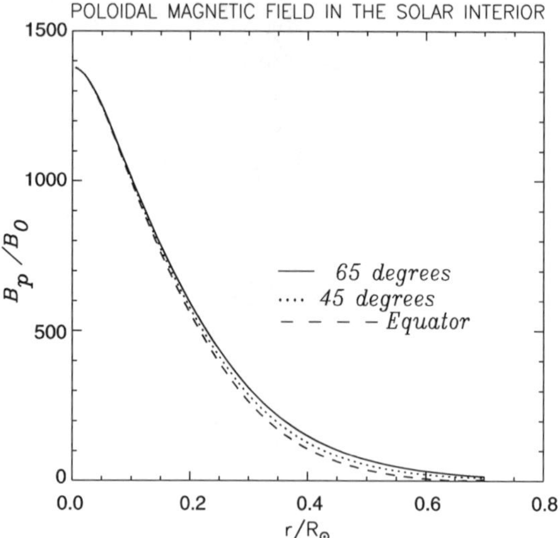

structure as $x \to \infty$. With appropriate boundary conditions at base of the convection zone and with helioseismically inferred rotation, one can uniquely estimate the unknown coefficients and eigen values respectively. In case of sun and with the available helioseismically inferred rotation [8, 42], geometry of the poloidal field structure is computed [64, 65] and is presented in Fig. 19.10. If one considers strength of interplanetary magnetic field structure ($\sim 10^{-5}$ G) as a scaling factor, strength of sun's poloidal magnetic field structure near its center is found to be $\sim 10^7$–10^8 G. From the dynamical constraints, one can rule out such a strong magnetic field structure near the center. However, if the 22 yr solar cycle and activity phenomenon is due to global slow MHD modes [65], B_0 is estimated to be $\sim 10^{-2}$ G and, in that case strength of poloidal magnetic field structure near the center turns out be $\sim 10^5$ G. Off course one would have found signature of such a strong field structure in the observations of p mode oscillations. Unfortunately low degree 'p' modes that penetrate deeply in the radiative core can not sense close to the center. Hence, one has to wait for the discovery of elusive g modes that are highly sensitive to the central structure of the sun and hence their characteristics are modified [22, 98–100] by such a strong magnetic field structure.

19.5 Concluding Remarks

Although seismology of the sun has provided some of the detailed global aspects of thermal and dynamic structures of the sun's interior, other global structures such as detailed informations regarding meridional flow (this information is needed to accept or reject so called flux transport dynamo models) and magnetic field structure are necessary. Hitherto solar seismologists came to the conclusion that the difference between the sun's thermal structure (especially the sound speed) and the modeled thermal structure is almost similar. However, recent revision of heavy-element abundances [2, 3, 10–12] that are reduced drastically, further worsened the difference between the sun's and modeled thermal structure leading to revisit of exotic astrophysical solutions that are used to explain the solar neutrino problem. Importantly, as noticed by the previous studies [16, 26], critical examination of modeling and method of computation in estimating the heavy-elemental abundance are necessary.

As for the dynamical helioseismic inferences, especially the rotational structure, 'p' modes are not the true representatives to sense the deep radiative core. For this purpose one has to wait for discovery of 'g' modes that have high sensitivity near the core. Hence knowing of magnitude and form (i.e., dependent or independent of radius and latitude) of thermal and rotational structures near the core are essential for understanding overall structure and temporal variations of the sun on short ($\sim$11 years) and long ($\sim$ billion years) time scales. Although helioseismic inferences of rotation rate of the convective envelope ruled out the increase of rotation rate (as is assumed to be in the earlier turbulent dynamo models) from surface to the interior and cylindrical isorotational contours (simulations by Gilman and Miller [45]), physics of isorotational contours, especially near the surface, is completely not understood. However, it should be appreciated that the simulated isorotational contours from base of convection zone to $0.935R_\odot$ [19, 21, 43, 57, 69, 74, 88, 102, 103] are almost similar to the rotational isocontours as inferred from helioseismology although none of the present simulations reproduce the near surface decreasing rotational profile from $0.935R_\odot$ to the surface. Even if numerical simulations achieve the goal of reproducing near surface rotational profile, from the MHD stability criterion [39, 85], $r^2\frac{d}{dr}[\frac{\Omega^2}{r^2}] - (\frac{1}{r^2})\frac{d}{dr}[\frac{(r^2 B_\phi^2)}{4\pi\rho}] > 0$ (where r is the radial coordinate, Ω is angular velocity, ρ is density and B_ϕ is toroidal magnetic field structure), one can show that such a decreasing rotational profile near the surface is unstable (independently, Miesch and Hindman [87], recently came to a similar conclusion) unless sun acquired near equipartition toroidal magnetic field structure. Interestingly, toroidal field structure estimated by theoretical [20, 57] and helioseismic [5, 13] inferences yield of similar magnitudes. Hence, near surface rotational profile is stable.

However, why the sun acquired such a decreasing near surface rotational profile and hence the toroidal magnetic field structure are remained to be explained. Although beyond the scope of this presentation, such a near surface decreasing rotational profile can be explained if one speculates that during the early history of the planetary formation, protoplanetary mass that rotated with Keplerian velocity might have accreted on to the sun that resulted in winding of threaded ambient large scale poloidal magnetic field structure into toroidal magnetic field structure. If this speculation is correct, such a protoplanetary accretion scenario might give some clues regarding absence of super earths near vicinity of the sun and peculiar high solar mass [51] compared to other stars in the sun's neighborhood in the Galaxy. Incase we accept the accretion scenario (during the early solar system formation), longstanding problems such as angular momentum and planetary formation of the solar system can be solved. With such an accretion scenario and hence high solar mass, unsolved faint young sun paradox [89] and recent puzzle of computation of low heavy elemental abundances [12] that are incompatible [16, 17] with the helioseismic inferences can also be alleviated upto some extent [52, 55, 84, 92, 130] and references there in; [110].

As for the secondary inversion of poloidal magnetic field structure, this method requires the rotational profile of the interior as inferred by the helioseismology for estimation of different magnetic moments. Instead, as the weak ($\sim$1 G) poloidal magnetic field structure isorotates with the internal rotation of the plasma and hence rotation is a function of magnetic flux, Eq. (19.14) can be used for simultaneously inverting rotational and poloidal magnetic field structures of the interior (Gokhale and Cristensen Dalsgaard, private communication).

References

1. R. Agnihotri, K. Dutta, W. Soon, J. Atmos. Sol.-Terr. Phys. **73**, 1980 (2011)
2. C. Allende Prieto, D.L. Lambert, M. Asplund, Astrophys. J. **556**, L63 (2001)
3. C. Allende Prieto, D.L. Lambert, M. Asplund, Astrophys. J. **573**, L137 (2002)
4. M. Anguera Gubau, P.L. Palle, P. Hernandez, T. Roca-Cortes, Sol. Phys. **128**, 79 (1990)
5. H.M. Antia, in *Proceedings of IAU Coll. 188*. ESA SP, vol. 505 (2002), p. 71
6. H.M. Antia, S. Basu, Astrophys. J. **720**, 494 (2010)
7. H.M. Antia, S.M. Chitre, Astron. Astrophys. **347**, 1000 (1999)
8. H.M. Antia, S. Basu, S.M. Chitre, Mon. Not. R. Astron. Soc. **298**, 543 (1998)
9. H.M. Antia, S.M. Chitre, M.J. Thompson, Astron. Astrophys. **360**, 335 (2000)
10. M. Asplund, A. Nordlund, R. Trampedach, R.F. Stein, Astron. Astrophys. **359**, 743 (2000)
11. M. Asplund, N. Grevesse, A.J. Sauval, C. Allende Prieto, D. Kiselman, Astron. Astrophys. **417**, 751 (2004)
12. M. Asplund, N. Grevesse, A.J. Sauval, Astron. Soc. Pac. Conf. Ser. **336**, 25 (2005)
13. C.S. Baldner, H.M. Antia, S. Basu, T.P. Larson, Astron. Nachr. **331**, 879 (2010)
14. S. Basu, Mon. Not. R. Astron. Soc. **288**, 572 (1997)
15. S. Basu, Astrophys. Space Sci. **328**, 43 (2010)

16. S. Basu, H.M. Antia, Phys. Rep. **457**, 217 (2008)
17. S. Basu, W.J. Chaplin, Y. Elsworth, R. New, A.M. Serenelli, G.A. Verner, Astrophys. J. **655**, 660 (2007)
18. S. Basu, W.J. Chaplin, Y. Elsworth, R. New, A.M. Serenelli, Astrophys. J. **699**, 1403 (2009)
19. A.S. Brun, J. Toomre, Astrophys. J. **570**, 865 (2002)
20. A.S. Brun, M.S. Miesch, J. Toomre, Astrophys. J. **614**, 1073 (2004)
21. A.S. Brun, M.S. Miesch, J. Toomre, Astrophys. J. **742**, 79 (2011)
22. C.P. Burgess, N.S. Dzhalilov, T.I. Rashba, V.B. Semikoz, J.W.F. Valle, Mon. Not. R. Astron. Soc. **348**, 609 (2004)
23. S. Chandrasekhar, Astrophys. J. **124**, 232 (1956)
24. S. Chandrasekhar, Astrophys. J. **124**, 244 (1956)
25. S. Chandrasekhar, Astrophys. J. **139**, 664 (1964)
26. W.J. Chaplin, S. Basu, Sol. Phys. **251**, 53 (2008)
27. A.R. Choudhuri, P.A. Gilman, Astrophys. J. **316**, 788 (1987)
28. T.G. Cowling, Mon. Not. R. Astron. Soc. **94**, 39 (1953)
29. J. Cuypers, Astron. Astrophys. **89**, 207 (1980)
30. C. Dalsgaard, Rev. Mod. Phys. **74**, 1073 (2002)
31. C. Dalsgaard, *Lecture Notes on Stellar Oscillations* (2003), p. 66
32. C. Dalsgaard, J. Schou, in *ESA SP*, vol. 286 (1988) p. 149
33. C. Dalsgaard, D.O. Gough, K.B. Libbrecht, Astrophys. J. **341**, L103 (1989)
34. C. Dalsgaard et al., Science **272**, 1286 (1996)
35. F.-L. Deubner, Astron. Astrophys. **44**, 371 (1975)
36. F.-L. Deubner, D. Gough, Annu. Rev. Astron. Astrophys. **22**, 593 (1984)
37. S. D'Silva, A.R. Choudhuri, Astron. Astrophys. **272**, 621 (1993)
38. S. D'Silva, R. Howard, Sol. Phys. **151**, 213 (1994)
39. B. Dubrulle, E. Knobloch, Astron. Astrophys. **274**, 667 (1993)
40. W.A. Dziembowski, P.R. Goode, Astrophys. J. **347**, 540 (1989)
41. W.A. Dziembowski, P.R. Goode, Astrophys. J. **376**, 782 (1991)
42. W.A. Dziembowski, P.R. Goode, K.G. Libbrecht, Astrophys. J. **343**, L53 (1989)
43. J.R. Elliott, M.S. Miesch, J. Toomre, Astrophys. J. **533**, 546 (2000)
44. J. Feynman, Adv. Space Res. **40**, 1173 (2007)
45. P.A. Gilman, J. Miller, Astrophys. J. Suppl. Ser. **46**, 211 (1981)
46. P. Goldreich, P. Kumar, Astrophys. J. **326**, 462 (1988)
47. D.O. Gough, P.P. Taylor, Mem. Soc. Astron. Ital. **55**, 215 (1984)
48. D.O. Gough, M.J. Thompson, in *Solar Interior and Atmosphere* (1991), p. 519
49. G. Grec, E. Fossat, M. Pomerantz, Sol. Phys. **82**, 55 (1983)
50. S. Grevesse, A. Noels, in *Origin and Evolution of the Elements* (1993), p. 15
51. B. Gustafsson, Phys. Scr. **130**, 014036 (2008)
52. J.A. Guzik, K. Mussack, Astrophys. J. **713**, 1108 (2010)
53. J.D. Haigh, Living Rev. Sol. Phys. **4**, 2 (2007)
54. C.J. Hansen, J.P. Cox, H.M. Van Horn, Astrophys. J. **217**, 151 (1977)
55. W.C. Haxton, A.M. Serenelli, Astrophys. J. **687**, 678 (2008)
56. G. Hill, F.L. Deubner, G. Isaak, in *Solar Interior and Atmosphere* (1991), p. 329
57. K.M. Hiremath, Bull. Astron. Soc. India **29**, 169 (2001)
58. K.M. Hiremath, J. Astrophys. Astron. **27**, 367 (2006)
59. K.M. Hiremath, J. Astrophys. Astron. **27**, 277 (2006)
60. K.M. Hiremath, in *The Proceedings of ILWS Workshop*, Goa (2006), p. 178
61. K.M. Hiremath, Astrophys. Space Sci. **314**, 45 (2008)
62. K.M. Hiremath, Sun & Geosphere **4**, 16 (2009)
63. K.M. Hiremath, Sun & Geosphere **5**, 17 (2010)
64. K.M. Hiremath, PhD Thesis, Bangalore University, 1994
65. K.M. Hiremath, M.H. Gokhale, Astrophys. J. **448**, 437 (1995)
66. K.M. Hiremath, M.R. Lovely, New Astron. **17**, 392 (2012)
67. K.M. Hiremath, P.I. Mandi, New Astron. **9**, 651 (2004)

68. R. Howe, Living Rev. Sol. Phys. **6**, 1 (2009)
69. L.L. Kitchatinov, G. Rüdiger, Astron. Astrophys. **299**, 446 (1995)
70. S.G. Korzennik, M.J. Thompson, J. Toomre, IAU Symp. **181**, 211 (1996)
71. A.G. Kosovichev, Astrophys. J. Lett. **469**, L61 (1996)
72. V.A. Kotov, A.B. Severnyi, T.T. Tsap, Mon. Not. R. Astron. Soc. **183**, 61 (1978)
73. J.R. Kuhn, Astrophys. J. **331**, L131 (1988)
74. M. Kuker, G. Rudiger, L.L. Kitchatinov, Astron. Astrophys. **530**, 48 (2011)
75. P. Kumar, P. Goldreich, Astrophys. J. **343**, 558 (1989)
76. P. Kumar, J. Franklin, P. Goldreich, Astrophys. J. **328**, 879 (1988)
77. D. Layzer, R. Rosner, H.T. Doyle, Astrophys. J. **229**, 1126 (1979)
78. J.W. Leibacher, R.F. Stein, Astrophys. J. Lett. **7**, 191L (1971)
79. R.B. Leighton, IAU Symp. **12**, 321 (1960)
80. R.B. Leighton, R.W. Noyes, G.W. Simon, Astrophys. J. **135**, 474 (1962)
81. K.G. Libbrecht, Astrophys. J. **334**, 510 (1988)
82. K.G. Libbrecht, Astrophys. J. **336**, 1092 (1989)
83. D. Lynden-Bell, J.P. Ostriker, Mon. Not. R. Astron. Soc. **136**, 293 (1967)
84. J. Meléndez, M. Asplund, B. Gustafsson, D. Yong, Astrophys. J. **704**, L66 (2009)
85. L. Mestel, in *Stellar Magnetism* (Oxford University Press, London, 1999), p. 395
86. L. Mestel, N.O. Weiss, Mon. Not. R. Astron. Soc. **226**, 123 (1987)
87. M.S. Miesch, B.W. Hindman, Astrophys. J. **743**, 79 (2011)
88. M.S. Miesch, A.S. Brun, J. Toomre, Astrophys. J. **641**, 618 (2006)
89. D.A. Minton, R. Malhotra, Astrophys. J. **660**, 1700 (2007)
90. S. Musman, D.M. Rust, Sol. Phys. **13**, 261 (1970)
91. J. Nishikawa, S. Hamana, K. Mizugaki, T. Hirayama, Publ. Astron. Soc. Jpn. **38**, 277 (1986)
92. A. Nordlund, arXiv:0908.3479 (2009)
93. P.L. Palle, C. Regulo, T. Roca-Cortes, Astron. Astrophys. **224**, 253 (1989)
94. C.A. Perry, Adv. Space Res. **40**, 353 (2007)
95. J.H. Piddington, Astrophys. Space Sci. **90**, 217 (1983)
96. F.P. Pijpers, M.J. Thompson, Astron. Astrophys. **262**, 33 (1992)
97. F.P. Pijpers, M.J. Thompson, Astron. Astrophys. **281**, 231 (1994)
98. T. Rashba, J. Phys. Conf. Ser. **118**, 012085 (2008)
99. T.I. Rashba, V.B. Semikoz, J.W.F. Valle, Mon. Not. R. Astron. Soc. **370**, 845 (2006)
100. T.I. Rashba, V.B. Semikoz, S. Turck-Chièze, J.W.F. Valle, Mon. Not. R. Astron. Soc. **377**, 453 (2007)
101. G.C. Reid, J. Atmos. Sol.-Terr. Phys. **61**, 3 (1999)
102. M. Rempel, Astrophys. J. **622**, 1320 (2005)
103. F.J. Robinson, K.L. Chan, Mon. Not. R. Astron. Soc. **321**, 723 (2001)
104. D. Ryu et al., Space Sci. Rev. **166**, 1 (2012)
105. N. Scafetta, B.J. West, Phys. Today **March**, 50–51 (2008)
106. P.H. Scherrer, J.M. Wilcox, Sol. Phys. **82**, 37 (1983)
107. P.H. Scherrer, J.M. Wilcox, V.A. Kotov, A.B. Severny, T.T. Tsap, Nature **277**, 635 (1979)
108. J. Schou, J. Christensen-Dalsgaard, M.J. Thompson, Astrophys. J. **433**, 389 (1994)
109. T. Sekii, IAU Symp. **181**, 189 (1996)
110. A.M. Serenelli, W.C. Haxton, C. Pena-Garay, arXiv:1104.1639 (2011)
111. A.B. Severnyi, V.A. Kotov, T.T. Tsap, Nature **259**, 87 (1976)
112. A.B. Severnyi, V.A. Kotov, T.T. Tsap, Sov. Astron. **23**, 641 (1979)
113. H. Shibahashi, M. Takata, Publ. Astron. Soc. Jpn. **48**, 377 (1996)
114. H. Shibahashi, M. Takata, IAU Symp. **181**, 167 (1997)
115. H. Shibahashi, M. Takata, S. Tanuma, in *Proceedings of the 4th Soho Workshop*. ESA SP (1995), p. 9
116. H. Shibahashi, K.M. Hiremath, M. Takata, in *ESA SP*, vol. 418 (1998), p. 537
117. H. Shibahashi, K.M. Hiremath, M. Takata, IAU Symp. **185**, 81 (1998)
118. H. Shibahashi, K.M. Hiremath, M. Takata, Adv. Space Res. **24**, 177 (1999)
119. K.P. Shine, Space Sci. Rev. **94**, 363 (2000)

120. W. Soon, Geophys. Res. Lett. **32**(16), L16712 (2005)
121. H.C. Spruit, in *Inside the Sun* (1990), p. 415
122. J.O. Stenflo, in *Solar Surface Magnetism* (1993)
123. M. Takata, H. Shibahashi, Astrophys. J. **504**, 1035 (1998)
124. M.J. Thompson, J. Christensen-Dalsgaard, M.S. Miesch, J. Toomre, Astron. Astrophys. Rev. **41**, 599 (2003)
125. M. Tiwari, R. Ramesh, Curr. Sci. **93**, 477 (2007)
126. S. Turck-Chièze, S. Couvidat, Rep. Prog. Phys. **74**, 086901 (2011)
127. S. Turck-Chièze, A. Palacios, J.P. Marques, P.A.P. Nghiem, Astrophys. J. **715**, 1539 (2010)
128. R.K. Ulrich, Astrophys. J. **162**, 99 (1970)
129. W. Unno, Y. Osaki, H. Ando, H. Sao, H. Shibahashi, in *Noradial Oscillations of Stars* (1989), p. 108
130. R.A. Winnick, P. Demarque, S. Basu, D.B. Guenther, Astrophys. J. **576**, 1075 (2002)

Index

M. Mohan (ed.), *New Trends in Atomic and Molecular Physics*,
Springer Series on Atomic, Optical, and Plasma Physics 76,
DOI 10.1007/978-3-642-38167-6, © Springer-Verlag Berlin Heidelberg 2013

Printed by Printforce, the Netherlands